HANDBOOK OF
PHYCOLOGICAL METHODS

PHYSIOLOGICAL
AND BIOCHEMICAL METHODS

EDITED BY

JOHAN A. HELLEBUST
PROFESSOR OF BOTANY
UNIVERSITY OF TORONTO, TORONTO, CANADA

AND

J. S. CRAIGIE
SENIOR RESEARCH OFFICER
ATLANTIC REGIONAL LABORATORY
NATIONAL RESEARCH COUNCIL OF CANADA, HALIFAX, CANADA

SPONSORED BY THE
PHYCOLOGICAL SOCIETY OF AMERICA INC.

CAMBRIDGE UNIVERSITY PRESS
CAMBRIDGE
LONDON · NEW YORK · MELBOURNE

Published by the Syndics of the Cambridge University Press
The Pitt Building, Trumpington Street, Cambridge CB2 1RP
Bentley House, 200 Euston Road, London NW1 2DB
32 East 57th Street, New York, NY 10022, USA
296 Beaconsfield Parade, Middle Park, Melbourne 3206, Australia

First published 1978

Printed in the United States of America
Typeset by Progressive Typographers, Inc.,
York, Pennsylvania
Printed by Hamilton Printing Company,
Rensselaer, New York
Bound by Payne Edition Bindery, Inc.,
Chester, New York

Library of Congress Cataloging in Publication Data
Main entry under title:
Handbook of phycological methods.
CONTENTS: [1] Culture methods and growth measurements,
edited by J. R. Stein. – v. 2. Physiological and biochemical methods,
edited by J. A. Hellebust and J. S. Craigie.
Includes bibliographies.
1. Algology – Technique.
I. Stein, Janet R. II. Hellebust, J. A. III. Craigie, J. S.
IV. Phycological Society of America.
QK565.2.S73   539'.31   73-79496
ISBN 0 521 20049 0

Contents

Contributors *page* ix   Editors' preface *page* xiii

Introduction   *page* 1

I   ISOLATION OF ORGANELLES AND   3
    MEMBRANES

   1   Chloroplasts from *Euglena gracilis*   5
      C. A. Price
   2   Nuclei and chromatin from *Euglena gracilis*   15
      D. E. Buetow
   3   Mitochondria from *Chlorella*   25
      Neil G. Grant
   4   Microbodies from *Chlorogonium* and   31
      Spirogyra
      Helmut Stabenau
   5   Isolation and characterization of   39
      diatom membranes
      Cornelius W. Sullivan

II   ANALYSIS OF CHEMICAL   57
    CONSTITUENTS

   6   Chlorophylls and carotenoids   59
      Arne Jensen
   7   Algal biliproteins   71
      Harold W. Siegelman and J. Helen Kycia
   8   Nucleic acids   81
      Rose Ann Cattolico
   9   Protein determination by dye binding   91
      Gary Kochert
  10   Carbohydrate determination by the   95
      phenol–sulfuric acid method
      Gary Kochert

11 Glycerolipids and fatty acids of algae   99
    J. P. Williams
12 Carrageenans and agars   109
    J. S. Craigie and C. Leigh
13 β-1,3-Glucans in diatoms and brown   133
    seaweeds
    Sverre Myklestad
14 Alginic acid   143
    Bjørn Larsen
15 Fucoidan   151
    Bjørn Larsen
16 Phenolic compounds in brown and   157
    red algae
    Mark A. Ragan and J. S. Craigie
17 Brown seaweeds: analysis of ash,   181
    fiber, iodine, and mannitol
    Bjørn Larsen
18 Quantitation of the macromolecular   189
    components of microalgae
    Gary Kochert
19 ATP, ADP, and AMP determinations   197
    in water samples and algal cultures
    *David M. Karl and Osmund
    Holm-Hansen*

III ENZYMES   207

20 Ribulose-bis-phosphate carboxylase in   209
    Chlamydomonas
    S. S. Badour
21 Nitrate reductase in marine   217
    phytoplankton
    Richard W. Eppley
22 Threonine deaminase in marine   225
    microalgae
    N. J. Antia and R. S. Kripps
23 Phosphoenolpyruvate carboxylase of   233
    blue-green algae
    Brian Colman
24 Aldolases of multicellular marine   239
    algae
    Kazutosi Nisizawa
25 The glucosyltransferases of red algae   245
    Jerome F. Fredrick

26 Anion-activated adenosine   255
    triphosphatases
    Paul G. Falkowski
27 Extracellular acid phosphatase of   263
    Ochromonas danica
    N. J. Patni and S. Aaronson

IV PHYSIOLOGICAL AND
    BIOCHEMICAL PROCESSES   267

28 Determination of photosynthetic rates   269
    and ^{14}C photoassimilatory products of
    brown seaweeds
    Bruno P. Kremer
29 Polarographic measurements of   285
    photosynthesis and respiration
    Alan D. Jassby
30 Polarographic measurements of   297
    respiration following light-dark
    transitions
    Alan D. Jassby
31 Hill reaction and   305
    photophosphorylation with
    chloroplast preparations from
    Chlamydomonas reinhardi
    Stephen Lien
32 Respiratory assimilation of ^{14}C-labeled 317
    substrates by a microalga
    Keith E. Cooksey
33 Respiratory processes in mitochondria   329
    Neil G. Grant
34 Determination of protein synthesis   337
    during synchronous growth
    Dennis J. O'Kane and Raymond F. Jones
35 Nucleic acid synthesis   353
    Rose Ann Cattolico
36 Nitrogen fixation   363
    Timothy H. Mague
37 Uptake of organic substrates   379
    Johan A. Hellebust and Yi-Hung Lin
38 Release of organic substances   389
    C. Nalewajko

V NUTRIENTS   399

39 Nitrate uptake   401
Richard W. Eppley
40 Phosphate uptake   411
F. P. Healey
41 Sulfate uptake   419
*Monica Lik-Shing Tsang, Robert C.
Hodson, and Jerome A. Schiff*
42 Uptake of silicic acid   427
Farooq Azam and Sallie W. Chisholm
43 Uptake of micronutrients and toxic   435
metals
Anthony G. Davies

VI ION CONTENT AND TRANSPORT   447

44 Measurement of ionic content and   449
fluxes in microalgae
J. Barber
45 Ion transport and ion-stimulated   463
adenosine triphosphatases
Cornelius W. Sullivan

VII INHIBITORS   477

46 Inhibitors used in studies of algal   479
metabolism
S. S. Badour

VIII APPENDIX   489

List of suppliers   491

IX INDEXES

Subject   495
Author   500
Taxonomic   511

Contributors

Aaronson, S., Biology Department, Queens College, City University of New York, Flushing, New York 11367 (Chapter 27)

Antia, N. J., Pacific Environment Institute, Environment Canada, West Vancouver, British Columbia V7V 1N6, Canada (Chapter 22)

Azam, Farooq, Institute of Marine Resources, A-018, Scripps Institute of Oceanography, University of California, San Diego, La Jolla, California 92093 (Chapter 42)

Badour, S. S., Department of Botany, University of Manitoba, Winnipeg, Manitoba R3T 2N2, Canada (Chapters 20, 46)

Barber, J., Department of Botany, Imperial College of Science and Technology, Prince Consort Road, London, SW7 2BB, England (Chapter 44)

Buetow, D. E., Department of Physiology and Biophysics, 524 Burrill Hall, University of Illinois, Urbana, Illinois 61801 (Chapter 2)

Cattolico, Rose Ann, Botany Department AK-10, University of Washington, Seattle, Washington 98195 (Chapters 8, 35)

Chisholm, Sallie W., Department of Civil Engineering 48-425, Massachusetts Institute of Technology, 77 Massachusetts Avenue, Cambridge, Massachusetts 02139 (Chapter 42)

Colman, Brian, Department of Biology, York University, Downsview, Ontario M3J 1P3, Canada (Chapter 23)

Cooksey, Keith E., Rosenstiel School of Marine and Atmospheric Science, University of Miami, 4600 Rickenbacker Causeway, Miami, Florida 33149 (Chapter 32)

Craigie, J. S., Atlantic Regional Laboratory, National Research Council of Canada, 1411 Oxford Street, Halifax, Nova Scotia B3H 3Z1, Canada (Chapters 12, 16)

Davies, Anthony G., Marine Biological Association, The Laboratory, Citadel Hill, Plymouth PL1 2PB, England (Chapter 43)

Eppley, Richard W., Institute of Marine Resources, A-018, Scripps Institute of Oceanography, University of California, San Diego, La Jolla, California 92093 (Chapters 21, 39)

Falkowski, Paul G., Oceanographic Sciences Division, Brookhaven National Laboratory, Upton, New York 11973 (Chapter 26)

Fredrick, Jerome F., The Research Laboratories, Dodge Chemical Company, 3425 Post Road, Bronx, New York 10469 (Chapter 25)

Grant, Neil G., Department of Biology, The City College of the City University of New York, New York, New York 10031 (Chapters 3, 33)

Healey, F. P., Department of the Environment, Fisheries and Marine Service, Freshwater Institute, 501 University Crescent, Winnipeg, Manitoba R3T 2N6, Canada (Chapter 40)

Hellebust, Johan A., Department of Botany, University of Toronto, Toronto, Ontario M5S 1A1, Canada (Chapter 37)

Hodson, Robert C., Department of Biology, University of Delaware, Newark, Delaware 19711 (Chapter 41)

Holm-Hansen, Osmund, Institute of Marine Resources, A-018, Scripps Institute of Oceanography, University of California, San Diego, La Jolla, California 92093 (Chapter 19)

Jassby, Alan D., Lawrence Berkeley Laboratory, University of California, Berkeley, California 94720 (Chapters 29, 30)

Jensen, Arne, Institute of Marine Biochemistry, University of Trondheim, N-7034 Trondheim-NTH, Norway (Chapter 6)

Jones, Raymond F., Division of Biological Sciences, State University of New York at Stony Brook, Stony Brook, New York 11794 (Chapter 34)

Karl, David M., Institute of Marine Resources, A-018, Scripps Institute of Oceanography, University of California, San Diego, La Jolla, California 92093 (Chapter 19)

Kochert, Gary, Botany Department, University of Georgia, Athens, Georgia 30602 (Chapters 9, 10, 18)

Kremer, Bruno P., Botanisches Institut der Universität Köln, Gyrhofstrasse 15, D-5000 Köln 41, Germany (Chapter 28)

Kripps, R. S., Division of Human Nutrition, School of Home Economics, University of British Columbia, Vancouver, British Columbia V6T 1W5, Canada (Chapter 22)

Kycia, J. Helen, Biology Department, Brookhaven National Laboratory, Upton, New York 11973 (Chapter 7)

Larsen, Bjørn, Institute of Marine Biochemistry, University of Trondheim, N-7034 Trondheim-NTH, Norway (Chapters 14, 15, 17)

Leigh, C., Atlantic Regional Laboratory, National Research Council of Canada, 1411 Oxford Street, Halifax, Nova Scotia B3H 3Z1, Canada (Chapter 12)

Lien, Stephen, Department of Plant Sciences, Indiana University, Bloomington, Indiana 47401 (Chapter 31)

Lin, Yi-Hung, Department of Botany, University of Toronto, Toronto, Ontario M5S 1A1, Canada (Chapter 37)

Mague, Timothy H., Bigelow Laboratory for Ocean Sciences, West Boothbay Harbor, Maine 04575 (Chapter 36)

Myklestad, Sverre, Institute of Marine Biochemistry, University of Trondheim, N-7034 Trondheim-NTH, Norway (Chapter 13)

Nalewajko, C., Scarborough College, University of Toronto, West Hill, Ontario M1C 1A4, Canada (Chapter 38)

Nisizawa, Kazutosi, Department of Fisheries, College of Agriculture and Veterinary Medicine, Nihon University, Shimouma-3, Setagaya-ku, Tokyo, 154 Japan (Chapter 24)

O'Kane, Dennis J., Division of Biological Sciences, State University of New York at Stony Brook, Stony Brook, New York 11794 (Chapter 34)

Patni, N. J., Biology Department, Queens College, City University of New York, Flushing, New York 11367 (Chapter 27)

Price, C. A., Waksman Institute of Microbiology, Rutgers University, New Brunswick, New Jersey 08903 (Chapter 1)

Ragan, Mark A., Institute of Marine Biochemistry, University of Trondheim, N-7034 Trondheim-NTH, Norway (Chapter 16)

Schiff, Jerome A., Institute for Photobiology of Cells and Organelles, Brandeis University, Waltham, Massachusetts 02154 (Chapter 41)

Siegelman, Harold, W., Biology Department, Brookhaven National Laboratory, Upton, New York 11973 (Chapter 7)

Stabenau, Helmut, Pflanzenphysiologisches Institut der Universität Göttingen, Untere Karspüle 2, 3400 Göttingen, Germany (Chapter 4)

Sullivan, Cornelius W., Department of Biological Sciences, University of Southern California, Los Angeles, California 90007 (Chapters 5, 45)

Tsang, Monica Lik-Shing, Institute for Photobiology of Cells and Organelles, Brandeis University, Waltham, Massachusetts 02154 (Chapter 41)

Williams, J. P., Department of Botany, University of Toronto, Toronto, Ontario M5S 1A1, Canada (Chapter 11)

Editors' preface

In 1967 a special editorial committee was struck by the Phycological Society of America to consider the publication of a source book on methods in experimental phycology. The editorial committee eventually proposed the publication of four volumes to include methods from four main subdivisions of experimental phycology: culture techniques and growth measurements, cytological methods, biochemical and physiological methods, and field-oriented methods. The first volume (although not formally called Volume I) of the proposed *Handbook of Phycological Methods – Culture Methods and Growth Measurements* – was edited by Professor Janet R. Stein, University of British Columbia, and published by Cambridge University Press in 1973. It was received by experimental phycologists as a valuable source of techniques.

The editing of the present volume – Volume II, *Physiological and Biochemical Methods* – was undertaken in 1975 and has progressed smoothly, although not as rapidly as we had hoped. We gratefully acknowledge the practical and timely advice of Dr. Janet Stein. Her cooperation together with that of other members of the editorial committee for this volume lightened our editorial task considerably.

We were pleasantly surprised by the alacrity of the response of phycologists from whom we solicited contributions to this volume. Our authors made sincere efforts to meet the relatively early deadline and for this we are most appreciative. Several excellent phycologists volunteered to contribute additional chapters on interesting and useful techniques. We regret that the majority of these could not be included owing to severe space limitations.

Cambridge University Press has been most cooperative during the entire editorial process. We particularly thank Mr. Kenneth I. Werner for this help.

University of Toronto　　　　　　　　　　　　　　Johan A. Hellebust
Toronto
Atlantic Regional Laboratory　　　　　　　　　　　　J. S. Craigie
National Research Council of Canada
Halifax

Editorial Committee

J. S. Craigie, Atlantic Regional Laboratory, National Research Council of Canada, Halifax, Nova Scotia, Canada.

Richard W. Eppley, Institute of Marine Resources, University of California, La Jolla, California

Johan A. Hellebust, University of Toronto, Toronto, Ontario, Canada

Osmund Holm-Hansen, Institute of Marine Resources, University of California, La Jolla, California

Janet R. Stein, University of British Columbia, Vancouver, British Columbia, Canada

Introduction

We have attempted to present as many useful methods as possible for physiological and biochemical investigations with algae. In view of the extreme diversity of the algae, ranging from blue-green bacteria-like organisms to large complex seaweeds, it is clearly impossible to include in a single volume methods applicable to all situations. We have therefore selected methods that cover a broad range of experimental procedures and as wide a variety of algal types as possible. Most of these are described in sufficient detail to be useful to scientists with relatively little previous experience in experimental phycology. We realize that methods which apply to all types of algae or to algae in different physiological conditions are not yet available. In such cases the authors have described the limitations of their techniques and have included brief discussions of how procedures may be modified to suit different algae or other conditions. References to more advanced or specialized techniques, or to methods for other algal species, are frequently included in the various chapters. The brief list of references at the end of this introduction will provide the advanced student with information and special techniques, many of which may be applied in phycological research.

Names of the suppliers of most of the specialized materials and equipment necessary for the various methods are cited, and the reader should consult the appendix for addresses. Sources of equipment and materials are published annually in the United States in *Science* (American Association for the Advancement of Science, 1515 Massachusetts Ave., N.W., Washington, D.C. 20005) and in Canada in *Research and Development* (Maclean Hunter, 418 University Ave., Toronto 101, Ontario M5W 1A7) and *Laboratory Products News* (Southam Business Publications Ltd., 1450 Don Mills Rd., Don Mills, Ontario M3B 2X7).

In most cases the sources of the algae described in this handbook are given. The clone number of the species is provided when known, followed by the acronym for the algal culture collection from which the clone was obtained (e.g., UTEX, see Appendix: List of Suppliers).

We hope this handbook will find a wide variety of users. It should serve as a useful source of techniques for graduate students and scientists who may occasionally use algae as experimental material, as well as for specialists in experimental phycology. Many of the procedures described in this volume are suitable for courses in experimental phycology. Several chapters have been included for those interested in the industrially important chemical components of seaweeds. Although field techniques will be treated in a subsequent handbook, the present volume includes several chapters of interest to scientists concerned with field investigations as well as with the effects of toxic metals on algae.

General references

Bergmeyer, H. U. 1974. *Methods of Enzymatic Analysis,* 2nd English ed., vols. 1–4. Academic Press, New York.

Colowick, S. P., and Kaplan, N. O. (eds.). 1955–75. *Methods in Enzymology,* vols. 1–43. Academic Press, New York. (Especially vols. 23 and 24 on photosynthesis and nitrogen fixation are of interest as they deal extensively with algal material.)

Dawson, R. M. C., Elliot, D. C., Elliot, W. H., and Jones, K. M. (eds.). 1969. *Data for Biochemical Research,* 2nd ed. Oxford University Press, London. 654 pp.

Herbert, D., Phipps, P. J., and Strange, R. E. 1971. Chemical analysis of microbial cells. *Methods Microbiol.* 5B, 209–344.

Lowry, O. H., and Passonneau, J. V. 1972. *A Flexible System of Enzymatic Analysis.* Academic Press, New York. 291 pp.

Prescott, D. M. (ed.). 1975. Yeast cells. *Methods Cell Biol.* 12, 1–395.

Umbreit, W. W., Burris, R. H., and Stauffer, J. F. 1972. *Manometric and Biochemical Techniques,* 5th ed. Burgess, Minneapolis. 387 pp.

Whistler, R. L., and BeMiller, J. N. (eds.). 1964–76. *Methods in Carbohydrate Chemistry,* vols. 4–7. Academic Press, New York.

Whistler, R. L., and Wolfrom, M. L. (eds.). 1962–3. *Methods in Carbohydrate Chemistry,* vols. 1–3. Academic Press, New York.

Section I

Isolation of organelles and membranes

1: Chloroplasts from Euglena gracilis

C. A. PRICE

Waksman Institute of Microbiology,
Rutgers University, New Brunswick, New Jersey 08903

CONTENTS

I	**Introduction**	*page*	**6**
II	**Test organism, materials, and equipment**		**7**
	A. Test organism		7
	B. Materials		7
	C. Equipment		8
III	**Procedures**		**8**
	A. Preparation of crude chloroplasts		8
	B. Density gradient centrifugation		8
IV	**Properties of isolated chloroplasts**		**9**
	A. Purity and integrity		9
	B. Composition		10
	C. Photosynthesis		11
	D. Protein synthesis		11
V	**Acknowledgments**		**13**
VI	**References**		**13**

I. Introduction

Students of chloroplast development who hope to exploit modern microbiological methods are faced with a dilemma in choosing a test organism: *Chlamydomonas reinhardi* has sex, but does not yield intact isolated chloroplasts; *Euglena gracilis* seems devoid of sex, but one can recover its chloroplasts in an intact and at least superficially functional state.

A successful isolation of chloroplasts should yield plastids that are morphologically intact, including the envelope and all the stromal contents; physiologically intact, as shown by normal rates of CO_2 fixation and CO_2-dependent O_2 evolution; and free of other subcellular components. In general, density gradient centrifugation is the method of choice for the purification of subcellular organelles, and this has been true for studies requiring chloroplasts scrupulously free of other components of the cell. Nonetheless, students of photosynthesis or protein synthesis have been obliged to accept the lower level of purity offered by differential centrifugation. Chloroplasts obtained from density gradients were inactive or only feebly active in CO_2 fixation or in the incorporation of amino acids into polypeptides.

For a number of years *Euglena* chloroplasts were obtained by centrifugal flotation in 2 M sucrose (Eisenstadt and Brawerman 1963). Although Leech (1964) had shown that morphologically intact chloroplasts from higher plants could be obtained from tissue breis of higher plants by isopyknic sedimentation in sucrose gradients, exposure of *Euglena* chloroplasts to high osmotic pressures invariably resulted in the loss of the chloroplast envelope and, consequently, most of the stroma. Although the resulting systems of thylakoids obtained by flotation in sucrose were not strictly chloroplasts, they did contain, a fortiori, most of the chloroplast DNA and relatively little nuclear DNA. Flotation thus provided a convenient method for the enrichment of chloroplast DNA.

We achieved the first true isolation of chloroplasts from *Euglena* by employing rate zonal sedimentation in isosmotic gradients of Ficoll (Vasconcelos et al. 1971). The plastids were morphologically intact, but did not photosynthesize and manifested other physiological activi-

[6]

ties only on even-numbered days of the week. The high viscosity and high cost of Ficoll, the low capacity of rate zonal sedimentation, and the erratic behavior of the chloroplasts obtained prompted us to seek improved methods.

We subsequently found that intact and fully functional chloroplasts could be obtained by isopyknic sedimentation of spinach breis in gradients of the silica sol Ludox AM (Morgenthaler et al. 1974, 1975). This recipe was not immediately applicable to *Euglena,* but was adapted by combination with the gradient conditions employed in the earlier Ficoll gradients (Salisbury et al. 1975). This is the method I shall describe here.

II. Test organism, materials, and equipment

A. Test organism

Euglena gracilis (Klebs) strain Z (Pringsheim) UTEX 753 was grown photoheterotrophically on a modified Hutner medium (Vasconcelos et al. 1971).

B. Materials

1. Breaking mix. Sucrose, 0.15 M; sorbitol, 0.15 M; Ficoll (Pharmacia), 1% w/v; polyvinylsulfate, 2 μg ml^{-1}; NaCl, 15 mM; mercaptoethanol, 5 mM; HEPES-NaOH buffer, 5 mM, pH 6.8.

2. Ludox AM. A commercial silicate sol (Du Pont de Nemours). It is essential that Ludox be purified before exposing chloroplasts to it. We adjust the pH to neutrality by the slow addition with stirring of the hydrogen form of a sulfonic acid type of cation-exchange resin, such as Dowex 50W-X8. The resin is easily removed by filtration. The sol is then passed slowly through a column of granular, activated charcoal. We employ a column 50 $\times$ 2.5 cm with a flow rate of 5 ml min^{-1} (Price and Dowling 1977).

3. Ludox AM-PEG-BSA. Immediately prior to use, the Ludox is made to contain 10% w/v polyethylene glycol, PEG 6000 (Carbowax, Union Carbide) and 1% w/v bovine serum albumin (BSA, from Sigma).

4. Starting solution. This solution contains the following components (final concentration): Ludox AM-PEG-BSA, 10% v/v; sucrose, 0.15 M; sorbitol, 0.15 M; Ficoll, 1% w/v; polyvinylsulfate, 2 μg ml^{-1}; NaCl, 15 mM; glutathione, 5 mM; HEPES-NaOH buffer, 5 mM, pH 6.8.

5. Final solution. The same composition as the starting solution (II.B.4), except that the final solution contains 70% v/v Ludox AM-PEG-BSA.

6. Gradients. Linear, 10–70% v/v in Ludox AM-PEG-BSA prepared from the starting (II.B.4) and final (II.B.5) solutions. The gradients may be prepared in a two-cylinder gradient former (e.g., IEC No. 3651, Damon/IEC).

C. Equipment

Because our separation of chloroplasts is by isopyknic sedimentation, any swinging bucket rotor capable of ca. 5000 rpm and containing 10 –30 ml of gradient is adequate for samples containing 4 mg of chlorophyll or less. For processing tens of milligrams of chlorophyll, the CF-6, RJ, or K types of continuous-flow rotors (Damon/IEC) are suitable (Price et al. 1973; Morgenthaler et al. 1975). Most other zonal rotors are not suitable because chloroplast envelopes are stripped upon passage through the narrow bores of the seals.

III. Procedures

A. Preparation of crude chloroplasts

Exponentially growing *Euglena* cells are harvested by centrifugation. All subsequent operations are carried out at 0–4 C as rapidly as possible. The cells are then washed twice in deionized H_2O, once in the breaking mix (II.B.1), and the cells weighed. The cells are then resuspended in 2 ml of breaking mix per gram wet weight and disrupted in a French pressure cell (American Instrument Co.) at ca. 100 kg cm^{-2}, and the brei collected in a chilled vessel containing 4 volumes of breaking mix.

The cell brei is clarified by centrifugation at 1000 rpm for 1 min to pellet whole cells and aggregates. The resulting supernatant is centrifuged at 3000 rpm for 3 min and the green pellet resuspended by gentle aspiration to a concentration of ca. 1 mg chlorophyll per milliliter. The crude chloroplasts are then immediately layered onto waiting gradients.

Alternatively the clarified brei can be centrifuged directly on gradients. In either case the chloroplasts must not be exposed to the other components of the cell brei any longer than necessary.

B. Density gradient centrifugation

The gradients are centrifuged at 7000 rpm for approximately 15 min. We prefer to use a meter that measures the $\int \omega^2 dt$. Continue the centrifugation for 500 $rad^2 \cdot \mu sec^{-1}$.

Centrifugation resolves the chlorophyll of the sample into two discrete zones. Thylakoids are in the upper zone; intact chloroplasts in the lower, more dense zone (Fig. 1-1).

Fig. 1-1. Separation of *Euglena* chloroplasts by density gradient centrifugation. The photograph shows separation of intact chloroplasts (*I*, lower zone) from stripped thylakoids (*s*, upper zone) following sedimentation at 7000 rpm for 15 min in an SB287 rotor. Starting material was (*left*) clarified cell brei or (*right*) crude chloroplasts.

The chloroplasts may be recovered nonquantitatively by withdrawing the lower zone through a bent pipette. For quantitative recovery the gradient should be unloaded in fractions. Collecting from the bottom has the advantage that the intact chloroplasts are recovered before the tubing has a chance to become contaminated with other fractions of the gradient. However, fractionation of gradients from the top usually results in better resolution.

The chloroplasts can be recovered from the gradient by diluting the fractions with 2 volumes of breaking mix and centrifuging at 3000 rpm for 5–10 min.

IV. Properties of isolated chloroplasts

A. Purity and integrity

Electron photomicrographs of chloroplasts isolated as described show the organelles with their outer envelopes and stromal contents intact (Fig. 1-2). The appearance of the particles is identical with that of

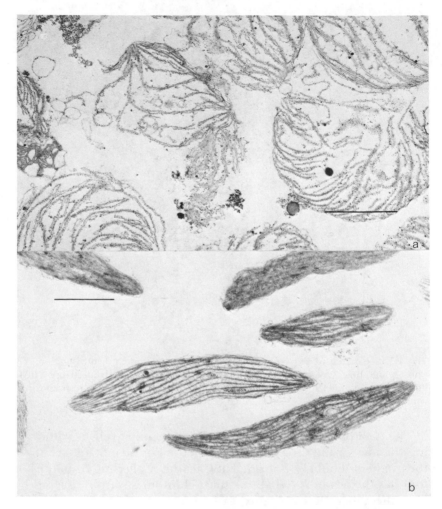

Fig. 1-2. Electron photomicrographs of *Euglena* chloroplasts separated in silica sol gradients. *a*. Thylakoid zone. *b*. Intact chloroplasts. Calibration bar is 1 μm. (From Salisbury et al. 1975.)

chloroplasts in situ. No other kinds of particles are visible in the electron photomicrographs.

B. *Composition*

The gradient zone containing intact chloroplasts is rich in ribulose-bis-phosphate carboxylase (RuBPC) and indeed is the only sedimenting zone of the enzyme (Fig. 1-3).

The polypeptides of various subfractions of the chloroplast have

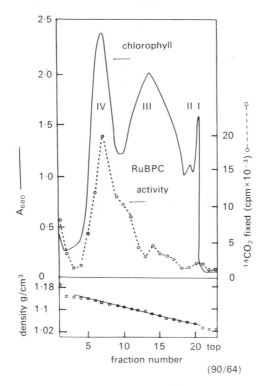

Fig. 1-3. Sedimentation profile of *Euglena* cell brei. The gradient was collected from bottom to top and fractions analyzed for RuBPC activity and chlorophyll. The peak of RuBPC coincides with the position of the intact plastids (IV). (From Salisbury et al. 1975.)

been analyzed (Vasconcelos et al. 1976) and correspond qualitatively to those observed in chloroplasts from higher plants.

C. Photosynthesis

Although the chloroplasts contain RuBPC and show Hill reaction activity, the critical capacity for CO_2-dependent O_2 evolution or fixation of CO_2 at physiologically significant rates has not yet been demonstrated.

D. Protein synthesis

Suspensions of the isolated chloroplasts are capable of vigorous incorporation of exogenous amino acids into polypeptides and proteins (Vasconcelos 1976). Gel electrophoresis of two of the chloroplast subfractions is shown in Fig. 1-4. Analysis of the stromal fraction shows that most of the label has gone into the large subunit of RuBPC. The

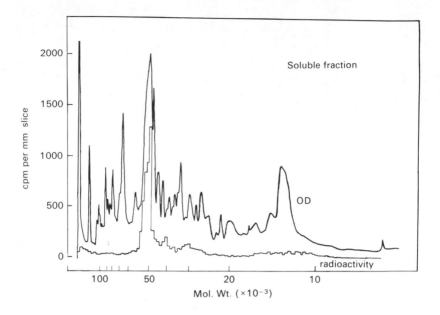

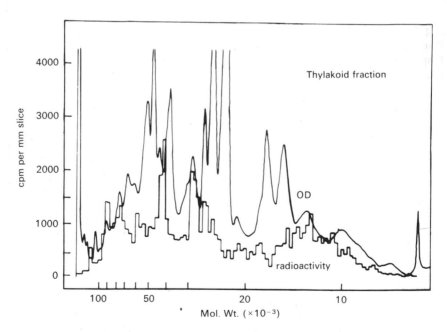

Fig. 1-4. Incorporation of ^{3}H-leucine into stromal proteins (*upper figure*) and thylakoids (*lower figure*) of isolated *Euglena* chloroplasts. Separation of polypeptides was obtained by electrophoresis on SDS-polyacrylamide gels. Solid lines show absorbance of Coomassie blue stain; histograms show radioactivity in 1-mm slices obtained from gels. (From Vasconcelos 1976.)

pattern of incorporation into polypeptides of other fractions is also similar to that of higher plant chloroplasts.

V. Acknowledgments

This work was supported in part by grants from the National Institutes of Health (HD-05602) and a Biomedical Research Support Grant to Rutgers University.

VI. References

Eisenstadt, J., and Brawerman, G. 1963. The incorporation of amino acids into the protein of chloroplasts and chloroplast ribosomes of *Euglena gracilis. Biochim. Biophys. Acta* 76, 319–21.

Leech, R. M. 1964. The isolation of structurally intact chloroplasts. *Biochim. Biophys. Acta* 79, 637–9.

Morgenthaler, J.-J., Marsden, P. F., and Price, C. A. 1975. Factors affecting the separation of photosynthetically competent chloroplasts in gradients of silica sols. *Arch. Biochem. Biophys.* 168, 289–301.

Morgenthaler, J.-J, Price, C. A., Robinson, J. M., and Gibbs, M. 1974. Photosynthetic activity of spinach chloroplasts after isopycnic centrifugation in gradients of silica. *Plant Physiol.* 54, 532–4.

Price, C. A., Breden, E. N., and Vasconcelos, A. C. 1973. Isolation of intact spinach chloroplasts in the CF-6 continuous-flow zonal rotor: Implications for membrane-bound organelles. *Anal. Biochem.* 54, 239–46.

Price, C. A., and Dowling, E. L. 1977. On the purification of the silica sol Ludox AM. *Anal. Biochem.* 82, 243–5.

Salisbury, J. L., Vasconcelos, A. C., and Floyd, G. L. 1975. Isolation of intact chloroplasts of *Euglena gracilis* by isopycnic sedimentation in gradients of silica. *Plant Physiol.* 56, 399–403.

Vasconcelos, A. C. 1976. Synthesis of proteins by isolated *Euglena gracilis* chloroplasts. *Plant Physiol.* 58, 719–21.

Vasconcelos, A. C., Mendiola-Morgenthaler, L. R., Floyd, G. L., and Salisbury, J. L. 1976. Fractionation and analysis of polypeptides of *Euglena gracilis* chloroplasts. *Plant Physiol.* 58, 87–90.

Vasconcelos, A., Pollack, M., Mendiola, L. R., Hoffmann, H.-P., Brown, D. H., and Price, C. A. 1971. Isolation of intact chloroplasts from *Euglena gracilis* by zonal centrifugation. *Plant Physiol.* 47, 217–21.

2: Nuclei and chromatin from Euglena gracilis

D. E. BUETOW

Department of Physiology and Biophysics,
University of Illinois, Urbana, Illinois 61801

CONTENTS

I	Introduction	*page* 16
II	Materials and test organism	17
	A. Equipment	17
	B. Test organism	17
	C. Growth medium	17
III	Methods	17
	A. Isolation of nuclei	17
	B. Isolation and fractionation of chromatin	19
IV	Sample data	19
V	Discussion	21
VI	Acknowledgments	23
VII	References	23

I. Introduction

Any technique for isolating intact, functional subcellular particles from *Euglena* encounters a major problem in the breaking of the tough pellicle without, at the same time, damaging the particles. Leedale (1958) isolated the nucleus from individual *Euglena* cells following the softening of the pellicle by overnight incubation in a saturated solution of pepsin. Parenti et al. (1969) reported a mass isolation of *Euglena* nuclei achieved by initially weakening cell pellicles by freezing and thawing the cells and treating them with Triton or by exposing the cells to pancreatic protease, but gave no estimates of final yields of nuclei and did not present any microscopic evidence for purity or intactness of the final nuclear preparations. McLennan and Keir (1975) isolated nuclei from *Euglena* in 80% yield by initially homogenizing the cells with glass beads on a vortex mixer. This latter method works best with a relatively small number of cells (60×10^6 in a 2-ml sample), and the nuclei are contaminated with pellicular fragments and paramylon granules. Active DNA polymerase, however, can be isolated from these nuclear preparations. Aprille and Buetow (1973) isolated *Euglena* nuclei in a final average yield of 29% by first weakening the cell pellicle with Triton followed by long incubation of the cells in saturated pepsin at a low pH. This method yields nuclei that are free of significant cytoplasmic contamination and are capable of incorporating amino acids (Aprille and Buetow 1974); however, these nuclei are not well preserved ultrastructurally (Aprille and Buetow 1973).

Several other methods that either did not lyse *Euglena* or did not yield pure intact nuclei were tried in this laboratory. Procedures that did not break the cells included low pressures (less than 3200 lb in.$^{-2}$, *Euglena* in 0.25 M sucrose, 10% dextran) in a French pressure cell (American Instrument Co.), hypotonic shock in distilled water, and homogenization in a Potter-Elvehjem homogenizer in up to 10% Triton X-100 solutions. Procedures that lysed the cells as well as most of the nuclei included pressures higher than 3200 lb in.$^{-2}$ in the French pressure cell and shearing in a Waring Blendor (Aprille and Buetow

1973). A method combining freezing and thawing, citric acid (pH 2.6), and sonication yielded a fraction enriched in nuclei, but not pure (Lynch 1973; Lynch and Buetow 1975). Also, as is the case in higher cell nuclei isolated with citric acid procedures (Busch and Smetana 1970), these latter *Euglena* nuclei show an altered ultrastructure. Other attempts that failed to preserve nuclear structure and reduce contamination of the final nuclear preparation included the above citric acid method with the cell homogenate sedimented through 2.5 M sucrose, freezing and thawing of the cells followed by sonication or homogenization in 0.25 M sucrose, and initial swelling of the cells in hypotonic solution (no sucrose) followed by freezing and thawing and sonication (Lynch 1973; Lynch and Buetow 1975).

The method presented in this chapter yields ultrastructurally intact, enzymatically active nuclei from *Euglena* (Lynch and Buetow 1975). A method is also presented for the isolation and fractionation of chromatin from these nuclei (Lynch et al. 1975).

II. Materials and test organism

A. Equipment

Sharples Super Centrifuge (The Chemical Rubber Co.); Bronwill Bio-sonik III sonicator (Bronwill Scientific); 12-l carboys.

B. Test organism

Euglena gracilis var. *bacillaris* (strain SM-L1), a streptomycin-bleached strain UTEX 945.

C. Growth medium

Cultures are grown in the dark at 27 C in 12-l carboys containing 6 l of defined medium (Buetow 1965) with 0.2 M ethanol as carbon source (Buetow and Padilla 1963). When cultures reach densities of 3×10^5 to 8×10^5 cells per milliliter (midlogarithmic phase of growth), they are harvested at 2 C with a Sharples Super Centrifuge. The concentrated cells are washed once by resuspension in 0.9% NaCl with centrifugation at $1000 \times g$ for 10 min.

III. Methods

A. Isolation of nuclei

Steps 2–7 are done at 0–2 C. MES is 2-[N-morpholino]ethane sulfonic acid (Sigma). Yields of isolated nuclei are estimated by staining them with 2% methyl green in 1% acetic acid (Aprille and Buetow

1973) and counting them with a hemacytometer or by measuring the DNA content of the final pellet (Lynch and Buetow 1975).

1. Freeze cell pellet in a dry ice–acetone bath and thaw it under running tap water at 20 C.

2. Wash cells once by resuspension in buffer containing 50 mM MES (pH 5.55), 7 mM $MgCl_2$, and 1.25% Triton X-100. Centrifuge at 1000 × g for 10 min.

3. Resuspend pellet in 10 volumes buffer as step 2 and incubate for 30 min at 0–2 C.

4. Sonicate the incubation mixture for 20–40 sec with 10-sec bursts at 238 W in.$^{-2}$ until 75–80% of the cells are broken. Sonication is done with a Bronwill Biosonik III sonicator equipped with a 1.9-cm stainless-steel probe. The sample for sonication is divided into 10-ml aliquots and sonicated for 10-sec bursts in a 40-ml plastic beaker cooled in an ice bath. Following each 10-sec burst, the probe and solution are allowed to cool for 20 sec.

5. Layer the sonicated mixture over 10% dextran (40 000 MW, Pharmacia) in buffer containing 50 mM MES (pH 5.55), 7 mM $MgCl_2$, and 0.5% Triton X-100. Centrifuge at 1000 × g for 15 min. The resultant crude nuclear pellet contains nuclei, paramylon, and occasional unbroken cells.

6. Resuspend pellet by homogenization in 1 volume buffer containing 50 mM MES (pH 5.55), 7 mM $MgCl_2$, 1.25% Triton X-100, and 0.1 M freshly made sodium metabisulfite ($Na_2S_2O_5$). Layer over 4 volumes 2.4 M sucrose in the same buffer. Mix top one-half of gradient with a glass rod. Centrifuge at 21 000 × g for 1 h.

7. Collect interface material and add to it an equal volume of buffer containing 50 mM MES (pH 5.55), 7 mM $MgCl_2$, and 1.25% Triton X-100. Mix gently with a glass rod. Layer over 2.4 M sucrose in the same buffer. Stir top one-half of gradient with a glass rod. Centrifuge at 21 000 × g for 1 h. Pellet contains nuclei in a yield of 25–37%.

The only contaminant in the final nuclear pellet consists of occasional paramylon granules. Pure intact nuclei are not isolated at pH 2.6 or at 7.4–8.5. Over the pH range 5.5–7.0 with MES buffer, good cell breakage is combined with little apparent nuclear damage only at pH 5.55. Sodium citrate (100 mM) as buffer at pH 5.5 causes more aggregation of cellular debris than does MES buffer. Sonication (step 4) is continued until 75–80% of the cells are broken as monitored by light microscopy. Further sonication produces nuclear damage. The density of nuclei and undamaged whole cells is similar at step 5. The addition of 0.1 M $Na_2S_2O_5$ (but not sodium bisulfite) to the buffer (step 6), however, results in a decreased density of the nuclei, enabling them to be separated from the whole cells. The change in nuclear density is readily reversible, as when nuclei are subsequently suspended in buffer lacking metabisulfite, they then sediment through 2.4 M sucrose (step 7).

B. Isolation and fractionation of chromatin

Whole chromatin can be isolated from either the final purified nuclear pellet (III.A.7) or the crude nuclear pellet obtained in III.A.5. Chromatin is fractionated (Lynch et al. 1975) by sonication followed by differential centrifugation (Leake et al. 1972). All steps are done at 0–2 C.

1. Swell nuclei by suspension in 5 volumes buffer containing 100 mM TRIS-HCl (pH 7.4).

2. Sonicate swollen nuclei with a Bronwill Biosonik III sonicator for a total of 60 sec, as in III.A.4.

3. Centrifuge the sonicated nuclei at 500 × g for 10 min to sediment paramylon granules and large debris; the resulting supernatant contains whole chromatin in yields of 73–84% on a DNA basis.

4. Centrifuge supernatant at 3000 × g for 20 min to obtain a pellet containing mainly heterochromatin. Free this pellet of any small amounts of trapped euchromatin by resuspending it in 5 volumes of 100 mM TRIS-HCl (pH 7.4) by means of a tight-fitting Dounce homogenizer (Wheaton Scientific) and by centrifuging the mixture again at 3000 × g for 20 min. The resultant pellet is the heterochromatin fraction.

5. Centrifuge the supernatant from step 3 at 75 000 × g for 1 h in an SB-206 rotor in an International B-60 centrifuge (International Equipment Co.); the pellet contains a mixture of heterochromatin and euchromatin ("interchromatin" fraction), and the supernatant contains euchromatin.

The use of TRIS buffer in the above technique permits the measurement of RNA polymerase activity in whole chromatin and its subfractions. Continued centrifugation of nuclei or chromatin in TRIS buffer, however, causes artificial aggregation. For other purposes, it may be necessary to avoid such aggregation. If so, a pH 7.4 buffer containing 0.25 M sucrose and 10 mM TES (N-tris[hydroxymethyl]-methyl-2-aminomethane sulfonic acid; Sigma) can be substituted for the TRIS buffer in steps 1 and 4 (Lynch et al. 1975).

IV. Sample data

An electron micrograph of a typical isolated interphase nucleus is shown in Fig. 2-1. Isolated nuclei appear intact and contain condensed chromosomes and a distinct nucleolus as typically found in the intact interphase cell (Fig. 2-2). The DNA and protein contents of the

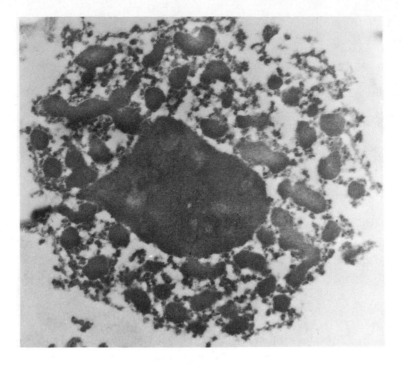

Fig. 2-1. Electron micrograph of isolated nucleus of *Euglena gracilis*. × 17 300. (From Lynch and Buetow 1975.)

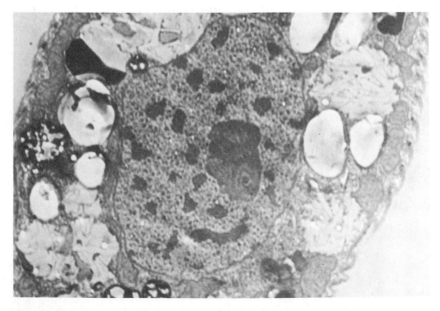

Fig. 2-2. Electron micrograph of nucleus of *Euglena gracilis* in situ. × 10 900. (From Lynch and Buetow 1975.)

Table 2-1. *DNA and protein content of Euglena nuclear fractions (pg/nucleus)*

Extract	DNA	Protein
Whole nucleus	4.4	19.8
0.14 M NaCl	0.1	1.2
0.10 M TRIS (pH 7.6)	<0.1	1.7
0.25 N HCl	0.15	4.4
Residual	4.1	12.5

Note: Isolated nuclei were extracted successively with NaCl, TRIS, and HCl. At each extraction step, nuclei were stirred gently for 30 min at 0–2 C and then centrifuged at 0–2 C at 6000 × *g* for 15 min.
Source: Lynch and Buetow (1975).

Table 2-2. *DNA content and RNA polymerase activity of Euglena chromatin fractions*

Fraction	Designation	Percent of total chromatin DNA	RNA polymerase activity[a]
Pellet (3000 × *g*)	Heterochromatin	60	0.43
Pellet (75 000 × *g*)	Interchromatin (hetcroeuchromatin)	26	0.66
Supernatant (75 000 × *g*)	Euchromatin	14	4.45

[a] Picomoles of CTP incorporated (100 μg DNA)$^{-1}$ (10 min)$^{-1}$; enzyme assay and table as in Lynch et al. (1975).

average *Euglena* nucleus are given in Table 2-1. The mass ratio of acid-soluble protein to DNA is 1.0, as also found in higher cell nuclei (Bonner et al. 1968). Even though the chromosomes always appear condensed (Figs. 2-1, 2-2), three distinct chromatin fractions with different degrees of condensation can be isolated from these nuclei (Fig. 2-3). As expected, the euchromatin fraction contains most of the total endogenous RNA polymerase activity (Table 2-2).

V. Discussion

The nuclear isolation method (III.A) also works for a heat-bleached mutant of *E. gracilis* strain Z grown as in II.C if sonication is increased to 50 sec, but yields are low. The method appears to depend in part

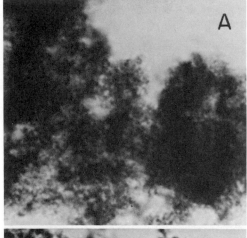

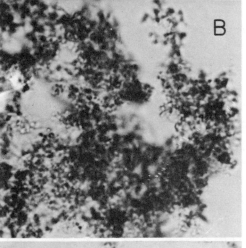

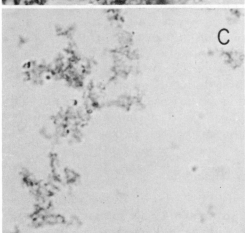

Fig. 2-3. Chromatin fractions were stained with 1% toluidine blue and photographed by means of a light microscope equipped with a Zeiss camera. × 1550. (Lynch *et al.* 1975.)

A heterochromatin

B interchromatin

C euchromatin

on the density of the cells and/or the nuclei and, thus, may need to be modified for various strains of *E. gracilis*.

VI. Acknowledgments

These studies were supported by NIH grants HD 03163 and GM 19641.

VII. References

Aprille, J. R., and Buetow, D. E. 1973. A method for isolating nuclei from *Euglena gracilis. Arch. Mikrobiol.* 89, 355–62.

Aprille, J. R., and Buetow, D. E. 1974. Incorporation of labelled amino acids by nuclei isolated from *Euglena gracilis. Arch. Mikrobiol.* 97, 195–201.

Bonner, J., Dahmus, M. E., Fambrough, D., Huang, R. C., Marushige, K., and Tuan, D. Y. H. 1968. The biology of isolated chromatin. *Science* 159, 47–56.

Buetow, D. E. 1965. Growth, survival and biochemical alteration of *Euglena gracilis* in medium limited in sulfur. *J. Cell. Comp. Physiol.* 66, 235–42.

Buetow, D. E., and Padilla, G. M. 1963. Growth of *Astasia longa* on ethanol. I. Effects of ethanol on generation time, population density and biochemical profile. *J. Protozool.* 10, 121–3.

Busch, H., and Smetana, K. 1970. *The Nucleolus.* Academic Press, New York. 626 pp.

Leake, R. E., Trench, M. E., and Barry, J. M. 1972. Effect of cations on the condensation of hen erythrocyte nuclei and its relation to gene activation. *Exp. Cell Res.* 71, 17–26.

Leedale, G. F. 1958. Nuclear structure and mitosis in the Euglenineae. *Arch. Mikrobiol.* 32, 32–64.

Lynch, M. J. 1973. The isolation and partial characterization of nuclei, chromatin and histones from *Euglena gracilis.* Ph.D. thesis, University of Illinois, Urbana. 114 pp.

Lynch, M. J., and Buetow, D. E. 1975. Isolation of intact nuclei from *Euglena gracilis. Exp. Cell Res.* 91, 344–8.

Lynch, M. J., Leake, R. E., O'Connell, K. M., and Buetow, D. E. 1975. Isolation, fractionation and template activity of the continuously condensed chromatin of *Euglena gracilis. Exp. Cell Res.* 91, 349–57.

McLennan, A. G., and Keir, H. M. 1975. Subcellular location and growth stage dependence of the DNA polymerases of *Euglena gracilis. Biochim. Biophys. Acta* 407, 253–62.

Parenti, F., Brawerman, G., Preston, J. F., and Eisenstadt, J. M. 1969. Isolation of nuclei from *Euglena gracilis. Biochim. Biophys. Acta* 195, 234–43.

3: Mitochondria from Chlorella

NEIL G. GRANT

*Department of Biology, The City College of the
City University of New York, New York, New York 10031*

CONTENTS

I	**Introduction**	*page* **26**
II	**Materials and test organism**	**26**
	A. Glassware and supplies	26
	B. Chemicals and solutions	26
	C. Equipment	27
	D. Test organism	27
III	**Methods**	**27**
	A. Growth	27
	B. Breakage of cells	27
	C. Centrifugation	28
	D. Protein assay	28
	E. Criteria of purity and intactness	29
IV	**Sample data**	**29**
V	**Discussion**	**29**
VI	**Acknowledgments**	**29**
VII	**References**	**30**

I. Introduction

The study of mitochondria from *Chlorella* utilizes methods developed for yeast (Schatz 1967), other microbes, and higher plants (Bonner 1967). Several problems must be kept in mind when working with mitochondria from algae. There is usually a thick, durable, complex cell wall that makes gentle release of mitochondria difficult; chloroplasts and chlorophyll complicate the purification and study of algal mitochondria by contaminating the mitochondrial fraction with membrane fragments and obscuring the spectral characteristics of the respiratory pigments. Nevertheless, *Chlorella* is a rewarding organism to study because of the work that has been done on ultrastructure, life cycle, nutrition, and biochemistry.

Higher plant mitochondria have been satisfactorily studied using etiolated or storage tissues. A similar situation occurs with *Chlorella prototheoides*, which, when grown on 1% glucose and relatively low concentrations of organic nitrogen, has diminished chloroplast structure and very little chlorophyll (Matsuka and Hase 1965).

II. Materials and test organism

A. Glassware and supplies

(1) Fernbach or low-form flasks, Pyrex, 2.8 l, with stainless-steel or cotton-wool plugs. (2) Erlenmeyer flasks, Pyrex, 250 ml, with stainless-steel or cotton-wool plugs. (3) Flasks for mechanical cell homogenizer, Pyrex, 75 ml. (4) Beakers, Pyrex, 100 ml. (5) Beads, glass, 0.45–0.55 mm diam., acid-washed and exhaustively rinsed with distilled water. (6) Funnel, plastic. (7) Funnel, glass, ca. 100 mm diam. (8) Flask, filter, 250 ml. (9) Centrifuge tubes, polycarbonate, 50 ml. (10) Miracloth (Calbiochem). (11) Ice buckets, pipettes, etc.

B. Chemicals and solutions

1. Growth medium. Per liter of distilled water: KH_2PO_4, 0.7 g; K_2HPO_4, 0.3 g; $MgSO_4 \cdot 7\ H_2O$, 0.3 g; $FeSO_4 \cdot 7\ H_2O$, 3 mg; thiamine·HCl, 10 μg; glucose, 10 g; glycine, 0.5 g; Arnon's A_5 solution, 1 ml, pH 6.3.

Arnon's A_5 solution contains, per liter: H_3BO_3, 2.9 g; $MnCl_2 \cdot 4 H_2O$, 1.8 g; $ZnSO_4 \cdot 7 H_2O$, 0.22 g; $CuSO_4 \cdot 5 H_2O$, 0.08 g; MoO_3, 0.018 g. The medium should be autoclaved for 15 min at 15 lb in.$^{-2}$.

Glucose may be autoclaved separately in solution and added aseptically or filtered through a Millipore filter.

2. *Breakage medium.* Sucrose, 0.4 M; EDTA, 1.0 mM; cysteine 0.05%; HEPES buffer, 10 mM; pH 7.2 adjusted with 1.0 N KOH at 0 C.

3. *Wash medium.* As above, but with cysteine omitted.

C. *Equipment*

Oxygen electrode apparatus; chart recorder, 10 mV; high-speed refrigerated centrifuge or centrifuge in cold room; cell homogenizer, Braun, MSK.

D. *Test organism*

Chlorella protothecoides Krüger UTEX 25.

III. Methods

A. *Growth*

Maintain the alga on agar slants in normal room light. For preparation of mitochondria, start cultures on Petri dishes containing 1.5% agar in culture medium and then transfer a loop of alga to 100 ml of culture medium in 250-ml Erlenmeyer flasks. After about 4 days of growth in the dark at 23 C, use the contents to inoculate 1 l of culture medium in 2.8-l flasks. Place these on a shaker in a dark room or wrap to exclude light.

Monitor the growth of the culture and harvest by centrifugation when it is in the late exponential phase. Mitochondria may be extracted now, but more reliable results are obtained by rinsing the algae and transferring to "greening medium," which is the same as the growth medium with the omission of glucose and the addition of 1 g glycine per liter. If the culture is incubated in this medium for 12–18 h, the respiratory activity and level of cytochrome aa_3 increases, as presumably does the amount of mitochondria (Grant and Hommersand 1974a). The conditions and pattern of greening are variable, so some experience may be necessary to obtain a culture that has maximum respiratory activity and has not yet started to green (Matsuka and Hase 1966).

B. *Breakage of cells*

After the cells have been centrifuged, rinse them with breakage medium and adjust the suspension so that it is about 25% packed cell volume. Place 25 ml of the cell suspension into cold 100-ml beakers containing 50 g of glass beads. Pour this slurry into the MSK flasks and place in the MSK. The flask is easier to withdraw from the MSK if it is wiped with a little light mineral oil before insertion. Modulate the flow of CO_2 and the operation of the MSK so that the flasks will not freeze or become too warm. Shake the MSK at 4000 cycles per minute for about 5–15 sec. Check with a microscope for amount of cell breakage. It is necessary to work as rapidly and gently as possible. After satisfactory breakage is obtained, check and adjust the pH to 7.2 (this is seldom necessary).

Decant the fluid from the beads and through the Miracloth that lines the funnel inserted in the filter flask. The Miracloth should have been wetted with the breakage medium, and the whole apparatus should be cold. Rinse out the beads in the flasks with more medium and filter as before. The filtering serves to remove glass beads and some bead fragments.

C. Centrifugation

Distribute the filtrate into ice-cold polycarbonate centrifuge tubes and centrifuge in the cold according to this schedule:

1. Centrifuge $1000 \times g$, 10 min, to sediment unbroken cells, cell walls, and large cytoplasmic fragments.

2. Carefully decant the supernatant fluid and centrifuge this at $10\ 000 \times g$, 10 min.

3. Decant and discard the supernatant fluid.

4. Suspend the pellet in 20 ml of wash medium and gently mix with a pipette. During these stages, carefully separate the tan mitochondrial layer, the yellow oily material that floats on the surface of the medium, and the heavy white material under the mitochondria.

5. Finally centrifuge at $6000 \times g$, 10 min.

6. Discard supernatant fluid and suspend mitochondrial pellet in 1 ml wash medium.

Yield varies from 5 to 20 mg protein from about 5 g fresh weight of *Chlorella*.

D. Protein assay

The Lowry protein assay is not satisfactory for use with *Chlorella* mitochondria because of interference from HEPES and other buffers and from mannitol, which is sometimes used instead of sucrose in the wash medium. I recommend the use of the semimicro Kjeldahl method described in Umbreit et al. (1972). The value for total nitrogen is then multiplied by 6.25, a factor that has been used for higher plant mitochondria (Lance and Bonner 1968).

Table 3-1. *Characteristics of Chlorella mitochondria*

Substrate (mM)	State 3 respiration (nmol O_2 min^{-1} mg^{-1} protein)	Respiratory control	ADP/O ratio
TRIS-succinate (8.0)	57	2.2	0.71
TRIS-malate (30)	41	1.6	1.32
NADH (1.0)	73	Not observed	—

E. Criteria of purity and intactness

Because mitochondria are complex structures, it is difficult to judge how much change has occurred during their isolation. Electron micrographs showing the structure of isolated mitocondria can be compared with the structure of mitochondria in situ. Biochemical criteria include respiratory control (ADP acceptor control), stability for an appreciable length of time, high rates of O_2 uptake, and increase in O_2 uptake rate upon addition of an uncoupler of oxidative phosphorylation in the presence of substrate and absence of ADP.

IV. Sample data

Results with mitochondria isolated from *Chlorella* are variable, owing in part to changes occurring during isolation, the variability of the respiration of a batch of cells, and the presence of two mitochondrial terminal oxidase systems, which may vary in their characteristics and proportions.

Some representative data are shown in Table 3-1.

V. Discussion

The respiratory control of the *Chlorella* mitochondria I have isolated from *Chlorella* is rather low. It appears that this is caused partly by damage during isolation, as the mitochondria oxidize added reduced horse heart cytochrome *c*, which indicates that the membranes have become more permeable (Grant and Hommersand 1974b).

VI. Acknowledgments

This work was done while I was a doctoral candidate in the Department of Botany, University of North Carolina, Chapel Hill, supported by a National Aeronautics and Space Administration Predoctoral Fellowship and an American Cancer Society Institutional Subgrant.

VII. References

Bonner, W. D., Jr. 1967. A general method for the preparation of plant mitochondria. *Methods Enzymol.* 10, 126–33.

Grant, N. G., and Hommersand, M. H. 1974a. The respiratory chain of *Chlorella protothecoides.* I. Inhibitor responses and cytochrome components of whole cells. *Plant Physiol.* 54, 50–6.

Grant, N. G., and Hommersand, M. H. 1974b. The respiratory chain of *Chlorella protothecoides.* II. Isolation and characterization of mitochondria. *Plant Physiol.* 54, 57–9.

Lance, C., and Bonner, W. D., Jr. 1968. The respiratory chain components of higher plant mitochondria. *Plant Physiol.* 43, 756–66.

Matsuka, M., and Hase, E. 1965. Metabolism of glucose in the process of "glucose-bleaching" of *Chlorella protothecoides. Plant Cell Physiol.* 6, 721–41.

Matsuka, M., and Hase, E. 1966. The role of respiration and photosynthesis in the chloroplast regeneration in the "glucose-bleached" cells of *Chlorella protothecoides. Plant Cell Physiol.* 7, 149–62.

Schatz, G. 1967. Stable phosphorylating submitochondrial particles from bakers' yeast. *Methods Enzymol.* 10, 197–202.

Umbreit, W. W., Burris, R. H., and Stauffer, J. F. 1972. *Manometric and Biochemical Techniques,* 5th ed., pp. 259–60. Burgess, Minneapolis.

4: Microbodies from Chlorogonium and Spirogyra

HELMUT STABENAU

*Pflanzenphysiologisches Institut der Universität Göttingen,
Untere Karspüle 2, 3400 Göttingen, Germany*

CONTENTS

I	Introduction	*page* 32
II	Test organisms and materials	32
	A. Test organisms	32
	B. Materials	33
III	Methods	33
	A. Homogenization and first purification step	33
	B. Isolation of microbodies	34
IV	Sample data	34
V	Discussion	36
VI	References	37

I. Introduction

Electron microscope studies have firmly established that algae possess microbodies similar in structure to those of higher plants (peroxisomes, glyoxysomes) in having a single limiting membrane and a fine granular matrix. They normally appear spherical or oval in section, 0.4–0.8 μm in diameter. As demonstrated biochemically and by the cytochemical test with diaminobenzidine, catalase is also a characteristic enzyme of the algal microbodies.

The method and success of isolation of algal microbodies depend on the alga involved, as will be discussed with respect to a filamentous and a unicellular organism. Furthermore, with *Spirogyra* and *Chlorogonium* dissimilar data were obtained, indicating that the microbodies in algae can be of different types.

II. Test organisms and materials

A. Test organisms

1. *Chlorogonium elongatum* is a flagellate (Chlamydomonaceae) with a major axis of 20–25 μm. The cells grow well under autotrophic or heterotrophic conditions. A culturing medium is described by Stabenau (1971).

As demonstrated by electron micrographs, the microbodies from *Chlorogonium* are located in the peripheral cytoplasm between the elongated mitochondria. In general they exhibit a spherical or oval profile in section with a diameter of about 0.4 μm. No inclusions are observed in the microbodies from this alga.

The cells also contain enzymes typical of peroxisomes and glyoxysomes from higher plants, though not all of them are constituents of the algal microbodies (Stabenau 1974a; Stabenau and Beevers 1974). Like several other algae (Frederick et al. 1973), *Chlorogonium* possesses a glycolate dehydrogenase that is different from the glycolate oxidase of higher plants (Nelson and Tolbert 1970).

2. *Spirogyra* can be cultivated as reported elsewhere (Stabenau

1976). Besides other peroxisomal enzymes, this alga contains a glycolate oxidase that is similar to the higher plant enzyme (Frederick et al. 1973; Stabenau 1975). Microbodies isolated from *Spirogyra* appear spherical, with a diameter between 0.4 and 0.8 μm. Inclusions were not seen.

B. Materials

1. Mortar, about 7 cm diam.
2. Sand, grains about 0.5 mm diam.
3. Two sintered glass filters, pore sizes 90–150 μm and 40–90 μm.
4. A device to prepare a linear sucrose gradient (Fig. 4-1).
5. Sucrose solutions, 0.98 M (30% w/w) and 2.09 M (60% w/w), with an addition of EDTA, 1 mM. The percentage of sugar is determined by a refractometer here and in the following step.
6. Isolation medium, containing the following: tricine buffer, 0.15 M, pH 7.5; KCl, 10 mM; MgCl$_2$, 1 mM; EDTA, 1 mM; sucrose, concentration depends on the algal material, e.g., 0.63 M (20% w/w) for *Chlorogonium* and 0.37 M (12% w/w) for *Spirogyra*.

III. Methods

A. Homogenization and first purification step

1. Chlorogonium. The algae in 1 l of suspension are collected by low-speed centrifugation (about 4000 × *g*). They are washed three times with distilled water and once in 10 ml of the isolation medium containing 20% sucrose. To break the cells, they are suspended in 4 ml of the same isolation medium and gently ground with sand in an ice-cold mortar for 2 min. Only a small amount of the sand should be used, so there still remains a thin layer of suspension above the sand pellet.

After the cells are disrupted, a further 6 ml of the isolation medium is added and the homogenate decanted. The crude homogenate is then centrifuged for 8 min at 2 C and 1400 × *g*. By this step whole cells and most of the chloroplast particles are removed from the homogenate. If necessary, centrifuge again.

2. Spirogyra. The filaments are collected on a sieve (about 1 mm diam. mesh size). Then they are washed several times with water and once in 20 ml of isolation medium containing 12% sucrose and finally transferred to 10 ml of the isolation medium. The cells are broken by cutting the mass of filaments with scissors for 10 min. The crude homogenate then has to be passed twice through sintered glass filters with average pore sizes of 120 and 65 μm, respectively. For this no suction should be used. During the second filtration through the 65-

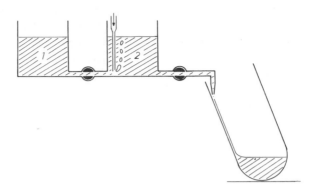

Fig. 4-1. Preparation of a sucrose gradient. At the beginning, chambers 1 and 2 contain solutions with 30% and 60% (w/w) of sucrose, respectively. Solutions are mixed in chamber 2 by aerating with air or N_2.

μm filter, most of the chloroplast particles are retained. The homogenate usually then appears brown. All steps should be performed at low temperature.

B. Isolation of microbodies

To separate the cell organelles, the crude homogenate must be layered on a linear gradient from 30% to 60% (w/w) sucrose, which may be prepared as shown in Fig. 4-1. After centrifugation in a swinging bucket for 4 h at 60 000 × g and 2 C, the different organelles are moved to the regions of their specific density in the gradient, whereas the top layer still contains all the soluble material of the homogenate. The gradient is then separated into 1-ml fractions. For this several commercial devices are available; however, the simplest successful method is to prick a small hole in the bottom of the centrifuge tube through which the solution slowly drains drop by drop and can be collected in different tubes.

After determination of the sugar concentration in each fraction by refractometry, distinctive enzymes may be tested to analyze the gradient as demonstrated by the following.

IV. Sample data

Because catalase is a marker of all microbodies, the distribution pattern of this enzyme is most suitable to show to which density the organelles moved during gradient centrifugation. The profile for catalase is also a good indicator of how much the microbodies were broken during the isolation procedure. Under good conditions, more than 50% of the total enzyme activity is found within the gradient

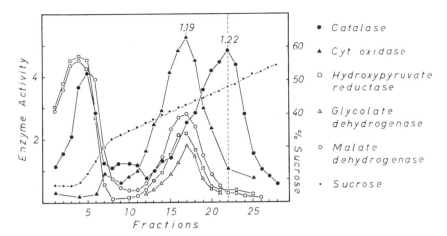

Fig. 4-2. Distribution of enzymes after separation of organelles from autotrophic *Chlorogonium* cells on a linear sucrose gradient. Unspecialized microbodies occur at density 1.22 g cm^{-3}, and mitochondria at density 1.19 g cm^{-3}. (Constructed from data in Stabenau 1974a.)

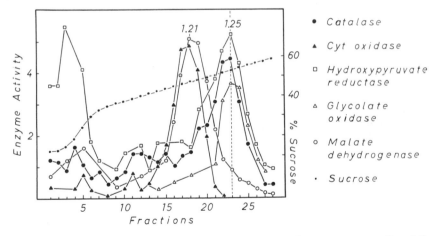

Fig. 4-3. Separation of cell organelles from *Spirogyra* on a linear sucrose gradient. Microbodies of the leaf peroxisomal type occur at density 1.25 g cm^{-3}; mitochondria at density 1.21 g cm^{-3}. (Constructed from data in Stabenau 1975, 1976.)

(Figs. 4-2 and 4-3). All microbodies presumably are broken if catalase activity can be measured only in the top of the gradient.

To evaluate the separation of microbodies from mitochondria, cytochrome oxidase, which is tightly bound to mitochondrial membranes, may be tested. Additionally, the determination of malate dehydrogenase is suggested, and this can be done by a simple test. As this enzyme is completely soluble, monitoring the activity is also an easy method to check whether the mitochondria were broken. With *Chlorogonium* a maximum of 40% of the total enzyme was found in the mitochondrial band (Fig. 4-2), whereas about 70% was located in the mitochondria from *Spirogyra* (Fig. 4-3).

To check whether the microbodies are of the leaf peroxisomal type, glycolate oxidase and especially hydroxypyruvate reductase may be tested, as these are distinctive markers for this organelle. However, enzymes characteristic for peroxisomes and glyoxysomes (Beevers 1969; Tolbert 1971) are not always located within microbodies of algae. As shown in Fig. 4-2 and Stabenau (1974a), enzymes of glycolate metabolism from *Chlorogonium* were identified in the mitochondrial region and the glyoxysomal marker isocitrate lyase was found not to be a constituent of the microbodies from this alga. Instead, this enzyme was located exclusively in the top fraction of the gradient and therefore appears to be soluble (Stabenau and Beevers 1974). The only enzymes of the microbodies from *Chlorogonium* so far identified are catalase and uricase. Therefore, these organelles differ from those of *Spirogyra,* which are of the leaf peroxisomal type.

All assays used for determination of algal enzyme activities have been described (Stabenau 1974a, 1974b, 1976; Stabenau and Beevers 1974).

V. Discussion

A major problem in the isolation of microbodies, especially from unicellular algae, is to break the cells without damaging the organelles. Gentle grinding with sand turned out to be a successful homogenization method for *Chlorogonium* as well as for *Chlamydomonas* (Stabenau 1974b). Disrupting the algae by a Potter-Elvehjem homogenizer is a suitable alternate procedure. However, with other algae different methods may be more convenient, and this should be checked in every case.

The specific activities of enzymes from algal microbodies are relatively low compared with those from higher plants. Therefore, monitoring the activity of the marker enzymes in the gradient fractions sometimes may be difficult. This is even more apparent when small amounts of homogenate are required to yield a good organelle separation. In relation to such problems, it is useful to know that the activities of the enzymes can be increased under certain conditions (Stabenau and Ulrich 1975). As could be shown for *Chlorogonium* and *Chlamydomonas,* reducing the CO_2 concentration in the aeration mixture from 2% to 0.03% during growth of the cells resulted in a 15-fold increase in the specific activity of catalase.

VI. References

Beevers, H. 1969. Glyoxysomes of castor bean endosperm and their relation to gluconeogenesis. *Ann. N.Y. Acad. Sci.* 168, 313–24.

Frederick, S. E., Gruber, P. J., and Tolbert, N. E. 1973. The occurrence of glycolate dehydrogenase and glycolate oxidase in green plants: An evolutionary survey. *Plant Physiol.* 52, 318–23.

Nelson, E. B., and Tolbert, N. E. 1970. Glycolate dehydrogenase in green algae. *Arch. Biochem. Biophys.* 141, 102–10.

Stabenau, H. 1971. Die Regulation des Photosyntheseapparates bei *Chlorogonium elongatum:* Dangeard unter dem Einfluss von Licht und Acetat. *Biochem. Physiol. Pflanz.* 162, 371–85.

Stabenau, H. 1974a. Localization of enzymes of glycolate metabolism in the alga *Chlorogonium elongatum Plant Physiol.* 54, 921–4.

Stabenau, H. 1974b. Verteilung von Microbody-Enzymen aus *Chlamydomonas* in Dichtegradienten. *Planta* 118, 35–42.

Stabenau, H. 1975. Zur Lokalisation von Glykolsäure-Oxidase in *Spirogyra. Ber. Dtsch. Bot. Ges.* 88, 469–71.

Stabenau, H. 1976. Microbodies from *Spirogyra:* Organelles of a filamentous alga similar to leaf peroxisomes. *Plant Physiol.* 58, 693–5.

Stabenau, H., and Beevers, H. 1974. Isolation and characterization of microbodies from the alga *Chlorogonium elongatum. Plant Physiol.* 53, 866–9.

Stabenau, H., and Ulrich, F. 1975. Regulation of peroxisomal and glyoxysomal enzymes in the alga *Chlorogonium* by different CO_2 concentrations. In *XII International Botanical Congress, Leningrad* (Abstracts). USSR Academy of Sciences, p. 371.

Tolbert, N. E. 1971. Microbodies: Peroxisomes and glyoxysomes. *Annu. Rev. Plant Physiol.* 22, 45–74.

5: Isolation and characterization of diatom membranes

CORNELIUS W. SULLIVAN

Department of Biological Sciences,
University of Southern California, Los Angeles, California 90007

CONTENTS

I	**Introduction**	*page* **40**
II	**Materials and test organisms**	**40**
	A. Laboratory apparatus	40
	B. Media and reagents	40
	C. Organisms	41
III	**Methods**	**43**
	A. Cell growth and harvesting	43
	B. Cell disruption	44
	C. Differential centrifugation	44
	D. Discontinuous sucrose gradient centrifugation	45
	E. Characterization of membranes	47
IV	**Sample data**	**49**
	A. Yield	49
	B. Enrichment	52
	C. Purity	52
	D. Identification	52
V	**Discussion**	**53**
VI	**Acknowledgments**	**54**
VII	**References**	**55**

I. Introduction

This chapter describes methods that have proved useful for the isolation and characterization of plasma membranes (plasmalemma) and various subcellular membranes including those of the Golgi, chloroplasts, and mitochondria of two diatoms (Bacillariophyceae). These methods should not be used in "cookbook" fashion to prepare membranes from all diatoms or other microalgal species, but they should at least provide a basic guideline and possible starting point for the development of new and better techniques. I have endeavored to cite criteria for assessing how good a particular technique is so that your own research efforts will follow fruitful lines.

II. Materials and test organisms

A. Laboratory apparatus

(1) Magnetic stirrer, 6 × 6 in. (Cat. No. 4810, Cole-Parmer). (2) Yeda Pressure Press. (3) Sharples Super Centrifuge. (4) Membrane filters, pore size 0.45 μm (Millipore).

B. Media and reagents

All solutions are prepared with distilled deionized water, and aseptic conditions are rigorously maintained throughout.

1. Synthetic seawater. SSW contains in 1 l of water the following: 40 g Utility Marine Mix, 0.5 g $NaNO_3$, 27 mg $K_2HPO_4 \cdot 3H_2O$, 0.2 g $Na_2SiO_3 \cdot 9H_2O$, 0.5 mg thiamine·HCl, 1 mg cyanocobalamine, 12 mg disodium EDTA, 0.4 mg H_3BO_3, and 1 ml of the trace element solution of Burkholder and Nickell (1949). The medium is adjusted to pH 7.1 with 5 N HCl, then autoclaved at 120 C, 15 lb in.$^{-2}$, for 20 min. One addition to the medium contained in 12 ml of water (2 g D-glucose, 150 mg citric acid, and 596 mg Na_2HPO_4 pH 6.6) is prepared separately and filter-sterilized through 0.45-μm pore size membrane filters before combining.

2. Protoplast isolation medium. P medium contains in 1 l of water the following: 1 g Bacto-tryptone (Difco), 0.5 g Bacto-yeast extract (Difco), and 30 g NaCl. The medium is adjusted to pH 8.0 with NaOH or HCl.

3. Artificial seawater medium. ASW_{12} contains in 1 l of water the following: 23.6 g NaCl, 4.9 g $MgSO_4 \cdot 7H_2O$, 4.1 g $MgCl_2 \cdot 6H_2O$, 1.1 g $CaCl_2$, 0.75 g KCl, 303 mg KNO_3, 45.6 mg $K_2HPO_4 \cdot 3H_2O$, 2.3 mg H_3BO_3, 12 mg disodium EDTA, 0.5 mg thiamine·HCl, 120 mg $Na_2SiO_3 \cdot 9H_2O$, 0.66 mg glycylglycine, 1 ml of the trace element mix of Burkholder and Nickell (1949) adjusted to pH 8.0 with NaOH.

4. Organelle isolation medium. I medium consists of 0.4 M sucrose, 0.02 M TRIS-HCl pH 7.4, 0.01 M $MgCl_2$, 0.01 M KCl, 1 mM 2-mercaptoethanol, and 0.1% bovine serum albumin (BSA).

5. First wash medium. W-1 medium consists of 0.4 M sucrose, 0.01 M $MgCl_2$, 0.01 M TRIS-HCl pH 7.4, and 1 mM 2-mercaptoethanol.

6. Second wash medium. W-2 medium consists of 0.4 M sucrose, 0.01 M TRIS-HCl pH 7.4, 1 mM EDTA, and 0.1% BSA.

7. Buffer A. 10 mM TRIS-HCl pH 7.2.

8. Reagents. (BSA) fraction V (Sigma); TRIS, "Sigma 9" (Sigma); Utility Marine Mix; spectroquality glycerol (Curtin-Matheson); dithiothreitol (Calbiochem).

C. Organisms

1. Nitzschia alba. *N. alba* is an apochlorotic pennate diatom that grows well (4–5 h doubling time) at 25–30 C in SSW supplemented with 1% glucose as a carbon and energy source. Its special advantage for membrane studies was shown by Hemmingsen (1971), who demonstrated that *N. alba* can be induced to shed its cell wall to form protoplasts. The osmotically sensitive protoplasts are gently lysed in hypotonic solution. This procedure provides internal membranes, which appear as large sheets or vesicles, and intact mitochondria. Because severe grinding or pressure shearing procedures are not employed, some of the isolated membranes retain their characteristic shapes and can be identified on morphological criteria (Fig. 5-1).

2. Cylindrotheca fusiformis (Reimann and Lewin). *C. fusiformis* is a photoautotrophic marine pennate diatom that grows with a doubling time of 12 h in ASW_{12} medium. These cells apparently do not form protoplasts; they do possess a lightly silicified cell wall, however, and a physical method of cell disruption is required.

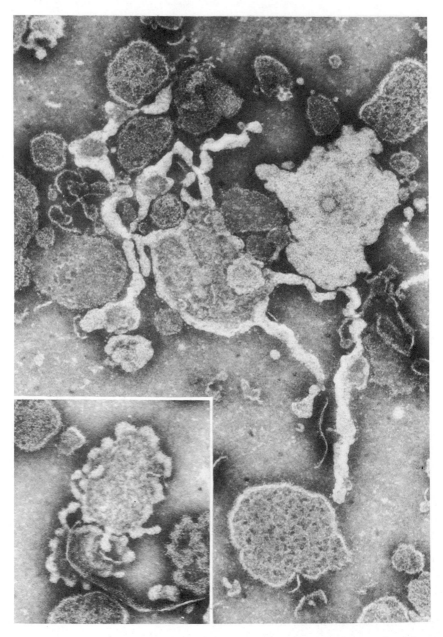

Fig. 5-1. Negatively stained Golgi vesicles and tubular membrane structures from fraction II of *Nitzschia alba*. ×120 000. (From Sullivan et al. 1974.)

III. Methods

A. Cell growth and harvesting

1. N. alba. Axenic clonal cultures of *N. alba* are grown in 250-ml Erlenmeyer flasks containing 100 ml SSW medium at 30 C for 36 h on a rotary shaker. When cells reach a density of 1×10^6 cells ml^{-1} the culture is transferred to 4 l of SSW medium contained in a 6-l Erlenmeyer flask and incubated at 30 C for 36 h with stirring provided by a 5-cm Teflon-coated magnetic stirring bar and a 6×6-in. Cole-Parmer stirrer. Stirring at 200 rpm is sufficient to provide aeration by inducing cavitation.

When cells are in the late exponential growth phase (7×10^5 to 9×10^5 cells ml^{-1}), the flasks are removed from stirrers and the cells are permitted to settle for 8 h. Approximately 3.5 l of supernatant fluid is decanted aseptically, and the cells are resuspended in the remaining medium, then transferred aseptically to sterile 250-ml plastic centrifuge bottles and centrifuged at $5000 \times g$, 5 min, at 20 C in a Sorvall refrigerated centrifuge. For routine large-scale isolation of membranes, cells are harvested from 24–48 l of medium by this procedure.

a. Protoplast formation. After centrifugation, the supernatant fluid is aseptically decanted and 100 ml sterile P medium is added to resuspend the cells by gentle shaking. Additional P medium is added to obtain a final density of 10^7–10^8 cells ml^{-1}, and 100-ml aliquots of the suspension are dispensed to sterile 250-ml Erlenmeyer flasks. Full protoplast formation occurs during 12–14 h of incubation in P medium under conditions described for cell growth. Protoplasts are harvested at the peak of their formation by centrifugation at $4000 \times g$ for 10 min at 0 C. Observation by phase-contrast microscopy serves to assess the percentage of cells that have formed protoplasts, usually 50–80%, and allows one to check for possible bacterial contamination. The presence of bacteria in the protoplast suspension *requires termination* of the experiment at this stage.

2. C. fusiformis. Axenic clonal cultures are grown to a density of 1.5×10^6 cells ml^{-1} in 9 l of ASW$_{12}$ medium contained in 12-l carboys and incubated at 20 C with magnetic stirring (200 rpm) and aeration (10 l min^{-1}) in the light (17 klux). The cells from 40–60 l of medium are harvested by centrifugation in a Sharples Super Centrifuge at 22 C, 1700 rpm and a flow rate of 1.7 l min^{-1}. The cells are washed once by resuspension in I medium, centrifuged, and resuspended in I medium at 20% cell w/v in preparation for cell disruption.

a. Precautions. Axenic cultures of logarithmic phase cells are employed because of their metabolically active state and because they usually contain less storage lipid than stationary phase cells. Non-membrane-associated lipids (free lipids) may interfere with chemical analysis of membranes by adsorption or intercalation with the membranes after cell disruption. Free lipids may also damage the functional properties and/or integrity of membranes and associated enzyme activities, as was found in mitochondria isolated from fatty tissues (Mehard et al. 1971).

Membrane isolation for chemical analysis generally requires relatively large amounts of starting cell material; 10 g wet weight cells (5000 × g, 5 min centrifuged pellet) yields approximately 100 mg of membrane protein.

B. Cell disruption

1. N. alba: hypotonic lysis. Protoplasts are gently disrupted by controlled hypotonic lysis. The protoplast pellets are pooled and resuspended in 1 l of medium A at 0 C by hand shaking the suspension 20 times. Phase-contrast microscopy confirms lysis of protoplasts.

2. C. fusiformis: pressure cell shearing. A 30-ml aliquot of the washed and resuspended cells is added to the prechilled N_2 pressure press and held at 400 lb in.$^{-2}$ for 10 min in an ice-water bath. The cell lysate is slowly dripped (120 drops min^{-1}) from the exit orifice into an ice-chilled beaker containing an equal volume of I medium. The procedure is repeated until the total volume of cell suspension has been passed through the pressure press. Cell breakage varies from approximately 75% to 95%, as determined by phase-contrast microscopy.

C. Differential centrifugation

1. N. alba. The *N. alba* lysate is immediately centrifuged according to the scheme shown in Fig. 5-2, to yield a maximum quantity of cell membrane material in the P-3 pellet fraction. This fraction contains all cellular membranes except intact mitochondria, which are enriched in the P-2 fraction. For isolation and further purification of diatom mitochondria by discontinuous sucrose gradient centrifugation, see detailed methods reported previously (Paul et al. 1975). Isolation and purification of diatom cell walls are accomplished by resuspending pellet P-1 in medium A, 20% w/v, followed by layering on and repetitive centrifugation through an equal volume of 60% sucrose in buffer A at 1000 × g for 10 min, at 0 C. Approximately eight passes through the 60% sucrose solution remove all residual membrane and free lipid material from the cell walls. The final product is translucent.

2. C. fusiformis. The cell lysate is subjected to successive differential centrifugation in the Sorvall centrifuge at 0 C to separate the crude

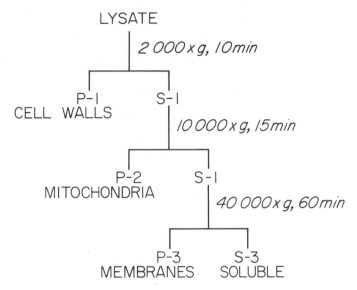

Fig. 5-2. Differential centrifugation scheme for the isolation of membrane fractions from *Nitzschia alba* protoplast lysate. (From Sullivan et al. 1974.)

organelle fractions. Unbroken cells are removed at 750 × g, 10 min. Five fractions are obtained enriched in: cell walls (750–1000 × g, 10 min), chloroplasts (1000–3500 × g, 15 min), and microsomes (30 000 –100 000 × g, 1 h). These crude fractions are washed three times with W-1 medium by centrifuging for 10 min at the upper g-force used to sediment the preceding fraction; the pellets are discarded and the supernatant fraction centrifuged to provide the fractions as indicated above.

It should be noted that the vesicle and microsomal fractions do not represent any particular subcellular membrane; they are based entirely on their sedimentation characteristics and potentially contain all cellular membranes. This procedure is especially useful for mitochondrial isolations, and when coupled with the discontinuous sucrose gradient centrifugation described previously, results in a six- to eight-fold purification of mitochondria based on succinate dehydrogenase and glycolate dehydrogenase enzyme enrichment (Paul et al. 1975).

D. Discontinuous sucrose gradient centrifugation

The 40 000 × g P-3 membrane pellet from *N. alba*, containing a total of 45–95 mg protein, is carefully resuspended in medium A containing 65% sucrose (w/v) by five strokes of a loose-fitting Teflon pestle homogenizer. The water associated with the pellet brings the final

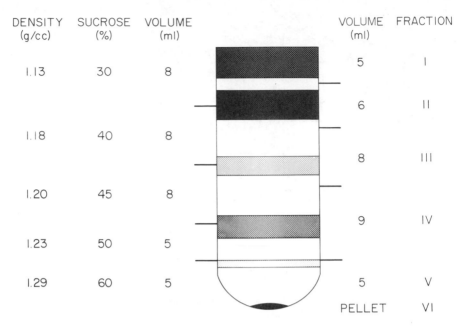

DENSITY (g/cc)	SUCROSE (%)	VOLUME (ml)	VOLUME (ml)	FRACTION
1.13	30	8	5	I
			6	II
1.18	40	8		
			8	III
1.20	45	8		
			9	IV
1.23	50	5		
1.29	60	5	5	V
			PELLET	VI

Fig. 5-3. Schematic representation of the discontinuous sucrose gradient and separated membrane bands formed by isopyknic centrifugation of P-3 membrane material. The stipple density is representative of the relative amounts of material in each of the membrane bands visualized after centrifugation. To the left of the gradient appears the approximate density, volume, and percentage of sucrose solutions that were layered onto the gradient before centrifugation. To the right of the gradients are listed the volumes removed and their respective designations. (From Sullivan et al. 1974.)

sucrose solution to 60%. A 5-ml aliquot is placed in the bottom of a 35-ml cellulose nitrate centrifuge tube, and over it the following medium-A-containing sucrose solutions are carefully layered without disrupting the interface: 5 ml, 50%; 8 ml, 45%; 8 ml, 40%; and 8 ml, 35%. The discontinuous gradient and membrane payload are then subjected to isopyknic flotation by centrifugation at 63 000 × g (25 000 rpm) at 5 C for 16 h in a Beckman model L preparative ultracentrifuge using a Beckman 25.1 swinging bucket (SW) rotor.

The resulting membrane "bands" are visible on examination of the tube by transmitted light; the relative density of the bands is shown schematically in Fig. 5-3. These fractions are carefully collected by a Pasteur pipet with a rubber bulb and removed to a prechilled 25-ml graduated cylinder for consistent volume measurement. The respective fractions are pooled from three tubes and diluted to 60 ml with medium A, then centrifuged at 100 000 × g (30 000 rpm, Beckman 30 rotor) at 5 C for 1 h. This procedure is repeated three times if the membranes are to be used for chemical analysis, as traces of sucrose from the gradient would otherwise interfere with the carbohydrate analysis. If the membranes are to be used for enzyme analysis they are suspended in 1–2 ml of medium A containing 20% spectroquality glycerol and 1 mM dithiothreitol.

Although these additions should be tested for their effects on each individual enzyme activity being assayed, enzyme activities generally appear to be stabilized in buffered media containing 20–50% glycerol, 10^{-4}–10^{-3} M dithiothreitol, and 5–10 mg membrane protein per milliliter. Storage on ice is generally preferred to freezing, but under the best conditions the activation of latent enzymes and/or inactivation of labile enzymes can, and often does, occur during storage. Therefore, great caution must always be exercised when one compares the enzyme activities among samples of stored material.

E. Characterization of membranes

Membranes can generally be characterized and identified on the basis of morphology, chemical composition, and associated enzyme activities. Newer techniques involve specific labeling with certain tracers (Scarborough 1975) or stains (Sundberg and Lembi 1976).

1. Morphological analysis by electron microscopy. Two general methods of observing membrane morphology are most useful. Negative staining techniques allow one to determine the gross surface morphology of membrane sheets and vesicles, and thin sectioning allows high-resolution observation of surface coat properties, membrane thickness, and presence or absence of "double track" OsO_4 staining characteristics.

a. Surface morphology. Negative staining is performed on freshly prepared membrane fractions removed directly from the sucrose gradient. Sucrose is removed by gently dropping the membrane suspension onto the surface of the negative staining solution by the method of Sturgess et al. (1973).

b. Morphology of thin sections. Membranes from individual pellets are fixed at room temperature for 1 h with 3% glutaraldehyde, in 0.1 M phosphate buffer, pH 7.0, containing 1 mM $CaCl_2$ according to Ulsamer et al. (1971). (If intact mitochondria or chloroplasts are being fixed, make the buffer to 0.4 M sucrose.) The samples are rinsed three times with phosphate buffer and then postfixed with phosphate buffer containing 1% OsO_4 for 1 h at room temperature. Samples then are dehydrated in ethanol series, followed by propylene oxide and embedded in Epon 812. The sections are stained in uranyl acetate and lead citrate as described by Reynolds (1963).

2. Chemical analysis. Any quantitative spectrometric or fluorometric method of analysis capable of detecting microgram or lower quantities of specific compounds is potentially useful. Analysis is performed

Table 5-1. *Total chemical composition (mg/10^{10} cells) of membrane fractions obtained during fractionation of protoplast lysate*

Component	Fractions					Gradient fractions					
	Lysate	P-1	P-2	P-3	S-3	I	II	III	IV	V	VI
Protein	1300	364	64.7	53.7	748	3.49	5.80*	4.24	4.66	3.16	1.72
Lipid	1952	1019	70.1	149	810	5.33	7.88*	6.45	5.16	2.95	2.29
Phospholipid	186	43.1	18.4	16.0	73.3	1.50	2.12*	1.14	0.713	0.499	0.186
Sterol	26.5	8.55	1.85	4.32	3.43	0.275	0.441*	0.336	0.276	0.254	0.069
Neutral lipid[a]	1716	946	53.7	53.4	493	4.05	5.56*	4.79	3.57	2.34	1.97
Carbohydrate A[b]	647	182	14.0	8.97	420	1.03	1.12	0.683	0.297	0.228	1.23*
Carbohydrate B	584	141	8.54	6.12	436	0.747	0.730	0.538	0.461	0.227	1.09*
Hexosamine	3.69	2.93	0.157	0.171	1.48	0.006	0.016*	0.012	0.009	0.007	0.010
Sialic acid	42.9	120	27.8	24.2	157	2.16	3.48*	1.53	1.30	0.98	0.69
RNA	20.2	4.22	0.355	0.558	12.6	0.054*	0.027	0.015	0.019	0.012	0.023
Total organic[c]	3919	1685	149	212	1990	9.90	14.82*	11.38	10.13	6.39	5.26
Dry weight	5361	3785	164.0	168.5	2670	11.51	15.73*	13.82	12.70	9.17	7.94

[a] Determined by subtraction of phospholipid + sterol from total lipid.
[b] Carbohydrates A and B determined by phenol-sulfuric and anthrone reaction, respectively.
[c] Sum of protein, lipid, and carbohydrate A.
* Highest value.
Source: Sullivan et al. (1974).

at two or more dilutions of sample in duplicate or triplicate. For details of analysis for diatom membrane protein, carbohydrate, RNA, phospholipid, sterol, sialic acid, and hexosamine, see Sullivan et al. (1974) and Table 5-1. Other physical characteristics, such as gravimetric analysis of dried membranes and membrane density as determined by migration on the sucrose gradient, may be of limited use.

3. Enzyme analysis. Enzymes are often used as markers for identifying various subcellular organelles and membranes. These are generally identified by isolation of morphologically recognizable membranes and determination of associated enzymes, or by histochemical staining of enzyme products in intact cells and visual localization by electron microscopic techniques. Well-established markers for animal cell membranes are: for plasma membrane, (Na^+, K^+)-ATPase and 5′-nucleotidase; for Golgi membrane, thiamine pyrophosphatase; for mitochondria, succinate dehydrogenase and cytochrome c oxidase. Except for 5′-nucleotidase, the same markers appear to apply to the diatom and to some extent to higher plants such as oats and barley (Leonard et al. 1973).

All enzyme assays should be carried out under optimal conditions determined for each enzyme in each species examined. The enzyme activity must be linear with amount of protein added. This generally falls into the range of 10–200 μg protein per assay. A variety of total and specific activities of enzymes associated with *N. alba* membranes is shown in Tables 5-2 and 5-3. Specific assay conditions can be found in Sullivan et al. (1974).

IV. Sample data

In membrane isolation studies, clear and concise data presentation is a matter of good bookkeeping (see Tables 5-1 to 5-3). It is most important to point out that the total activity and specific activity of all chemical and enzyme markers must be analyzed and reported for the total lysate, as all further determinations are based on these values.

A. Yield

One must depend on chemical or enzyme markers to estimate the total amount of membrane (in relative units) present in the total cell lysate. When this value is compared with the total units of enzyme activity in any desired fraction, a yield can be determined. Yield indicates how much of the total starting material has been isolated.

$$\text{yield} = \frac{\text{enzyme units in isolated membrane fraction}}{\text{enzyme units in total lysate}} \qquad (1)$$

Table 5-2. *Total enzyme activities ($\mu mol\ h^{-1}$) present in membrane fractions derived from 10^{10} cells of Nitzschia alba*

Enzyme	Fractions					Gradient fractions					
	Lysate	P-1	P-2	P-3	S-3	I	II	III	IV	V	VI
Mg^{2+}-ATPase	1196	222	173	157	456	4.74	10.8	12.9	24.2	16.2	4.04
(Na^+, K^+, Mg^{2+})-ATPase	2639	797	395	1084	359	26.2	112	132	179	104	39.0
Thiamine pyrophosphatase	1404	531	99.6	120	104	17.6	42.6	20.8	4.70	1.61	2.51
Acid phosphatase	637	179	122	97.2	164	8.13	16.1	8.99	5.36	3.13	2.13
UDPase	533	51.0	25.2	63.9	277	8.17	30.2	12.9	5.03	4.57	2.17
Phosphodiesterase	13.0	0	1.29	3.22	7.48	0.42	0.58	0.25	0.18	0.19	0.07
Glucose-6-phosphatase	104	10.9	1.94	20.4	59.8	1.88	3.82	1.95	1.02	0.66	0.17
Alkaline phosphatase	91.0	29.1	5.18	4.83	29.9	0.38	0.52	0.25	0.14	0.09	0.03
5'-Nucleotidase	312	—	3.88	21.5	37.4	1.92	1.86	1.19	0.56	0.38	0.10
Succinic dehydrogenase	1508	589	529	227	643	0.28	0.87	7.58	8.34	18.9	27.4
Cytochrome *c* oxidase[a]	2821	629	471	56.9	0	0.84	2.90	5.08	9.08	8.37	4.36
NADH cytochrome *c* reductase	5200	1092	1488	320	2992	10.5	23.2	42.2	156	85.3	12.0
NADPH cytochrome *c* reductase	—	—	—	—	—	12.5	12.2	13.1	2.79	5.68	0.00

[a] Relative units.

Source: Sullivan et al. (1974).

Table 5-3. *Specific activities (μmol h⁻¹ mg⁻¹ protein) of enzymes associated with membrane fractions derived from 10¹⁰ cells of Nitzschia alba*

Enzyme	Fractions					Gradient fractions					
	Lysate	P-1	P-2	P-3	S-3	I	II	III	IV	V	VI
Mg²⁺-ATPase	0.92	0.61	2.67	2.92	0.61	1.36	1.86	3.04	5.19	5.13	2.35
(Na⁺, K⁺, Mg²⁺)-ATPase	2.03	2.19	6.11	20.2	0.48	7.51	19.4	31.2	38.4*	32.8	22.7
Thiamine pyrophosphatase	1.08	1.46	1.54	2.25	0.14	5.03	7.35*	4.90	1.01	0.51	1.46
Acid phosphatase	0.490	0.493	1.88	1.81	0.22	2.33	2.77*	2.12	1.15	0.99	1.24
UDPase	0.41	0.14	0.39	1.19	0.37	2.43	5.20*	3.04	1.08	0.98	1.26
Phosphodiesterase	0.01	0.00	0.02	0.06	0.01	0.12*	0.10	0.06	0.04	0.06	0.04
Glucose-6-phosphatase	0.08	0.03	0.03	0.38	0.08	0.54	0.66*	0.46	0.22	0.21	0.10
Alkaline phosphatase	0.07	0.08	0.08	0.09	0.04	0.08	0.09*	0.06	0.03	0.03	0.02
5'-Nucleotidase	0.24	—	0.06	0.40	0.05	0.55*	0.32	0.28	0.12	0.12	0.06
Succinic dehydrogenase	1.16	1.62	8.18	4.23	0.86	0.08	0.15	1.79	4.47	6.00*	1.62
Cytochrome c oxidase[a]	2.17	1.73	7.29	1.06	0	0.24	0.50	1.20	1.95	2.65*	2.54
NADH cytochrome c reductase ($\times10^{-3}$)	4.0	3.0	23	6.0	4.0	3.0	4.0	10	37*	27*	7.0
NADPH cytochrome c reductase ($\times10^{-3}$)	—	—	—	—	—	3.6	2.1	3.1	0.6	1.8	0.0

[a] Relative units. * Highest value.
Source: Sullivan et al. (1974).

The yield of (Na^+, K^+, Mg^{2+})-ATPase in membrane fraction P-3 (Table 5-2) is 100 (1084/2639) = 41%.

B. Enrichment

By comparing the specific activity of a given enzyme from a partially purified fraction with the specific activity of the same enzyme in the total lysate, one obtains an estimate of enrichment.

$$\text{enrichment} = \frac{\text{enzyme specific activity in fraction}}{\text{enzyme specific activity in total lysate}} \qquad (2)$$

Thus the enrichment of the (Na^+, K^+, Mg^{2+})-ATPase in fraction P-3 (Table 5-3) is:

$$\text{enrichment} = \frac{20.2 \text{ units (mg protein)}^{-1}}{2.0 \text{ units (mg protein)}^{-1}} = 10.1\text{-fold} \qquad (3)$$

Note that the maximum possible enrichment decreases with increasing amount of a membrane in a cell. Thus if the plasma membrane constitutes 2% of the cell protein, the maximum enrichment for pure plasma membrane is 50-fold; if mitochondria make up 20% of the total cell protein, the maximum enrichment for pure mitochondria is 5-fold.

C. Purity

Estimation of the purity of a membrane preparation is most difficult. Having determined yield and enrichment based on *several* reliable chemical and/or enzyme markers for a specific membrane, one must then test for the presence of other markers that are indicative of other membranes; if they are present, the preparation is contaminated with some extraneous membrane material. If there is no evidence of other marker enzymes, one must consider the possibility that extraneous membranes contaminate the preparation but are unrecognized because their associated enzymes have not been tested for.

D. Identification

Based on the combined data, the membrane fractions isolated from the discontinuous sucrose gradient are characterized and identified as follows. The light component of the smooth membranes (fraction I) is isolated at the top of the gradient as large vesicles and sheets. It is distinguished chemically by the highest specific content of lipid, phospholipid, neutral lipid, carbohydrate, sialic acid, and RNA, and enzymatically by the highest specific activity of NADPH, cytochrome *c* reductase, phosphodiesterase, and 5'-nucleotidase. This fraction contains the lowest sterol/phospholipid molar ratio and the lowest Na^+

plus K^+ stimulation of Mg^{2+}-ATPase, and is probably enriched in endoplasmic reticulum.

The heavy component of the smooth membranes (fraction II) shows the highest total content of all chemical components analyzed except RNA and is similar to the light fraction in specific chemical content. It is enzymatically distinguished by the highest specific activities of thiamine pyrophosphatase, uridine diphosphatase, acid phosphatase, alkaline phosphatase, and glucose-6-phosphatase. Negatively stained samples prepared for electron microscopy reveal the presence of large sacs and vesicles with tubular structures typical of Golgi apparatus isolated from plant and animal cells. The heavy component appears to be enriched in Golgi membranes approximately 13-fold.

The plasma membranes (fractions IV and V) are distinguished by their high sterol content, high sterol/phospholipid molar ratio, and a low carbohydrate specific content as compared with the smooth membranes. They contain the highest specific and total activities of both the Mg^{2+}-ATPase and the (Na^+, K^+, Mg^{2+})-ATPase activities. Based on these enzymes, the plasma membranes are purified approximately 20-fold.

V. Discussion

An understanding of algal membrane structure and function ultimately depends on the complete chemical and enzymatic characterization of the membranes. A major aim of this chapter is to demonstrate two approaches to the problem of disruption of algal cells and to illustrate morphological, chemical, and enzymatic techniques useful in determining the fate of various membranes during their subsequent fractionation.

It is difficult to isolate and identify the membrane systems of microalgae for several reasons:

1. The tough cell walls are not easily disrupted by methods that leave the membranes in a morphologically recognizable form. We have taken advantage of the unique property of *N. alba,* which sheds its cell wall to form protoplasts when grown in a nutrient-rich medium lacking Ca^{2+} and Mg^{2+}, to avoid this troublesome barrier to algal membrane studies. Similar advantages might also be gained by using cell-wall-less mutants of algae such as *Chlamydomonas* CW 15 (Hyams and Davies 1972).

2. Standards for determining the identity and purity of algal or plant membrane preparations have not been fully established. Identification of algal membranes is difficult because of the paucity of specific membrane markers and because of the ubiquity of membrane-associated ATPase in both higher plants (Leonard et al. 1973) and

algae (Sullivan et al. 1975; Okita et al. 1976). The multiplicity of membrane-bound ATPases from higher plants and algae is presumably responsible for much of the confusion and apparently contrasting reports concerning ATP hydrolyzing activities among them. The complexity of the problem, as illustrated in the reports mentioned above, will, it is hoped, lead to some clarification by stimulating further research.

In addition to the methods already discussed for determining membrane purity, a promising staining technique using phosphotungstic acid recently reported by Sundberg and Lembi (1976) allows a visual as well as a quantitative appraisal of the plasmalemma of algal cells. An even better stabilization and specific labeling technique for plasma membranes using concanavalin A and diazotized ^{35}S-sulfanilic acid was recently reported by Scarborough (1975). This technique should be most useful for localization of plasma membranes of cell-wall-less cells or mutants and algal protoplasts.

3. Major limitations in the methods reported here include relatively low membrane yields, especially at the first step of fractionation. However, the yield might be improved by increasing the lysate volume and by addition of deoxyribonuclease to the lysate to lower the viscosity of the nucleoprotein gel caused by nuclear disruption. Also, if one is attempting to isolate a particular membrane, the use of a continuous sucrose gradient might be more productive than the discontinuous gradient because the former would allow finer resolution of membranes of similar density.

Elucidation of chemical or enzymatic markers for nuclear membrane, tonoplast, and silicalemma would allow their localization on the gradient. Presumably these membranes are on the gradient and present as unrecognized contaminants of the identified-membrane-enriched fractions. Further studies of the tentatively identified membranes and histochemical stains of whole cells might offer confirmation of the properties and identities of the diatom membranes.

VI. Acknowledgments

Research was supported by the National Institutes of Health grant GM-08229-13-14 to B. E. Volcani. The author was supported by the Department of Biological Sciences, University of Southern California, during preparation of the manuscript. The figures and tables are reproduced with the permission of Academic Press, New York.

VII. References

Burkholder, P. R., and Nickell, L. G. 1949. Atypical growth of plants. I. Culture of virus tumors of *Rumex* on nutrient agar. *Bot. Gaz.* 110, 426–37.

Hemmingsen, B. B. 1971. A mono-silicic acid stimulated adenosine triphosphatase from protoplasts of the apochlorotic diatom *Nitzschia alba*. Ph.D. thesis, University of California, San Diego. 132 pp.

Hyams, J., and Davies, D. R. 1972. The induction and characterisation of cell wall mutants of *Chlamydomonas reinhardi. Mutat. Res.* 14, 381–9.

Leonard, R. T., Hansen, D., and Hodges, T. K. 1973. Membrane-bound adenosine triphosphatase activities of oat roots. *Plant Physiol.* 51, 749–54.

Mehard, C. W., Packer, L., and Abraham, S. 1971. Activity and ultrastructure of mitochondria from mouse mammary gland and mammary adenocarcinoma. *Cancer Res.* 31, 2148–60.

Okita, T., Sullivan, C. W., and Volcani, B. E. 1976. Gel electrophoresis of ion-stimulated ATPases from protoplast membranes of the diatom *Nitzschia alba. Plant Sci. Lett.* 6, 129–34.

Paul, J. S., Sullivan, C. W., and Volcani, B. E. 1975. Photorespiration in diatoms: Mitochondrial glycolate dehydrogenase in *Cylindrotheca fusiformis* and *Nitzschia alba. Arch. Biochem. Biophys.* 169, 152–9.

Reynolds, E. S. 1963. The use of lead citrate at high pH as an electron-opaque stain in electron microscopy. *J. Cell Biol.* 17, 208–12.

Scarborough, G. A. 1975. Isolation and characterization of *Neurospora crassa* plasma membranes. *J. Biol. Chem.* 250, 1106–11.

Sturgess, J. M., Katona, E., and Moscarello, M. A. 1973. The Golgi complex. I. Isolation and ultrastructure in normal rat liver. *J. Membr. Biol.* 12, 367–84.

Sullivan, C. W., Volcani, B. E., Lum, D., and Chiappino, M. L. 1974. Isolation and characterization of plasma and smooth membranes of the marine diatom *Nitzschia alba. Arch. Biochem. Biophys.* 163, 29–45.

Sullivan, C. W., Volcani, B. E., and Lum, D. 1975. Multiple ion-stimulated adenosine triphosphatase activities associated with membranes of the diatom *Nitzschia alba. Arch. Biochem. Biophys.* 167, 437–43.

Sundberg, I., and Lembi, C. A. 1976. Phosphotungstic acid–chromic acid: A selective stain for algal plasma membranes. *J. Phycol.* 12, 48–54.

Ulsamer, A. G., Wright, P. L., Wetzel, M. G., and Korn, E. D. 1971. Plasma and phagosome membranes of *Acanthamoeba castellanii. J. Cell Biol.* 51, 193–215.

Section II

Analysis of chemical constituents

6: Chlorophylls and carotenoids

ARNE JENSEN

Institute of Marine Biochemistry,
University of Trondheim, N-7034 Trondheim-NTH, Norway

CONTENTS

I	**Introduction**	*page* **60**
II	**Chlorophylls**	**61**
	A. General precautions	61
	B. Extraction	61
	C. Measurement in crude extracts	62
	D. Separation by chromatography	63
	E. Quantitative measurement of separated pigments	64
III	**Carotenoids**	**64**
	A. Extraction	64
	B. Measurement in crude extracts	65
	C. Fractionation into polar and nonpolar pigments	65
	D. Saponification	66
	E. Separation by chromatography	66
	F. Quantitative measurement of separated pigments	68
	G. Problems	68
IV	**References**	**69**

I. Introduction

Chlorophylls are key compounds in plants for trapping light energy for photosynthesis; thus, their quantitative determination is of great importance in studies of photosynthesis, primary production, and related subjects. Chlorophylls other than chlorophyll *a* are distributed among the algal classes in a highly systematic way, making these compounds of great value in chemotaxonomic studies. Furthermore, these pigments are rapidly degraded during starvation, in stress periods, and upon death of the cells, rendering knowledge of presence and concentration of the various pigments and their breakdown products valuable indicators of the physiological state of the plant cells (Jensen and Sakshaug 1973).

These are some reasons for the need for specific and correct methods of qualitative and quantitative analysis of chlorophylls. Despite the many spectrophotometric equations available (Richards with Thompson 1952; Strickland and Parsons 1968) and the improvements introduced (Scor-Unesco 1966; Jeffrey and Humphrey 1975), the determination of chlorophylls in mixture gives nothing more than a rough estimate. Accurate and reliable determinations can be obtained only after separation of the compounds by suitable chromatographic procedures.

Carotenoids are ubiquitous polyenes studied mainly for their chemotaxonomic importance (Goodwin 1976), although some (e.g., fucoxanthin and peridinin) are active in collecting light for photosynthesis. They share with the chlorophylls a high sensitivity to oxygen, light, and acids. Most carotenoids are, however, relatively stable toward bases and can be worked up after saponification, a step that helps to reduce the problems caused by other lipids often present in the extracts. Some carotenoids, such as fucoxanthin and the carotenol esters, are altered by treatment with alkali, and it is recommended that extracts of unknown composition be tested by qualitative chromatography before and after saponification to detect possible changes occurring during the treatment with alkali.

Generally the precautions given for analysis of chlorophylls are necessary also when working with carotenoids. These include dim light, peroxide-free solvents, protection against exposure to acids, an oxygen-free atmosphere, and low temperature.

II. Chlorophylls

A. General precautions

The chlorophylls are labile compounds that may be easily degraded by both chemical and biochemical reactions. They are easily oxidized, especially in strong light, and they are immediately attacked by bases and acids. It is, therefore, necessary to work in dim light, to reduce the oxygen content in solvents, and to make use of protecting agents such as CO_2, carbonates, H_2S, or buffers. Good procedures are based on rapid workup of fresh material with no storage of raw material, extracts, or separated pigments if this can be avoided. If raw material must be stored, this should take place at -20 C or lower. It is especially important to ensure that all solvents used are free from peroxides, freshly distilled, and neutral.

If average values for the pigment content of algal populations are sought, it is important to secure representative samples. For the common seaweeds at least 20 individual plants must be collected (Baardseth and Haug 1953), rapidly chopped, or ground in a suitable mill (e.g., a Wiley Laboratory Mill fitted with a 2-mm perforated screen of stainless steel). The sample for extraction is apportioned after thorough mixing of the ground material.

B. Extraction

The best results are obtained with fresh material extracted directly with acetone or methanol. Usually chopped or ground seaweed samples (1–3 g) are best extracted with small portions of acetone in a mortar. The portions of extracts are transferred by a Pasteur pipette to a medium-porosity sintered glass filter, and the residual material is broken down by the pestle until a fine powder is obtained and no more colored material can be extracted. Some red seaweeds need special treatment because of their high content of lipoxidase. The algal material may be treated for 20 sec in hot steam immediately prior to extraction or dipped for 20 sec in boiling water; or the material may be crushed with dry ice in the mortar and extracted with a mixture of dry ice and acetone or methanol.

Planktonic algae are best extracted after collection on a glass fiber filter (Whatman GF/C covered with a thin layer of $MgCO_3$) with methanol that contains H_2S (Jensen and Sakshaug 1973). The filter is transferred to a 10-ml centrifuge tube, and the algal layer is scraped off together with the $MgCO_3$ and a thin layer of the glass fiber filter. The rest of the filter contains relatively few pigmented cells and is transferred to another centrifuge tube. Methanol (8 ml) containing H_2S is added to both tubes, and the algal material is thoroughly ground together with the glass fiber by means of a glass stirring rod. The debris is now centrifuged down and the supernatants from both tubes are transferred to a graduated cylinder by means of a small pipette. The total volume of the chlorophyll extracts must be recorded.

C. Measurement in crude extracts

For total chlorophyll *a* determination the absorbancy of the pigment extract at specified wavelengths is recorded. For brown and red seaweeds as well as most of the marine planktonic algae, the absorption at 666 nm (in acetone) corrected for turbidity by subtraction of the reading at 730 nm may be used for the estimation of the content of chlorophyll *a* according to the following formula:

$$c = \frac{(D_{666} - D_{730}) \times V \times 10}{890} \tag{1}$$

where c is the total amount of chlorophyll *a* (in milligrams) in the volume V (in milliliters) and D_{666} and D_{730} are the absorbancies at 666 and 730 nm, respectively, read in a 1.0-cm cell. The equation is based on the specific extinction coefficient, $E_{1\,cm}^{1\%} = 890$ (Stoll and Wiedemann 1959; Jensen 1966). If the absorbancy at 730 nm exceeds 0.010, the extract should be refiltered or discarded.

An estimation of the sum of chlorophylls c_1 and c_2 for the same plants may be obtained by determining the light absorption in 90% acetone at 664 and 630 nm and applying the equation of Jeffrey and Humphrey (1975):

$$\text{chlorophylls } c_1 + c_2 = 24.36\,D_{630} - 3.73\,D_{664} \tag{2}$$

A "trichromatic" method is needed to measure chlorophylls *a*, *b*, and $c_1 + c_2$ in extracts of mixed phytoplankton that contain green algae (Jeffrey and Humphrey 1975):

$$\text{chlorophyll } a = 11.85\,D_{664} - 1.54\,D_{647} - 0.08\,D_{630} \tag{3}$$

$$\text{chlorophyll } b = -5.43\,D_{664} + 21.03\,D_{647} - 2.66\,D_{630} \tag{4}$$

$$\text{chlorophylls } c_1 + c_2 = -1.67\,D_{664} - 7.60\,D_{647} + 24.52\,D_{630} \tag{5}$$

All absorbancies should be corrected for turbidity (by subtraction of the reading at 730 nm), and the solvent system again is 90% acetone. The values obtained by the equations must be multiplied by the volume of the extract (in milliliters) and divided by the volume of the water sample filtered (in liters). This will give the chlorophyll content in micrograms per liter (or milligrams per cubic meter).

It must be realized that the values obtained for the various chlorophylls, including those for chlorophyll *a*, are only approximate. The presence of degradation products such as pheophytins, chlorophyllides, and pheophorbides will interfere. This can be corrected for in part (Lorenzen 1967). For the determination of extremely small amounts of pigments found in ocean water, fluorimetric methods have been developed (Yentsch and Menzel 1963; Holm-Hansen et al. 1965; Loftus and Carpenter 1971). These methods require less than one-tenth the quantities needed for ordinary spectroscopy.

D. Separation by chromatography

For quantitative determination of chlorophylls, circular chromatography on kieselguhr paper is the method of choice, thanks to its rapidity, selectivity, high recovery, and eminent resolving properties. Chlorophylls *c* may be successfully separated on thin layers, but quantitative recovery is not so easily obtained by this method. Several authors (Jeffrey 1968; Holden 1976) still prefer two-dimensional thin-layer chromatography for quantitative determination of chlorophylls and carotenoids. The reason may be the lack of experience with circular chromatography on kieselguhr paper.

Prior to the chromatographic determination, an aliquot (containing 0.1–1 mg total chlorophylls) of the plant extract is rapidly taken to dryness in vacuo at room temperature and the pigments transferred quantitatively to a volumetric flask of 2- to 5-ml capacity using acetone or acetone-methanol to dissolve the pigments. A small amount of the concentrate (10–50 μl, containing 20–50 μg of total chlorophylls) is applied to the center of the chromatographic paper No. 287, 18 cm diameter (Schleicher and Schüll) by means of a micropipette, forming a spot 10–15 mm in diameter. The solvent is continuously removed by a stream of N_2 directed from below. A wick (ca. 5 mm wide) is cut in the paper to the center of the spot, and the paper is placed on a Petri dish (16 cm diameter) that contains 10% acetone in petroleum ether (boiling range 60–80 C). The dish is covered by a glass plate. The solvent is allowed to creep up the wick and to migrate to the rim of the Petri dish. During this development the dish should be covered by a piece of black cloth.

The developed chromatogram is removed from the dish, the solvent front and the positions of the pigmented zones are rapidly marked under dim light with a pencil and measured for later calculations of relative migration rates (R_f values).

The pigmented zones (chlorophyll *a*, deep green; chlorophyll *b*,

Table 6-1. *Wavelengths of maximum light absorption and specific extinction coefficients for algal chlorophylls*

Chlorophyll	λ_{max}(nm)	$E_{1\,cm}^{1\%}$	Reference
a	666 (100% acetone)	890	Jensen (1966)
b	647 (90% acetone)	514	Jeffrey and Humphrey (1975)
c_1	629 (100% acetone)	392	Jeffrey and Humphrey (1975)
c_2	630 (100% acetone)	372	Jeffrey and Humphrey (1975)

grass green; and chlorophylls c, yellowish green) are rapidly cut out with scissors and packed tightly into glass tubes that have been drawn out to a capillary in one end. The chlorophylls are eluted with acetone directly into 5-ml volumetric flasks, which then are filled to the mark with acetone.

The R_f values of the different chlorophylls are characteristic properties of the compounds. They may vary, however, depending on the moisture content of the paper, trace impurities of alcohol, and with the content of lipids in the extract. The sequence of the pigments, chlorophyll a followed by b and with chlorophylls c_1 and c_2 near the center, is, however, always the same as long as Schleicher and Schüll No. 287 paper and 10% acetone in petroleum ether are used.

E. Quantitative measurement of separated pigments

The pigments eluted from the colored zones of the paper chromatograms are usually pure enough to give nearly ideal absorption spectra. Recording of the absorbancy at one wavelength is, therefore, possible and sufficient. The absorption at 730 nm should be checked routinely for turbidity caused by cellulose fibers or kielselguhr particles. The characteristic wavelengths and the corresponding extinction values are given in Table 6-1. From the $E_{1\,cm}^{1\%}$ (absorbancy of a 1% solution in a cuvette of 1 cm pathlength at the specified wavelength) values, and the absorbancies obtained for the eluted samples, their pigment concentrations can be calculated.

III. Carotenoids

A. Extraction

Small-scale extraction (1–3 g of wet material) is best carried out on fresh material in a mortar, using acetone and/or methanol. Larger samples may be treated in a Waring Blendor with acetone and dry ice. To facilitate the pigment extraction from dried (and finely ground) material, samples should be moistened with twice their weight of water for 2–5 min before the extracting solvent is added.

The extraction in the mortar is carried out as described (II.B). Neutral or neutralized (by addition of $CaCO_3$) extracts may be stored overnight at -20 C in the dark after thorough purging with N_2 gas. Sometimes xanthophylls occur as protein complexes, and this is probably the case for fucoxanthin in brown algae and peridinin in dinoflagellates. Such complexes can be extracted by water or buffer solutions. They may be purified by gel filtration on Sephadex and will give very pure carotenoid pigments upon treatment with acetone (Thommen 1971, p. 656).

B. Measurement in crude extracts

In many cases an estimate of the total carotenoid content is desired and may be obtained from the total extract provided negligible amounts of chlorophylls are present. This can be checked by measuring the absorbancy of the suitably diluted extract at approximately 450 and 670 nm in the spectrophotometer. If the former absorption is 10 times the latter, the content of carotenoids may be estimated from the maximum absorbancy measured in the 450-nm region as follows:

$$C = \frac{D \times V \times f \times 10}{2500} \tag{6}$$

The total amount of carotenoids is then C (in milligrams), the absorbancy D in a 1.0-cm cell, the volume (milliliters) of the original extract is V, and the dilution factor is f. It is assumed that the pigments have an average extinction coefficient of 2500.

Chlorophylls, when present, may be removed by saponification. The total amount of carotenoids is then determined in the ether extract obtained (see III.D).

If alkali-labile carotenoids are involved, the content of polar carotenoids may be measured in the hypophase after phase separation as described in III.C.

C. Fractionation into polar and nonpolar pigments

To facilitate the chromatographic separation of the individual carotenoids, and to gain rapid information on the proportion of polar to nonpolar pigments, a partitioning of the compounds between two immiscible solvents may be introduced. For this purpose, either the extract must be taken to dryness in vacuo at room temperature and dissolved in equal volumes of petroleum ether and aqueous (4–40% water) methanol, or the pigments of the extract may be transferred to diethyl ether by the addition of ether and water (containing 2–5% NaCl to reduce the formation of emulsions). The ether phase must be washed with more water to remove acetone or methanol and dried over anhydrous Na_2SO_4. The solvent is finally removed at reduced pressure, and the residual pigments are dissolved in the two-solvent system mentioned above. The nonpolar carotenes, and the carotenol esters as well as chlorophylls *a* and *b*, will enter the petroleum ether phase (epiphase); the xanthophylls (oxygenated carotenoids) and chlorophylls *c* will remain in the methanolic lower phase (hypophase).

D. Saponification

Chlorophylls and interfering lipids may be removed by a saponification step. This will also cleave any carotenoid esters present and allow identification of the component carotenol. The treatment can be used only for carotenoids that are stable to alkali.

The crude pigment extracts must be taken to dryness before saponification. It is especially important that all the acetone is removed. Use a good vacuum pump. The residual lipids are dissolved in a small volume of ether, and an equal volume of 10% methanolic KOH solution is added. The saponification takes place at room temperature for 1–2 h in the dark; then water is added (containing 2–5% NaCl) to a separatory funnel, and the ether layer is separated from the aqueous phase. The latter is extracted several times with more ether and the combined extracts washed three times with water to remove alkali and methanol. The ether may then be dried over anhydrous Na_2SO_4.

E. Separation by chromatography

In nature the carotenoids almost always occur in mixtures, and separation into individual pigments is needed both for accurate identification and especially for quantitative determination. For larger samples and for preparative purposes (more than 0.1 mg of pigment), column chromatography is more convenient, although thick-layer plates are preferable for small-scale isolation of pure pigments for mass spectroscopy.

Column chromatography is so well known that only special adsorbents and solvent systems will be mentioned. A detailed description of the application of column chromatography to the carotenoid field is given by Liaaen Jensen and Jensen (1965). Al_2O_3 or $Ca(OH)_2$ are the most useful adsorbents for the separation of the epiphasic carotenoids (carotenes); MgO, $CaCO_3$, or silica gel are more suitable for the polar carotenoids of the hypophase (xanthophylls). In most cases petroleum ether containing varying amounts of acetone or diethyl ether is used as the first eluant. Acetone should be avoided when working with Al_2O_3 columns. Elution of the more polar pigments requires increasing concentrations of methanol. It is important that the columns are not overloaded with pigments. Usually a 1×25-cm column will carry up to 1 mg of total carotenoids, and a 4×30- to 4×40-cm column will carry 10–20 mg of pigments.

Thin-layer chromatography is especially well suited for two-dimensional separation and identification of very complex mixtures of carotenoids. Glass, plastic, or aluminum plates precoated with layers of Al_2O_3, silica gel, or cellulose are commercially available. Davies (1976) discussed in detail the application of thin-layer chromatography to the separation of carotenoids. The plates (usually 20×20 cm) should be activated by heating to 110 C for 1 h prior to use. Cooling must be carried out in a desiccator. The pigment mixture (10–30 μg, preferably dissolved in diethyl ether or acetone) is applied as a small spot in one corner of the plate, and the development is carried out in a closed glass tank containing the first solvent system. When the solvent front has moved 15–17 cm up the plate, the chromatogram is removed, dried rapidly in a stream of N_2, turned 90°, placed in a second tank, and developed with the second solvent system. Known carotenoids may be added to the mixture to demonstrate identity with components of the extract. Colored spots may also be scraped off the plates and extracted in a small funnel fitted with a sintered glass filter. The light absorption spectrum of the compound may thus be determined.

A versatile system based on thin layers of cellulose and 25% chloroform in petroleum ether (boiling range 60–80 C) for the first and 2% *n*-propanol in petroleum ether for the second development has been used for algal carotenoids by Jeffrey (1974).

The relative migration rates (R_f values) of the carotenoids, although indicative of identity, vary quite considerably with the activity of the adsorbent (moisture content) and with the quality of the solvent (traces of alcohol in the diethyl ether may increase R_f values significantly). Co-chromatography is therefore required for more reliable identification. Useful lists of R_f values are given by Davies (1976) and Bolliger and König (1969).

Paper chromatography is also the method of choice for quantitative determination of carotenoids when adsorbent-loaded papers and radial development are used (Jensen and Liaaen Jensen 1959; Jensen 1963; Davies 1976). The analytical procedure has been described in II.D. Also for qualitative work, chromatography on kieselguhr paper (Schleicher and Schüll No. 287) gives rapid and good resolution of many carotenoids. Extraction is easy and makes determination of visible absorption spectra a simple task. The R_f values are more repro-

Table 6-2. *Principal absorption maxima and specific extinction coefficients for some algal carotenoids*

Carotenoid	λ_{max}(nm)	$E_{1\ cm}^{1\%}$	Reference
β-Carotene	453 (petroleum ether)	2592	Schwieter et al. (1965)
γ-Carotene	462 (petroleum ether)	3100	Schwieter et al. (1965)
Diatoxanthin	453 (acetone)	ca. 2300	
Fucoxanthin	449 (acetone)	1600	Jensen (1966)
Lutein	445 (acetone)	ca. 2500	
Neoxanthin	442 (acetone)	ca. 2240	
Peridinin	466 (acetone)	1340	Jeffrey and Haxo (1968)
Violaxanthin	442 (acetone)	2400	Jensen (1966)
Zeaxanthin	452 (acetone)	2340	Aasen and Liaaen Jensen (1966)

ducible than those obtained on thin layers. In addition, the separation of *cis-trans*-isomers of many carotenoids is best achieved on kieselguhr paper. Co-chromatography of iodine-catalyzed isomerization mixtures of carotenoids on such paper forms a rather reliable indicator of identity (Jensen 1963). Lists of R_f values of carotenoids on various adsorbent-loaded paper have been compiled by Jensen and Liaaen Jensen (1959) and Jensen (1963, 1966).

F. Quantitative measurement of separated pigments

The solutions of single carotenoids obtained by extraction either from thin-layer plates or from paper chromatograms are, after adjustment to a known volume, measured in the spectrophotometer. Either the entire spectrum in the visible range, or the absorbancy at characteristic wavelengths, is obtained. Comprehensive lists of main absorption maxima of carotenoids in a number of solvents, together with an extensive list of specific extinction coefficients ($E_{1\ cm}^{1\%}$), are given in the review by Davies (1976). Table 6-2 gives the main absorption maxima and extinction values for a number of common algal carotenoids. For calculating the pigment concentration, use equation 1 after substituting the appropriate absorbancies and extinction coefficients. Recoveries from paper chromatograms are usually 95% or higher.

G. Problems

Identification of the chlorophylls presents an easy problem, but it is a formidable task to deliver a decisive proof of the identity of a carotenoid. R_f values in several chromatographic systems together with the visible light spectrum are not sufficient because the number of known carotenoids amounts to several hundreds and many are very closely related. Infrared and nuclear magnetic resonance spectra are needed, together with mass spectra, which, because of the specific fragmentation patterns, are very useful.

The *cis-trans*-isomerization that easily takes place around many of the carbon-carbon double bonds of the polyene chain may also constitute a confusing problem to the novice. The pigment in a single zone of a chromatogram may give rise to two or more zones upon rechromatography, and this formation of *cis*-isomers reduces the light extinction of the carotenoid in question and leads to an underestimation of pigment concentration.

Acid lability of the carotenoid epoxides is another complicating factor. It is very difficult to decide whether a furanoid isomer is a native pigment or an artifact formed by acid-catalyzed rearrangement of the corresponding epoxide. These and many other problems are discussed in the comprehensive treatises of Karrer and Jucker (1950), Isler (1971), and Goodwin (1976).

IV. References

Aasen, A. J., and Liaaen Jensen, S. 1966. Carotenoids of flexibacteria. IV. The carotenoids of two further pigment types. *Acta Chem. Scand.* 20, 2322–4.

Baardseth, E., and Haug, A. 1953. Individual variation of some constituents in brown algae, and reliability of analytical results. *Nor. Inst. Tang-Tareforsk. Rep.* 2, 1–23.

Bolliger, H. R., and König, A. 1969. K vitamins, including carotenoids, chlorophylls and biologically active quinones. In Stahl, E (ed.). *Thin Layer Chromatography: A Laboratory Handbook,* 2nd ed., pp. 259–311. Springer Verlag, Berlin.

Davies, B. H. 1976. Carotenoids. In Goodwin, T. W. (ed.). *Chemistry and Biochemistry of Plant Pigments,* vol. 2, 2nd ed., pp. 38–165. Academic Press, London.

Goodwin, T. W. 1976. Distribution of carotenoids. In Goodwin, T. W. (ed.). *Chemistry and Biochemistry of Plant Pigments,* vol. 1, 2nd ed., pp. 225–61. Academic Press, London.

Holden, M. 1976. Chlorophylls. In Goodwin, T. W. (ed.). *Chemistry and Biochemistry of Plant Pigments,* vol. 2, 2nd ed., pp. 1–37. Academic Press, London.

Holm-Hansen, O., Lorenzen, C. J., Holmes, R. W., and Strickland, J. D. H. 1965. Fluorometric determination of chlorophyll. *J. Cons., Cons. Int. Explor. Mer.* 30, 3–15.

Isler, O. (ed.). 1971. *Carotenoids.* Birkhäuser Verlag, Basel. 932 pp.

Jeffrey, S. W. 1968. Quantitative thin-layer chromatography of chlorophylls and carotenoids from marine algae. *Biochim. Biophys. Acta* 162, 271–85.

Jeffrey, S. W. 1974. Profiles of photosynthetic pigments in the ocean using thin-layer chromatography. *Marine Biol.* 26, 101–10.

Jeffrey, S. W., and Haxo, F. T. 1968. Photosynthetic pigments of symbiotic dinoflagellates (zooxanthellae) from corals and clams. *Biol. Bull.* (*Woods Hole, Mass.*) 135, 149–65.

Jeffrey, S. W., and Humphrey, G. F. 1975. New spectrophotometric equations for determining chlorophylls *a, b, c*₁ and *c*₂ in higher plants, algae and natural phytoplankton. *Biochem. Physiol. Pflanz.* 167, 191–4.

Jensen, A. 1963. Paper chromatography of carotene and carotenoids. In Lang, K. (ed.). *Carotine und Carotinoide,* pp. 119–28. Steinkopff Verlag, Darmstadt.

Jensen, A. 1966. Carotenoids of Norwegian brown seaweeds and of seaweed meals. *Nor. Inst. Tang-Tareforsk. Rep.* 31, 1–138.

Jensen, A., and Liaaen Jensen, S. 1959. Quantitative paper chromatography of carotenoids. *Acta Chem. Scand.* 13, 1863–8.

Jensen, A., and Sakshaug, E. 1973. Studies on the phytoplankton ecology of the Trondheimsfjord. II. Chloroplast pigments in relation to abundance and physiological state of the phytoplankton. *J. Exp. Mar. Biol. Ecol.* 11, 137–55.

Karrer, P., and Jucker, E. 1950. *Carotenoids.* Elsevier, New York. 384 pp.

Liaaen Jensen, S., and Jensen, A. 1965. Recent progress in carotenoid chemistry. In Holman, R. T. (ed.). *Progress in the Chemistry of Fats and Other Lipids,* vol. 8, pp. 129–212. Pergamon Press, Oxford.

Loftus, M. E., and Carpenter, J. H. 1971. A fluorometric method for determining chlorophylls *a, b,* and *c. J. Mar. Res.* 29, 319–38.

Lorenzen, C. J. 1967. Determination of chlorophyll and phaeo-pigments: Spectrophotometric equations. *Limnol. Oceanogr.* 12, 343–6.

Richards, F. A., with Thompson, T. G. 1952. The estimation and characterization of plankton populations by pigment analyses. II. A spectrophotometric method for the estimation of plankton pigments. *J. Mar. Res.* 11, 156–72.

Schwieter, U., Bolliger, H. R., Chopard-dit-Jean, L. H., Englert, G., Kofler, M., König, A., v. Planta, C., Rüegg, R., Vetter, W., and Isler, O. 1965. Synthesen in der Carotenoid-Reihe. 19. Physikalische Eigenschaften der Carotine. *Chimia* 19, 294–302.

Scor-Unesco, Paris, Working Group 17. 1966. Determination of photosynthetic pigments in sea-water. *Monogr. Oceanogr. Methodol.* 1, 9–18.

Stoll, A., and Wiedemann, E. 1959. Die kristallisierten naturlichen Chlorophylle *a* und *b. Helv. Chim. Acta* 42, 679–83.

Strickland, J. D. H., and Parsons, T. R. 1968. *A Practical Handbook of Seawater Analysis.* Bull. 167, Fishery Research Board of Canada, Ottawa. 311 pp.

Thommen, H. 1971. Metabolism. In Isler, O. (ed.). *Carotenoids,* pp. 637–68. Birkhäuser Verlag, Basel.

Yentsch, C. S., and Menzel, D. W. 1963. A method for the determination of phytoplankton chlorophyll and phaeophytin by fluorescence. *Deep-Sea Res.* 10, 221–31.

7: Algal biliproteins

HAROLD W. SIEGELMAN AND
J. HELEN KYCIA

Biology Department,
Brookhaven National Laboratory, Upton, New York 11973

CONTENTS

I	**Introduction**	*page*	**72**
II	**Experimental materials**		**73**
III	**Methods**		**73**
	A. Determination of algal biliprotein concentrations		73
	B. Extraction of biliproteins from algal cells		74
	C. Purification of algal biliproteins		75
	D. Isolation of algal bile pigments		77
IV	**Acknowledgment**		**78**
V	**References**		**78**

I. Introduction

The algal biliproteins (phycobiliproteins) function as important light-harvesting components for driving the photosynthetic reactions in blue-green, red, and cryptomonad algae. They are composed of algal bile pigments (phycobilins) covalently linked to a specific protein. The chromoproteins are intensely fluorescent, water-soluble, and of three general types: the phycoerythrins, which are red, and the phycocyanins and allophycocyanins, which are blue. They are organized into granules (phycobilisomes) localized on the outer surface of the thylakoids (Gantt 1975) and are now classified on the basis of absorbance spectra and taxonomic origin into many distinctive kinds (Chapman 1973; Glazer 1976; ÓCarra and ÓhEocha 1976).

Only two algal bile pigments have been isolated from biliproteins: phycocyanobilin (left) and phycoerythrobilin (right) (Siegelman *et al.* 1968). The specific linkage of the bile pigment to the protein is not yet firmly established for any biliprotein (Bogorad 1975; ÓCarra and ÓhEocha 1976).

This chapter describes simple procedures for the detection, extraction, purification, and quantitative estimation of the algal biliproteins. These proteins can provide valuable information for algal classification, physiological, biochemical, and ecological studies, and serve as excellent material for gaining experience in methods of protein chemistry. Field examination of the principal algal biliproteins in vivo can be conveniently achieved with a hand spectroscope. References to detailed studies of the algal biliproteins and bile pigments are found

[72]

in recent reviews (Chapman 1973; Bogorad 1975; Glazer 1976; ÓCarra and ÓhEocha 1976; Bennett and Siegelman 1977).

II. Experimental materials

A wide variety of biliprotein-containing algae can be used as a source of the proteins. Culture techniques provide a dependable and consistent source of algae, and algae can also be harvested from marine and freshwaters where dense algal growth occurs. Although fresh or frozen algae are preferred, biliproteins can be extracted from dried but uncooked algae. Nori, dried *Porphyra tenera* (Lemberg 1928; Svedberg and Katsurai 1929), is available in oriental food stores, and dulse, dried *Rhodymenia palmata* (Chapman et al. 1967), is commercially obtainable.

III. Methods

A. Determination of algal biliprotein concentrations

The relative concentrations of biliproteins in algal cells can be measured spectrophotometrically by the Shibata opal glass technique (Shibata et al. 1954). For example, Jones and Myers (1965) determined the ratios of phycocyanin, chlorophyll *a,* and carotenoids in *Anacystis nidulans* by measuring the absorbance spectrum of intact cells by the Shibata techniques. The spectrum was measured from 400 to 730 nm and the absorbance at 730 nm was subtracted from the spectrum to correct for scattering. Similar measurements can be made with algal cells collected on membrane filters. The Shibata technique is useful primarily for comparing pigments ratios of algae grown under varied experimental or environmental conditions. It can also be used with portions of a thallus of a macroalga. The principal deficiencies of the method are that quantitation of a particular pigment is not of high precision, and a trace constituent may not be detectable.

The algal biliproteins can be extracted from the algae and their concentration measured spectrophotometrically for a more quantitative determination. Myers and Kratz (1955) suspended *A. nidulans* in water and broke the cells by prolonged sonication. The solution was centrifuged at $7000 \times g$ for 15 min and absorbancies determined. The extinction coefficients of several biliproteins listed in Table 7-1 permit the quantitation of purified biliproteins. A procedure for the quantitative extraction of biliproteins by the combination of freezing and sonication was described by Bennett and Bogorad (1973). The concentrations of C-phycocyanin, C-phycoerythrin, and allophyco-

Table 7-1. *Properties of algal biliproteins*

Biliprotein	Absorption maxima (nm)	Molecular weight	$E_{1\,cm}^{1\%}$
C-phycocyanin	615	224 000	77
Allophycocyanin	650	90 000	65
R-phycocyanin	553; 615	273 000	66
Cryptomonad phycocyanin 615	588; 615	—	—
Cryptomonad phycocyanin 630	588; 630	37 300	—
Cryptomonad phycocyanin 645	583; 645	—	—
C-phycoerythrin	565	226 000	125
B-phycoerythrin	565	290 000	82
R-phycoerythrin	500; 565	290 000	81
Cryptomonad phycoerythrin 544	544	—	—
Cryptomonad phycoerythrin 555	555	27 800	—
Cryptomonad phycoerythrin 568	568	35 000	—

cyanin in crude extracts of *Fremyella diplosiphon* (in milligrams per milliliter) were calculated using a series of equations that corrects for spectral overlaps of the pigments:

$$\text{C-phycocyanin (PC)} = \frac{A615 - 0.474\,(A652)}{5.34} \quad (1)$$

$$\text{allophycocyanin (APC)} = \frac{A652 - 0.208\,(A615)}{5.09} \quad (2)$$

$$\text{C-phycoerythrin (PE)} = \frac{A562 - 2.41\,(PC) - 0.849\,(APC)}{9.62} \quad (3)$$

For maximum accuracy equivalent equations must be determined for quantitation of the algal biliproteins of each species (French 1960).

B. Extraction of biliproteins from algal cells

Several methods are available, and selection of a particular procedure is based on the ease of cell breakage. The algal biliproteins can sometimes be extracted by placing the algae in water and allowing the proteins to leach out for several days (Svedberg and Katsurai 1929). A general procedure that has proved useful for extracting many algae is repeated freezing and thawing in the presence of 0.05 M phosphate buffer, pH 6.7, made by mixing equal volumes of 0.1 M KH_2PO_4 and 0.1 M K_2HPO_4 solutions. Efforts should be made to keep all solutions below 5 C when thawing the cells. A mortar and pestle with sand as an abrasive or a blender can be used for easily fractured cells. Algae that are less easily ruptured can be ground to a fine powder with dry ice in a blender or in liquid N_2 in a large mortar and pestle and subsequently extracted with buffer. Some blue-green algae can be disrupted with lysozyme (Crespi et al. 1962); this is simply achieved on a small scale by adding uncooked egg white to the algae. The extraction process may have to be repeated several times for a quantitative yield. The extracted biliproteins are separated from the algae by filtration or centrifugation.

C. Purification of algal biliproteins

Purification can be achieved by several methods. A useful procedure that is generally applicable follows.

Blue-green algae such as *Phormidium luridum* or *Tolypothrix tenuis* are extracted by repeated (three times) freezing and thawing in 0.1 M potassium phosphate buffer, pH 6.7, containing 0.2 M NaCl, in a 1:1 algal fresh weight to buffer volume ratio. The resulting colored solution is clarified by centrifugation at 24 000 × g for 15 min or by filtration with the aid of diatomaceous earth. The volume of the solution is measured, and sufficient $(NH_4)_2SO_4$ is added with stirring to make the solution 25% saturated; the addition of 730 g of $(NH_4)_2SO_4$ to 1 l of solution is 100% saturation at 10 C. The solution is stirred for 15–30 min following the addition of $(NH_4)_2SO_4$ and then centrifuged at 10 000 × g for 10 min. The precipitate is dissolved in a minimal volume of 0.01 M potassium phosphate buffer, pH 6.7, containing 0.2 M NaCl. The supernatant solution is then made to 35% saturation by addition of more $(NH_4)_2SO_4$. The precipitate is removed by centrifugation, dissolved as above, and the supernatant is made to 60% saturation with $(NH_4)_2SO_4$. The resulting supernatant contains cytochromes and ferredoxin, which can be precipitated by making the solution 90% saturated. The absorption spectra of the various $(NH_4)_2SO_4$ fractions should be determined from 350 to 700 nm to assess purity. Further purification may be achieved by reprecipitation with $(NH_4)_2SO_4$ taking narrower $(NH_4)_2SO_4$ fractions (e.g., 10–20%, 20–25%, 25–35%).

Ammonium sulfate fractions can be further purified if necessary by chromatography on calcium phosphate. Calcium phosphate with excellent chromatographic properties is easily made (Siegelman et al. 1965). Prepare two solutions: 1.1 l of 2 M K_2HPO_4 and 1.0 l of 2.0 M $CaCl_2$ in distilled water. Place 150 ml of distilled water and 100 ml of $CaCl_2$ solution into a large beaker. While keeping the mixture stirred, add the K_2HPO_4 and the $CaCl_2$ solutions at about 5 ml min^{-1} each independently and simultaneously using pumps or Mariotte bottles. Check flow rates periodically. The calcium phosphate precipitate is brushite ($CaHPO_4 \cdot 2H_2O$). Hydroxylapatite, $Ca_{10}(PO_4)_6(OH)_2$, is prepared by adding 2.0 M KOH to a stirred sample of brushite until the

slurry reaches pH 11. Allow the mixture to stand until the pH drops
to about 8. Continue adding 2.0 M KOH until the pH does not drop
below 9. The hydroxylapatite is washed by decantation with 0.001 M
potassium phosphate buffer, pH 6.7, in 0.2 M NaCl to remove fine
particles and equilibrate for chromatographic use. The biliprotein
fraction to be purified is dialyzed against three to four changes of
0.001 M potassium phosphate buffer, pH 6.7, in 0.2 M NaCl for at least
4 h. A column about 2.5 × 7.5 cm is prepared by pouring a slurry of
hydroxylapatite into a 50-ml plastic syringe with a plug of glass-wool at
the bottom. A short piece of tubing with a hose clamp attached to the
bottom of the column will control the flow. Pass about 50 ml of 0.001
M potassium phosphate buffer, pH 6.7, in 0.2 M NaCl through the column. Then carefully pipette the dialyzed biliprotein sample onto the
column with a sample volume that will cause about one-third of the
column to be visibly colored. Wash the column with 50 ml of 0.001 M
potassium phosphate buffer, pH 6.7, in 0.2 M NaCl. Begin the elution
by carefully applying 25 ml of 0.005 M potassium phosphate buffer,
pH 6.7, in 0.2 M NaCl to the top of the column. If no fraction is removed, increase the buffer concentration in 25-ml increments stepwise by 0.025 M until a colored eluate is obtained. Allophycocyanin is
the most strongly absorbed biliprotein and may require 0.25 M potassium phosphate buffer in 0.2 M NaCl for its elution. The NaCl promotes absorption of the proteins to the hydroxylapatite; the phosphate promotes elution. The column can also be eluted with a linear
gradient of 0.001–0.1 M potassium phosphate buffer, pH 6.7, in 0.2
M NaCl.

Gel filtration of the $(NH_4)_2SO_4$ fractions on a column of coarse Sephadex G-100 (Pharmacia), 2.5–5 × 75–100 cm, may be useful in
further purification of the biliproteins. The elution solvent used is
0.001 M potassium phosphate buffer, pH 6.7, in 0.2 M NaCl. The
protein is applied to the column in a 2–10 ml volume, and the column
is developed with buffer. The fractions eluted from the column can
be absorbed directly on a hydroxylapatite column if further purification is required.

The ratio of absorbancies at 280 nm to the visible wavelength maximum of the biliprotein is a convenient measure of purity. A summary
of purification procedures and absorbance ratios for biliproteins is
shown in Table 7-2.

Additional methods suitable for the purification of the algal biliproteins are available. These include chromatography on cellulose ion
exchangers (Bennett and Bogorad 1971; Kobayashi et al. 1972), crystallization from $(NH_4)_2SO_4$ (Hattori and Fujita 1959), $(NH_4)_2SO_4$ gradient elution chromatography (Chapman et al. 1968), polyacrylamide
gel electrophoresis, and isoelectric focusing.

The algal biliproteins are multimeric and are composed of two dis-

Table 7-2. *Purification of algal biliproteins and absorbance ratios*

Species	Bili-protein	Purification steps 1	2	3	4	Absorbance ratio
Tolypothrix tenuis	PE	25% AS	G-100	HA	HA	542/278 = 8
	PC	35% AS	G-100	HA	HA	628/278 = 5
	APC	60% AS	G-100	HA	HA	650/278 = 6
Phormidium luridum	PC	30% AS	G-100	HA	HA	628/278 = 5
	APC	50% AS	G-100	HA	HA	650/278 = 6
Chroomonas sp.	CrPC	AS	G-100	G-100	HA	645/278 = 6.6
Phormidium sp. UTEX 485	PC	30% AS	G-100	—	—	615/278 = 3.3

Note: PC, phycocyanin; PE, phycoerythrin; APC, allophycocyanin; CrPC, cryptomonad phycocyanin; %AS, percent saturation with $(NH_4)_2SO_4$; G-100, Sephadex G-100; HA, hydroxylapatite.

similar subunits, each bearing one or more chromphores. These subunits can be dissociated and separated by sodium dodecyl sulfate (SDS)–polyacrylamide gel electrophoresis (Bennett and Bogorad 1971; Glazer and Cohen-Bazire 1971; Kobayashi et al. 1972). The algal biliproteins remain colored after treatment with SDS, and the separation of the subunits can thus be followed visually. Electrophoresis of native biliproteins on polyacrylamide gels can also be followed visually (Glazer and Cohen-Bazire 1971).

D. Isolation of algal bile pigments

The algal bile pigments can be cleaved from the protein by a number of reagents including conc. HCl, boiling methanol, and proteolytic enzymes (Bogorad 1975; ÓCarra and ÓhEocha 1976). The algal bile pigments are easily cleaved with methanol (Fujita and Hattori 1962; ÓCarra and ÓhEocha 1966; Cole et al. 1968) from either whole cells or isolated biliproteins. The cells are repeatedly extracted with methanol until free of chlorophyll and carotenoids. Fresh methanol is added and the mixture boiled under reflux for 16–24 h for maximum yield. Similarly, biliproteins can be boiled in methanol for 16–24 h. The cleaved diacid pigments are concentrated by evaporation of the methanol and converted to their dimethyl esters by addition of 7% BF_3 in methanol and boiling under reflux for 3 min (Cole et al. 1968; Beuhler et al. 1976). The solution is cooled and an equal volume of chloroform added; sufficient water is then added to form two phases. The chloroform phase, which contains the esterified pigments, is washed with water until neutral, centrifuged at 2000 × g for 5 min, and the chloroform layer evaporated to dryness. The dried residue is stored at -15 C in the dark. The dimethyl esters can be separated by thin-layer chromatography on silica gel with a solvent of benzene:ethyl acetate (65:35 v/v).

A simple procedure was devised for chromatographically demonstrating phycocyanobilin and phycoerythrobilin from nori (*Porphyra tenera*). A sample of dried nori (10 g) is ground with dry ice in a blender to a fine powder. The powder is extracted with 25 ml of 0.1 M potassium phosphate buffer, pH 6.7. The debris is removed by filtration through two layers of cheesecloth and reextracted with the fresh buffer and refiltered. The filtrate is clarified by centrifugation and the supernatant made to 5% acetic acid, which precipitates the biliproteins. The precipitate is washed repeatedly with methanol by centrifugation to remove acetic acid and other alcohol-soluble materials. The precipitated protein is boiled under reflux for 2 h with *n*-butanol. The *n*-butanol is filtered to remove the precipitated protein and evaporated. The residue is dissolved in 2 ml 7% BF_3 in methanol and boiled under reflux for 3 min and cooled. An equal volume of chloroform and 10 ml of water are added. The blue chloroform layer is transferred to a 12-ml centrifuge tube, and the chloroform layer is extracted three times with 8 ml H_2O. The chloroform layer is dried by evaporation and dissolved in 0.5 ml methanol. The solution is applied to silica gel thin-layer chromatography plates, and the phycobilin dimethyl esters are separated as described above.

IV. Acknowledgment

This research was carried out at Brookhaven National Laboratory under the auspices of the United States Energy Research and Development Administration.

V. References

Bennett, A., and Bogorad, L. 1971. Properties of subunits and aggregates of blue-green algal biliproteins. *Biochemistry* 10, 3625–34.

Bennett, A., and Bogorad, L. 1973. Complementary chromatic adaptation in a filamentous blue-green alga. *J. Cell Biol.* 58, 419–35.

Bennett, A., and Siegelman, H. W. 1977. Bile pigments of plants. In Dolphin, D. (ed.). *The Porphyrins.* Academic Press, New York (in press).

Beuhler, R. J., Pierce, R. C., Friedman, L., and Siegelman, H. W. 1976. Cleavage of phycocyanobilin from C-phycocyanin. *J. Biol. Chem.* 251, 2405–11.

Bogorad, L. 1975. Phycobiliproteins and complementary chromatic adaptation. *Annu. Rev. Plant Physiol.* 26, 369–401.

Chapman, D. J. 1973. Biliproteins and bile pigments. In Carr, N. G., and Whitton, B. A. (eds.). *The Biology of Blue Green Algae,* pp. 162–85. University of California Press, Berkeley.

Chapman, D. J., Cole, W. J., and Siegelman, H. W. 1967. Chromophores of allophycocyanin and R-phycocyanin. *Biochem. J.* 105, 903–5.

Chapman, D. J., Cole, W. J., and Siegelman, H. W. 1968. Cleavage of phycocyanobilin from C-phycocyanin. *Biochim. Biophys. Acta* 153, 692–8.

Cole, W. J., Chapman, D. J., and Siegelman, H. W. 1968. The structure and properties of phycocyanobilin and related bilatrienes. *Biochemistry* 7, 2929–35.

Crespi, H. L., Mandeville, S. E., and Katz, J. J. 1962. The action of lysozyme on several blue-green algae. *Biochem. Biophys. Res. Commun.* 9, 569–77.

French, C. S. 1960. The chlorophylls in vivo and in vitro. In Ruhland, W. (ed.). *Encyclopedia of Plant Physiology*, vol. V/1, pp. 252–97. Springer-Verlag, Berlin.

Fujita, Y., and Hattori, A. 1962. Preliminary note on a new phycobilin pigment from blue-green algae. *J. Biochem. (Tokyo)* 51, 89–91.

Gantt, E. 1975. Phycobilisomes: Light-harvesting pigment complexes. *Bioscience* 25, 781–8.

Glazer, A. N. 1976. Phycocyanins: Structure and function. In Smith, K. C. (ed.). *Photochemical and Photobiological Reviews*, vol. 1, pp. 71–115. Plenum, New York.

Glazer, A. N., and Cohen-Bazire, G. 1971. Subunit structure of the phycobiliproteins of blue-green algae. *Proc. Natl. Acad. Sci. U.S.A.* 68, 1398–401.

Hattori, A., and Fujita, Y. 1959. Crystalline phycobilin chromoproteins obtained from a blue-green alga, *Tolypothrix tenuis. J. Biochem. (Tokyo)* 46, 633–44.

Jones, L. W., and Myers, J. 1965. Pigment variations in *Anacystis nidulans* induced by light of selected wavelengths. *J. Phycol.* 1, 7–14.

Kobayashi, Y., Siegelman, H. W., and Hirs, C. H. W. 1972. C-phycocyanin from *Phormidium luridum:* Isolation of subunits. *Arch. Biochem. Biophys.* 152, 187–98.

Lemberg, R. 1928. Die Chromoproteiden der Rotalgen 1. *Justus Liebig's Ann. Chem.* 461, 46–89.

Myers, J., and Kratz, W. A. 1955. Relations between pigment content and photosynthetic characteristics in a blue-green alga. *J. Gen. Physiol.* 39, 11–22.

ÓCarra, P., and ÓhEocha, C. 1966. Bilins released from algae and biliproteins by methanolic extraction. *Phytochemistry* 5, 993–7.

ÓCarra, P., and ÓhEocha, C. 1976. Algal biliproteins and phycobilins. In Goodwin, T. W. (ed.). *Chemistry and Biochemistry of Plant Pigments*, pp. 328–76. Academic Press, London.

Shibata, J., Benson, A. A., and Calvin, M. 1954. The absorption spectra of suspensions of living microorganisms. *Biochim. Biophys. Acta* 15, 461–70.

Siegelman, H. W., Chapman, D. J., and Cole, W. J. 1968. The bile pigments of plants. *Biochem. Soc. Symp.* 28, 107–20.

Siegelman, H. W., Wieczorek, G. A., and Turner, B. C. 1965. Preparation of calcium phosphate for protein chromatography. *Anal. Biochem.* 13, 402–4.

Svedberg, T., and Katsurai, T. 1929. The molecular weights of phycocyanin and of phycoerythrin from *Porphyra tenera* and of phycocyanin from *Aphanizomenon flosaquae. J. Am. Chem. Soc.* 51, 3573–83.

8: Nucleic acids

ROSE ANN CATTOLICO

Botany Department AK-10,
University of Washington, Seattle, Washington 98195

CONTENTS

I Introduction   *page* 82
II Materials and test organisms   82
III Methods and discussion   82
   A. DNA isolation   82
   B. DNA quantitation   83
   C. RNA isolation   85
   D. RNA quantitation   86
IV References   88

I. Introduction

The purpose of this chapter is to present methods for the isolation and quantitation of DNA and RNA in algal cells. Problems and advantages of specific assay techniques used in the estimation of nucleic acid levels will be discussed.

II. Materials and test organisms

T_1 ribonuclease, pancreatic ribonuclease, and 3,5-diaminobenzoic acid (Sigma); Metricel Alpha-6 filters, pore size 0.45 μm (Gelman); polyethylene BEEM capsules, size 00, with standard pyramidal tips (Ladd Research Industries); acrylamide, bisacrylamide and NNN′N-tetramethylethylenediamine (TEMED) (BioRad); agarose (Marine Colloids); ammonium persulfate (Canalco). All other materials are of reagent grade.

Axenic cultures of *Chlamydomonas reinhardtii* Dangeard 137 C⁺, *Chlorella sorokiniana* Shihira and Krauss, *Ochromonas danica* Pringsheim, and *Olisthodiscus luteus* Carter were used in these studies.

III. Methods and discussion

A. DNA isolation

Olisthodiscus cells are collected at 3500 rpm in a Sorvall RC-5 centrifuge using an SS-34 rotor with a Szent-Györgyi-Blum continuous-flow attachment. SSC buffer (0.15 M NaCl, 0.15 M trisodium citrate dihydrate, pH 7.0) containing 2.5% sodium lauryl sarcosine is added to the pelleted cells at a ratio of 1.0 ml buffer solution per 2.2 × 10⁷ cells. The cells are lysed at ice temperature for 15 min followed by the addition of 1.5 volumes of phenol (redistilled) saturated with SSC buffer. The cell lysate and phenol solution are gently mixed using a rocking motion for 20 min at 5 C. The solution is centrifuged in an Hb-4 swinging bucket rotor for 15 min at 7500 rpm, and the supernatant is reextracted twice with saturated phenol as described above, except that a 1:1 volume of supernatant to phenol is maintained. The final supernatant is extracted three times with 2 volumes of ether. After ether treatment, 2 volumes of 95% ethanol (-20 C) are added to the supernatant. The mixture is rapidly shaken, and the nucleic acid is quickly spooled out. This spooling step is critical, for it eliminates contamination by coprecipitating polysaccharides. The spooled material is dissolved in 0.02 M TRIS-HCl buffer, pH 8.5, to a concentration of 1.0 mg nucleic acid per milliliter of buffer solution and treated with 10 units ml^{-1} of T_1 ribonuclease and 30 mg ml^{-1} pancreatic ribonuclease for 2 h at 37 C. Stock nuclease solutions, which are stored at -20 C, are made in a pH 8.0 buffer containing 0.01 M TRIS-HCl and 0.02 M NaCl, and are heated at 60 C for 30 min before use. After enzyme digestion, the DNA is precipitated at 25 C with 5 volumes of 2-propanol. The highly polymerized DNA product obtained by this method has a 230/260 ratio of 0.45, a 230/280 ratio of 1.86, and is recovered from the cells with 80–90% efficiency.

The presence of polysaccharide contaminants in nucleic acid preparations is a major technical problem (Segovia et al. 1965; Edelman 1975) in the isolation of DNA from algal cells (Edelman et al. 1967). This cellular component may seriously interfere with analysis and quantitation (Edelman et al. 1969; Guidice 1973) of an isolated DNA product. Edelman (1975) reports that concanavalin A chromatography has been useful in removing polysaccharides from DNA preparation, although the author warns (Edelman 1974) that this method provides no universal solution for algal systems. Removal of polysaccharide (also small amounts of RNA and protein) may also be accomplished by preparative CsCl centrifugation of the DNA product. Flamm et al. (1969) have written an excellent descriptive review of this method, and Brunk and Leick (1969) have introduced a two-step gradient technique that significantly reduces DNA centrifugation times.

B. DNA quantitation

1. Spectrophotometric. Absorbance is measured at 260 nm. For a DNA duplex, 20 absorbance units are equivalent to 1.0 mg of nucleic acid. Note that protein, phenol, trichloroacetic acid, and polysaccharides will interfere with nucleic acid estimation by this method. For quantitation of specific nucleic acid species (e.g., chloroplast, mitochondrial, ribosomal satellite), spectrophotometric measurement of DNA bands formed after analytical neutral CsCl equilibrium centrifugation has been frequently used (Stutz and Vandrey 1971; Bayen and Rode 1973; Lee and Jones 1973).

2. Colorimetric. The classic diphenylamine (Burton 1956) and indole (Ceriotti 1952) methods and their micromodifications (Hubbard et al.

1970; Abraham et al. 1972) are available. Investigators should note that the presence of trichloroacetic acid, protein, lipids, polysaccharides, and other cellular materials may interfere (Edelman et al. 1969; Hubbard et al. 1972; Guidice 1973, p. 201) with the color development in these techniques. The data of Hopkins et al. (1972) clearly demonstrate the danger of such interfering substances. These authors have shown that DNA accumulates in a stepwise manner during the synchronous growth of a thermophilic strain of *Chlorella,* even though initial studies of this cell system indicated that a linear increase in DNA occurred during the entire cell cycle. The anomalous result of a linear DNA increase was caused by non-DNA substances that gave a positive reaction in both indole and diphenylamine assay systems.

3. Fluorometric. This is an excellent method for quantitating DNA in whole cells (Holm-Hansen et al. 1968; Hinegardner 1971; Hesse et al. 1975). The assay technique is not only extremely sensitive, thereby requiring a small sample size, but also requires no separation of DNA from other cellular components. The difficulties encountered in the application of more classic methods of DNA extraction (Schmidt and Thannhauser 1945; Schneider 1945; Ogur and Rosen 1950) and the problems cited above resulting from the quantitation of DNA by spectrophotometric or colorimetric methods, therefore, may be entirely avoided (for detailed discussion see Hutchison and Munro 1961; Munro and Fleck 1966).

Cells are axenically removed from the culture flask and placed in a 12-ml conical test tube. A small quantity of formalin is added to kill the cells, and the sample is centrifuged at 2900 rpm. All but 0.5 ml of the growth medium is removed and the cell sample is transferred to a 6×50-mm Siliclad-coated test tube. The 12-ml conical test tube is rinsed twice with 0.2 ml of growth medium. The sample is then centrifuged in a Sorvall RC-5 centrifuge at 7000 rpm at 5 C using an Hb-4 swinging bucket rotor. All but 50 μl of the supernatant is aspirated, and the pellet is resuspended using a vortex mixer. The resuspended pellet is extracted with 80% acetone (*Chlamydomonas, Olisthodiscus, Ochromonas*) or 100% ethanol (*Chlorella*) until the pellet is free of pigments. The cells are then transferred onto Metricel filters using a modified Millipore apparatus (Cattolico and Gibbs 1975) and washed twice using approximately 0.15 ml of 0.6 N trichloroacetic acid (5 C) for each wash. Two washes are then made using 0.15 ml of 5 C ethanol:H_2O (2:1 v/v) followed by two washes using 0.15 ml of 60 C ethanol:H_2O (2:1 v/v). The filters are then removed from the apparatus, placed in BEEM polyethylene capsules, and allowed to dry at 20 C. DNA standards (0–0.75 μg DNA/0.01 ml made in 1.0 M NH_4OH) are added to a blank set of washed filters.

Purified 3,5-diaminobenzoic acid (50 μl) is added to each filter. This fluorescent agent is purified by adding 0.6 g of 3,5 diaminobenzoic acid to 2.0 ml of 4.0 N HCl at 5 C, and the solution is vortexed until completely dissolved. The fluorescent reagent is then extracted with 20 mg of charcoal by drawing the mixture in and out of a Pasteur pipette, and the mixture is centrifuged for 5 min in a clinical tabletop centrifuge. This extraction sequence is repeated seven times until a clear yellow solution is obtained. The filters, which have been saturated with the fluorescent agent, are incubated at 60 C for 30 min, then cooled for 5 min at room temperature; after this 0.5 ml 1 N HCl is added. The samples are immediately read in a Turner fluorometer equipped with a blue lamp and high-sensitivity sample holder, using an excitation maximum of 405 nm and an emission maximum of 520 nm. Values for total cellular DNA content of a variety of algal cells monitored by this method agree well with those values obtained using classic macromethods (Cattolico and Gibbs 1975).

C. RNA isolation

Chlamydomonas reinhardtii cells are harvested at 8000 rpm in a Sorvall RC-5 centrifuge using an SS-34 rotor with a Szent-Györgyi-Blum continuous-flow attachment at 5 C. The pellet is resuspended at a ratio of 3.5×10^7 cells per milliliter of 5 C TMK buffer (25 mM TRIS-HCl, 25 mM $MgCl_2 \cdot 6 H_2O$, and 25 mM KCl, pH 7.6). The cell suspension is then subjected to a total of 30–45 sec of sonication at 5 C, applied in three short bursts. Cell breakage is monitored by phase-contrast microscopy. Sodium dodecyl sulfate is added to the sonicate to produce a final concentration of 0.8%. Lysis of the cells without the sonication step results in a 60% loss of RNA during the isolation procedure (Cattolico and Jones 1972). The lysate is then extracted for 20 min at 10 C with redistilled phenol saturated with TMK buffer (1 volume of lysate to 1.5 volumes of phenol) by rapidly shaking the mixture on a New Brunswick G-2 rotary laboratory shaker. The supernatant is recovered by centrifugation for 15 min at 7000 rpm in an Hb-4 swinging bucket rotor. The supernatant is reextracted twice, as described above except that a supernatant/phenol ratio of 1:1 is maintained. The supernatant is then extracted three times with diethyl ether, after which the RNA is precipitated at a 1:2 volume ratio of supernatant to 95% ethanol (-20 C).

Olisthodiscus cells, which lack walls, may be lysed in 0.8% SDS-TMK buffer at 10 C at a ratio of 2.2×10^7 cells per milliliter of buffer solution, and the RNA extracted as described for the *Chlamydomonas* system.

Buffer solutions containing magnesium are most frequently used in the isolation of RNA from algal systems. The presence of Mg^{2+} has been reported to maintain hydrogen-bonded regions in the RNA molecule (Herbeck and Zundel 1976) and to offer protection against ribonuclease activity (Morrill and Reiss 1969). Our own observations indicate that this cation stimulates the removal (digestion) of cellular DNA during the isolation of RNA. It should be noted, however, that recent experiments (McIntosh and Cattolico, unpublished) have shown that RNA of good quality may be isolated by lysing *Olisthodiscus* cells at 5 C using 2.5% sodium lauryl sarcosine in a buffer system that contains high concentrations of sodium citrate. The primary nucleic acid product is heavily contaminated with DNA and, therefore, requires a DNase digestion step. The use of a buffer system containing EDTA or the presence of bentonite during the isolation procedure causes extracted RNA to be of poor quality. Diethylpyrocarbonate, a potent nuclease inhibitor (Solymosy et al. 1968; Wolf et al. 1970), may be used during RNA isolation, but caution is advised, for this reagent may cause modifications in both tRNA and rRNA species (Solymosy et al. 1971; Henderson et al. 1973). In all RNA isolations described above, the product obtained has a 230/260 nm absorbancy ratio of approximately 0.44, and a 260/280 ratio of 2.1.

D. RNA quantitation

1. Spectrophotometric. Absorbancy measurements are made at 260 nm. For the predominantly single-stranded RNA molecule, 25 absorbance units is equivalent to 1.0 mg of this nucleic acid. Protein, phenol, trichloroacetic acid, and other substances will affect nucleic acid measurement by this method. For the quantitation of specific RNA species, separation of RNA on 2.4% acrylamide–0.5% agarose gels and measurement by ultraviolet scanning is recommended. These agarose-acrylamide gels have the capacity to separate precursor rRNA, mature rRNA (both cytoplasmic and chloroplast species), and tRNA on the same gel.

Gels of 2.4% acrylamide–0.5% agarose are prepared by adding 17.2 ml of stock acrylamide solution (48.6 ml of 1.0 M TRIS-HCl, 4.8 g acrylamide, and 0.25 g bisacrylamide made to an 86-ml volume with distilled H_2O) to 2.5 ml of 3.2% TEMED. The solution is de-gassed by heating in a water bath for 15 min at 54 C. This TEMED-acrylamide mixture is added in a 1:1 volume ratio to a 1.0% agarose solution (which is refluxed for 20 min immediately before use and cooled to 54 C). This mixture was used to cast gels using 0.025 ml of 1.4% ammonium persulfate per milliliter as the polymerizing reagent. The polyacrylamide gels are cast in Lucite tubes (4 mm i.d. × 90 mm), each of which is capped on the bottom with Parafilm. Gels are polymerized for 7 min at 25 C, followed by 20 min at 5 C, and finally for 40 min at 25 C. After polymerization, the gels are extruded so that the Parafilm-formed base results in a perfectly flat loading platform. Then 1 cm of the extruded (meniscus) end of the gel is removed, and this end is capped with cotton bunting held in place with a rubber collar. This is done to prevent the loosened gel from slipping out of the tube. Gels are prerun 45 min in a pH 7.9 buffer containing 0.036 M TRIS-HCl,

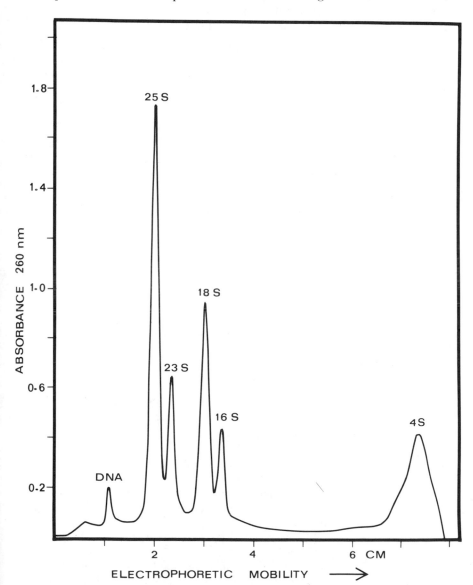

Fig. 8-1. Representative ribosomal RNA electrophoretic profile from *Chlamydomonas reinhardtii*. Cells were maintained on a 12-h light, 12-h dark regimen. The culture was harvested at h 1.0 of the synchronous cell cycle.

0.03 M NaH_2PO_4, and 0.001 M EDTA. The RNA samples, which are suspended in 50 μl NET buffer (0.1 M TRIS-HCl, 0.02 M EDTA, 0.01 M NaCl, pH 7.6), 20 μl of 50% sucrose made in NET buffer, and 1.0 μl of saturated bromphenol blue solution are subjected to electrophoresis at 8.2 V and 2.5 mA per tube for 2 h. It is of critical importance to maintain gel temperatures near 5 C during the electrophoresis. Algal rRNA species are extremely thermolabile and will be reduced to oligonucleotide subfractions by slight elevations in temperature (Doolittle 1973; Cattolico and Jones 1975). We have successfully maintained a temperature of 5.5 C in our electrophoresis system by simply passing the water from the water jacket of the electrophoresis bath through a copper coil immersed in ice water. A Cole-Parmer magnetic drive pump is used for this operation. After electrophoresis, gels are scanned at 260 nm in a Gilford spectrophotometer with a 2410 linear transport attachment at 1.0 cm min^{-1} using a chart speed of 1 in. min^{-1}. A representative scan of RNA extracted from *Chlamydomonas reinhardtii* is presented in Fig. 8-1.

2. Colorimetric. The classic orcinol reaction (Schneider 1957) may be used. It should be noted, however, that a large range of substances will interfere with the color development of this reaction, especially in plant systems (Smillie and Krotkov 1960; Hutchison and Munro, 1961; Munro and Fleck 1966).

3. Fluorometric. There is no assay for RNA that is totally specific. Measurement of RNA in tissue homogenates using ethidium bromide have been reported (Karsten and Wollenberger 1972; El-Hamalawi et al. 1975).

IV. References

Abraham, G. N., Scaletta, C., and Vaughan, J. H. 1972. Modified diphenylamine reaction for increased sensitivity. *Anal. Biochem.* 49, 547–9.

Bayen, M., and Rode, A. 1973. Heterogeneity and complexity of *Chlorella* chloroplastic DNA. *Eur. J. Biochem.* 39, 413–20.

Brunk, C. F., and Leick, V. 1969. Rapid equilibrium isopycnic CsCl gradients. *Biochim. Biophys. Acta* 179, 136–44.

Burton, K. 1956. A study of the conditions and mechanism of the diphenylamine reaction for the colourimetric estimation of deoxyribonucleic acid. *Biochem. J.* 62, 315–23.

Cattolico, R. A., and Gibbs, S. P. 1975. Rapid filter method for the microfluorometric analysis of DNA. *Anal. Biochem.* 69, 572–82.

Cattolico, R. A., and Jones, R. F. 1972. Isolation of stable ribosomal RNA from whole cells of *Chlamydomonas reinhardtii. Biochim. Biophys. Acta* 269, 259–64.

Cattolico, R. A., and Jones, R. F. 1975. An improved technique for the preparative electrophoresis and electroelution of high molecular weight ribosomal RNA. *Anal. Biochem.* 66, 35–46.

Ceriotti, G. 1952. A microchemical determination of desoxyribonucleic acid. *J. Biol. Chem.* 198, 297–303.

Doolittle, W. F. 1973. Postmaturational cleavage of 23s ribosomal ribonucleic acid and its metabolic control in the blue-green alga *Anacystis nidulans. J. Bacteriol.* 113, 1256–63.

Edelman, M. 1974. Carbohydrate exchange chromatography as a technique for nucleic acid purification. *Methods Enzymol.* 34B, 499–502.

Edelman, M. 1975. Purification of DNA by affinity chromatography: Removal of polysaccharide contaminants. *Anal. Biochem.* 65, 293–7.

Edelman, M., Hirsch, C. A., Hiatt, H. H., and Fox, M. 1969. Apparent changes in mouse liver DNA content due to interference by non-DNA diphenylamine-reacting cytoplasmic material. *Biochim. Biophys. Acta* 179, 172–8.

Edelman, M., Swinton, D., Schiff, J. A., Epstein, H. T., and Zeldin, B. 1967. Deoxyribonucleic acid of the blue-green algae (Cyanophyta). *Bacteriol. Rev.* 31, 315–31.

El-Hamalawi, A.-R., Thompson, J. S., and Barker, G. R. 1975. The fluorometric determination of nucleic acids in pea seeds by use of ethidium bromide complexes. *Anal. Biochem.* 67, 384–91.

Flamm, W. G., Birnstiel, M. L., and Walker, P. M. B. 1969. Preparation and fractionation and isolation of single strands of DNA by isopycnic ultracentrifugation in fixed-angle rotors. In Birnie, G. D., and Fox, S. M. (eds.). *Subcellular Components,* pp. 125–54. Butterworths, London.

Guidice, G. 1973. *Developmental Biology of the Sea Urchin Embryo.* Academic Press, New York. 469 pp.

Henderson, R. E. L., Kirkegaard, L. H., and Leonard, N. J. 1973. Reaction of diethyl pyrocarbonate with nucleic acid components: Adenosine-containing nucleotides and dinucleoside phosphates. *Biochim. Biophys. Acta* 294, 356–64.

Herbeck, R., and Zundel, G. 1976. Influence of temperature and magnesium ions on the secondary and tertiary structures of tRNA^{-Phe} and 23s RNA-infrared investigations. *Biochim. Biophys. Acta* 418, 52–62.

Hesse, G., Lindner, R., and Krebs, D. 1975. Schnelle fluorometrische desoxyribosespezifische DNS-Bestimmung an nicht desintegrierten Mikroorganismen mit 3,5-Diaminobenzoesäure (DABA). I. Chemischanalytische Methodik und Anwendung auf Hefepopulationen. *Z. Allg. Mikrobiol.* 15, 9–18.

Hinegardner, R. T. 1971. An improved fluorometric assay for DNA. *Anal. Biochem.* 39, 197–201.

Holm-Hansen, O., Sutcliffe, W. H., Jr., and Sharp, J. 1968. Measurement of deoxyribonucleic acid in the ocean and its ecological significance. *Limnol. Oceanogr.* 13, 507–14.

Hopkins, H. A., Flora, J. B., and Schmidt, R. R. 1972. Periodic DNA accumulation during the cell cycle of a thermophilic strain of *Chlorella pyrenoidosa. Arch. Biochem. Biophys.* 153, 845–9.

Hubbard, R. W., Matthew, W. T., and Dubowik, D. A. 1970. Factors influencing the determination of DNA with indole. *Anal. Biochem.* 38, 190–201.

Hubbard, R. W., Matthew, W. T., and Moulton, D. W. 1972. Factors influencing the determination of DNA with indole. II. Recovery of hydrolyzed DNA from protein precipitates. *Anal. Biochem.* 46, 461–72.

Hutchison, W. C., and Munro, H. N. 1961. The determination of nucleic acids in biological materials *Analyst* 86, 768–813.

Karsten, U., and Wollenberger, A. 1972. Determination of DNA and RNA in homogenized cells and tissues by surface fluorometry. *Anal. Biochem.* 46, 135–48.

Lee, R. W., and Jones, R. F. 1973. Induction of Mendelian and non-Mendelian streptomycin resistant mutants during the synchronous cell cycle of *Chlamydomonas reinhardtii. Mol. Gen. Genet.* 121, 99–108.

Morrill, G. A., and Reiss, M. M. 1969. Inhibition of enzymatic degradation of RNA by bound calcium and magnesium. *Biochim. Biophys. Acta* 179, 43–9.

Munro, H. N., and Fleck, A. 1966. The determination of nucleic acids. *Methods Biochem. Anal.* 14, 113–76.

Ogur, M., and Rosen, G. 1950. The nucleic acids of plant tissues. I. The extraction and estimation of desoxypentose nucleic acid and pentose nucleic acid. *Arch. Biochem.* 25, 262–76.

Schmidt, G., and Thannhauser, S. J. 1945. A method for the determination of desoxyribonucleic acid, ribonucleic acid, and phosphoproteins in animal tissues. *J. Biol. Chem.* 161, 83–9.

Schneider, W. C. 1945. Phosphorous compounds in animal tissues. I. Extraction and estimation of desoxypentose nucleic acid and of pentose nucleic acid. *J. Biol. Chem.* 161, 293–303.

Schneider, W. C. 1957. Determination of nucleic acids in tissues by pentose analysis. *Methods Enzymol.* 3, 680–4.

Segovia, Z. M. M., Sokol, F., Graves, I. L., and Ackermann, W. W. 1965. Some properties of nucleic acids extracted with phenol. *Biochim. Biophys. Acta* 95, 329–40.

Smillie, R. M., and Krotkov, G. 1960. The estimation of nucleic acids in some algae and higher plants. *Can. J. Bot.* 38, 31–49.

Solymosy, F., Fedorcsák, I., Gulyás, A., Farkas, G. L., and Ehrenberg, L. 1968. A new method based on the use of diethyl pyrocarbonate as a nuclease inhibitor for the extraction of undegraded nucleic acid from plant tissues. *Eur. J. Biochem.* 5, 520–7.

Solymosy, F., Hüvös, P., Gulyás, A., Kapovits, I., Gaal, O., Bagi, G., and Farkas, G. L. 1971. Diethyl pyrocarbonate, a new tool in the chemical modification of nucleic acids? *Biochim. Biophys. Acta* 238, 406–16.

Stutz, E., and Vandrey, J. P. 1971. Ribosomal DNA satellite of *Euglena gracilis* chloroplast DNA. *FEBS Lett.* 17, 277–80.

Wolf, B., Lesnaw, J. A., and Reichmann, M. E. 1970. A mechanism of the irreversible inactivation of bovine pancreatic ribonuclease by diethyl pyrocarbonate. *Eur. J. Biochem.* 13, 519–25.

9: Protein determination by dye binding

GARY KOCHERT

Botany Department,
University of Georgia, Athens, Georgia 30602

CONTENTS

I **Introduction**	*page*	**92**
II **Materials**		**92**
A. Reagents		92
B. Solutions		92
III **Methods**		**92**
A. Standard method		92
B. Micromethod		93
IV **Discussion**		**93**
V **References**		**93**

I. Introduction

A rapid and reproducible method for the determination of proteins by dye binding has been described by Bradford (1976). The assay is based on the fact that Coomassie Brilliant Blue G-250 exists in two different color forms. When the dye binds to protein, the red form is converted to blue. The protein-dye complex has a high extinction coefficient, which gives great sensitivity to the assay. It has further advantages in that it is less subject to interference from common laboratory reagents, is readily adaptable to a micromethod, and is about four times more sensitive than the standard Lowry (Lowry et al. 1951) method.

II. Materials

A. Reagents

Coomassie Brilliant Blue G-250 and bovine serum albumin (2 × crystallized) (Sigma).

B. Solutions

1. Protein reagent. Dissolve 100 mg of Coomassie Brilliant Blue G-250 in 50 ml of 95% ethanol. To this solution add 100 ml of 85% (w/v) H_3PO_4. Dilute the resulting solution to a final volume of 1 l with H_2O. Store at room temperature.

2. Protein standard solution. Dissolve 100 mg of bovine serum albumin in H_2O to a final volume of 100 ml. Store at 4 C.

III. Methods

A. Standard method

A standard curve is generated by pipetting a range (10–100 μg) of protein concentrations from the protein standard solution into a series of marked 12 × 100-mm test tubes. Adjust the volume of each tube to 0.1 ml with H_2O. Pipette 10, 50, and 100 μl of the unknown protein sample into three separate test tubes and adjust the volume of each tube to 0.1 ml with the buffer in which the unknown protein sample is dissolved. A reagent blank of 0.1 ml of the buffer should also be included. Add 5 ml of protein reagent to all tubes and mix immediately by vortex mixer or by inversion. Absorbance at 595 nm is measured after 2 min and before 1 h. Plot the weight of protein standard against the corresponding absorbance to generate a standard curve. The concentration of the unknown protein is determined graphically.

B. Micromethod

Dilute the protein standard solution 10-fold with water and pipette a range of protein concentrations (1–10 μg) into a series of marked 12 × 100-mm test tubes. Adjust the final volume of each tube to 0.1 ml with H_2O. Proceed as in the standard method except use only 1.0 ml of the protein reagent in all tubes. Absorbance at 595 nm is read in 1-ml cuvettes.

IV. Discussion

There is a slight amount of curvature in standard curves obtained by this method. This presents no problem if unknown protein concentrations are estimated graphically rather than by Beer's law.

Color development in this assay is essentially complete at 2 min. After about 20 min, aggregation of protein-dye complexes results in a gradual loss of color. For very precise determinations, read the samples between 5 and 20 min after addition of protein reagent.

Little or no interference in the binding assay was caused by $MgCl_2$, KCl, NaCl, ethanol, $(NH_4)_2SO_4$, TRIS, acetic acid, 2-mercaptoethanol, sucrose, glycerol, EDTA, and trace quantities of detergents. Quantities of detergents such as Triton X-100, sodium dodecyl sulfate, or Haemosol in concentrations higher than 0.1% cause serious interference.

The protein-dye complex generated in this assay has a tendency to bind to quartz cuvettes. The amount of colored complex bound is too small to interfere in the protein assay, but might interfere with other uses of the cuvettes. The complex can be removed by soaking the cuvettes in 0.1 M HCl. This problem can be avoided by using glass or plastic cuvettes, as the complex does not bind to these.

V. References

Bradford, M. 1976. A rapid and sensitive method for the quantitation of microgram quantities of protein utilizing the principle of protein-dye binding. *Anal. Biochem.* 72, 248–54.

Lowry, O. H., Rosebrough, N. J., Farr, A. L., and Randall, R. J. 1951. Protein measurement with the Folin phenol reagent. *J. Biol. Chem.* 193, 265–75.

10: Carbohydrate determination by the phenol–sulfuric acid method

GARY KOCHERT

Botany Department,
University of Georgia, Athens, Georgia 30602

CONTENTS

I	Introduction	*page* 96
II	Materials	96
	A. Reagents	96
	B. Solutions	96
III	Method	96
IV	Discussion	97
V	References	97

I. Introduction

Many colorimetric methods for reducing sugars and polysaccharides have been described. The commonly used anthrone technique (Spiro 1966) has several disadvantages. Anthrone reagent is relatively expensive, and solutions of it in H_2SO_4 are unstable. It is of limited utility for methylated sugars and pentoses, and many solvents commonly used for paper chromatographic separation of sugars interfere with the assay.

Phenol in H_2SO_4 can also be used for colorimetric determination of sugars, methylated sugars, and polysaccharides (Dubois et al. 1956). The assay is very simple, rapid, inexpensive, and highly sensitive. The color produced is very stable, and the assay is largely unaffected by the presence of proteins.

II. Materials

A. Reagents

1. A 90% phenol solution (e.g., from Fisher Scientific). It should be colorless when freshly opened.
2. H_2SO_4 should be of reagent grade (95.5%, specific gravity 1.48).

B. Solutions

1. Phenol reagent is made by adding 10 ml of H_2O to 90 ml of the 90% phenol solution. The resultant solution is colorless, but may develop a pale yellow color in time. The color does not interfere with the assay, and the solution is usable for many months when stored at room temperature.
2. Glucose standard solution consists of 50 mg of glucose dissolved in water to a final volume of 100 ml.

III. Method

To generate a standard curve, pipette a range (20–140 μl) of glucose standard solution into a series of marked 18 × 150-mm test tubes.

Adjust the volume of each tube to 2 ml with H_2O. Pipette two or three different volumes of the unknown carbohydrate solution into separate marked tubes, and adjust the volume of each to 2 ml with the buffer in which the unknown is dissolved. Include a reagent blank containing 2 ml of the buffer. Add 50 μl of phenol reagent to each tube and mix thoroughly. Then rapidly add 5 ml of H_2SO_4 to each tube. A 5-ml pipette with a portion of the tip cut off should be used to add the H_2SO_4 directly onto the surface of the liquid in the tubes. Delivery time should be 15–20 sec for 5 ml. This procedure promotes the mixing and heat development necessary for the assay. Allow the samples to stand at room temperature for 30 min and read the absorbance at 485 nm. The color is stable for several hours. Plot the weight of glucose standard against the corresponding absorbance to generate a standard curve. The concentration of the unknown carbohydrate is determined graphically.

IV. Discussion

Extinction coefficients vary somewhat for the different sugars detected by this assay. When a pure sugar is assayed, this presents no problem, but in a mixture of carbohydrates the values obtained are somewhat arbitrary.

Hexoses, disaccharides, oligosaccharides, polysaccharides, and methylated derivatives that possess a free or potentially free reducing group will react in this assay. Pentoses, methylpentoses, and uronic acids also react, but glucosamine and galactosamine are inert. These can be rendered reactive by deamination (see, however, Lee and Montgomery 1961). To avoid the use of micropipettes, phenol can be added as 1 ml of a 5% solution.

V. References

Dubois, M., Gilles, K. A., Hamilton, J. K., Rebers, P. A., and Smith, F. 1956. Colorimetric method for determination of sugars and related substances. *Anal. Chem.* 28, 350–56.

Lee, Y. C., and Montgomery, R. 1961. Determination of hexosamines. *Arch. Biochem. Biophys.* 93, 292–6.

Spiro, R. G. 1966. Analysis of sugars found in glycoproteins. *Methods Enzymol.* 8, 1–26.

11: Glycerolipids and fatty acids of algae

J. P. WILLIAMS

*Department of Botany,
University of Toronto, Toronto, Ontario M5S 1A1, Canada*

CONTENTS

I	**Introduction**	*page* **100**
II	**Methods**	**101**
	A. Extraction and purification of lipids	101
	B. Quantitative estimation of total lipids and pigments	102
	C. Separation by thin-layer chromatography	102
	D. Transesterification	104
	E. Gas-liquid chromatography of fatty acid methyl esters	105
III	**Discussion**	**106**
IV	**Acknowledgments**	**107**
V	**References**	**107**

I. Introduction

The term "lipids" represents a broad class of organic compounds that are generally soluble in organic solvents and insoluble in water. Included in this group may be sterols, quinones, some photosynthetic pigments, waxes, and acyl- or glycerolipids. Glycerolipids typically contain fatty acids esterified to glycerol molecules. They are major components of all membranes and perform an energy storage function in many plants. This chapter will be devoted to the extraction, separation, and quantitative determination of the major glycerolipids and their fatty acids in algae. The reader is referred to an excellent review on the fatty acids and saponifiable lipids in algae (Wood 1974) and to two general reviews on lipids and lipid metabolism in algae and higher plants (Kates 1970; Mazliak 1973).

The simplest forms of glycerolipids found in algae are the triglycerides, in which all three glycerol hydroxyl groups are esterified to fatty acids. Simple mono- and diglycerides may also be present, but these have not been identified with certainty. These lipids are usually regarded as energy storage products.

The "membrane" glycerolipids of algae are usually diglycerides with fatty acids esterified at the C-1 and C-2 hydroxyl groups of glycerol. The remaining C-3 hydroxyl group may be esterified either to a phosphatidyl or a glycosidic group.

The major phospholipids in algae are phosphatidylcholine (PC), phosphatidylethanolamine (PE), and phosphatidylglycerol (PG). Phosphatidylserine (PS), phosphatidylinositol (PI), and diphosphatidylglycerol (DPG) are also present in low levels, but are often difficult to detect. Three major glycosidic lipids are found in all algae: monogalactosyl diglyceride (MGDG), digalactosyl diglyceride (DGDG), and the plant sulfolipid (SL). Details of the structure of these lipids may be found in the review by Wood (1974). Of the above lipids, blue-green algae contain only PG, MGDG, DGDG, and SL.

[100]

II. Methods

A. Extraction and purification of lipids

The following technique is recommended for the extraction and purification of lipids from algal tissue of 1–5 g fresh weight (Williams and Merrilees 1970). The technique may be adapted to accommodate more material by increasing the quantities of solvent, etc. Quantities of algal material less than 1 g should be treated as equivalent to 1 g. However, the lipid recovered may not be sufficient for accurate analysis.

All glycerolipids are readily extracted from algal material with solvents that disrupt lipoprotein complexes. In some cases, breakage of the cells in the extracting solvent may be necessary to allow the solvent to penetrate the cell. This may be accomplished using one of the established cell-disruption techniques such as sonication, homogenization in a tissue grinder or blender, or freezing and grinding with a pestle and mortar. The cells from cultures or suspensions of algae may be collected on glass fiber, paper, or Millipore filters, and the filters and algae homogenized in a high-speed blender.

The algal cells are homogenized in 30–50 ml chloroform: methanol (2:1 v/v). The use of freshly distilled solvents in this and the subsequent chromotographic procedures is recommended. The cell wall material, precipitated proteins, and nucleic acids are removed by filtration on paper, glass fiber, or Millipore filters that are resistant to organic solvents. The lipid extract is collected in a 250-ml round-bottom flask. Before the extract is dried, the water and soluble nonlipid components must be removed. Sephadex G-25 (1 g) is added to the filtrate for each gram of water present in the extracted tissue. The mixture is partially dried on a rotary evaporator until the Sephadex begins to gel and adheres to the walls of the flask. Chloroform (10–20 ml) is added to the flask, and the partial drying process is repeated. The Sephadex is then resuspended in chloroform and the suspension poured into a fritted glass chromatography column (1 cm diam.). The flask is rinsed several times with chloroform and the washings added to the column. The Sephadex is then washed with chloroform until all the chlorophyll has been removed. The Sephadex retains the water and nonlipid materials from the cells; the chloroform contains the lipid and pigments. The eluate from the column is dried on a rotary evaporator, and the lipids are redissolved in a small volume of chloroform (1 ml g^{-1} of algal tissue). The lipid solution may be stored under N_2 in a freezer.

In place of the method described above, one of the methods of Folch et al. (1957) or Bligh and Dyer (1959) may be used. In these methods the tissue extract separates into an organic solvent (lipid) phase and an aqueous phase. The lipid phase is washed with water or salt solutions to remove nonlipid contaminants before being dried and redissolved in chloroform, as above.

B. Quantitative estimation of total lipids and pigments

The total lipid and pigment of the algal cells may be estimated from the extract by conventional gravimetric analysis, by the total lipid carbon estimation method of Holm-Hansen et al. (1967), or by a dichromate method (Chp. 18).

C. Separation by thin-layer chromatography

Two-dimensional chromatography provides the best separation of individual lipids. It is, however, time-consuming, and less lipid material can be applied to the plate. One-dimensional chromatography is, therefore, generally used for fatty acid analysis and quantitative estimates. The following thin-layer chromatography techniques are recommended, although many other systems have also been proved successful (Stahl 1969).

1. Two-dimensional separations. Silica Gel H (E. Merck) plates, prepared in the usual way, are used. The lipid spot is placed in one corner of the plate 2 cm from each edge. The plate is run first in chloroform:methanol:NH_4OH (65:25:4 v/v), dried quickly, and run in the second direction, using chloroform:acetone:methanol:acetic acid:water (50:20:10:10:5 v/v). An outline of a typical separation is shown in Fig. 11-1.

2. One-dimensional separations. The phospholipids and glycolipids are separated using the following thin-layer chromatography system. Silica Gel G (E. Merck) plates (20 × 20 cm) impregnated with $(NH_4)_2SO_4$ are prepared by spreading a slurry of 40 g silica gel in 100 ml 2% $(NH_4)_2SO_4$ solution. The plates are activated for 4 h at 95–100 C just prior to use. The lipids are applied to the plates as streaks at least 1 cm long. The applied streak is dried quickly with nitrogen, and the plate developed in acetone:benzene:water (91:30:8 v/v) (Pohl et al. 1970). The simple glycerides, free fatty acids, and other neutral lipids may be separated on Silica Gel G plates prepared without $(NH_4)_2SO_4$ using less polar solvents (e.g., hexane:diethyl ether, 80:20 v/v) (Stahl 1969).

3. Identification of lipids. The lipids separated by the above techniques may be identified by chromatography with authentic standards, by spray reagents, or, in some cases, by fatty acid content. With two-dimensional chromatography, a separate plate is necessary for each

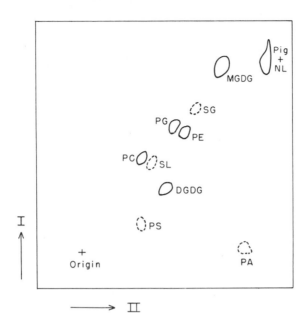

Fig. 11-1. Diagram of a two-dimensional thin-layer chromatograph of *Chlamydomonas* lipid extract separated on Silica Gel H in solvent I, chloroform:methanol:NH₄OH (65:25:4 v/v), and solvent II, chloroform:acetone:methanol:acetic acid:water (50:20:10:10:4 v/v). *Pig + NL*, pigments and neutral lipids; *MGDG*, monogalactosyl diglyceride; *PG*, phosphatidylglycerol; *SG*, sterol glycoside; *DGDG*, digalactosyl diglyceride; *SL*, sulfolipid; *PE*, phosphatidylethanolamine; *PC*, phosphatidylcholine; *PS*, phosphatidylserine; *PA*, phosphatidic acid.

spray reagent. With one-dimensional separations, a section of the plate may be exposed to the spray reagent while the remainder of the plate is protected from the spray by a glass plate.

The following spray reagents should allow the identification of the major glycerolipids in algae.

a. Phosphate spray. Spray lightly. All phosphate-containing lipids will be revealed immediately as blue spots or bands. On heating at 100 C, the galactolipids are charred brown.

Preparation: Dissolve 16 g (NH₄)₂MoO₄ in 120 ml water (solution I). Shake 80 ml of this solution with 40 ml conc. HCl and 10 ml mercury for 30 min and then filter (solution II). Add 200 ml conc. H₂SO₄ to the remaining 40 ml of solution I and then add solution II. Cool and dilute to 1 l (Vaskovsky and Kostetsky 1968).

b. Ninhydrin spray. Spray lightly. Phosphatidylethanolamine and phosphatidylserine react on heating with a typical purple ninhydrin–amino acid reaction.

Preparation: Dissolve ninhydrin (0.5%) in ethanol.

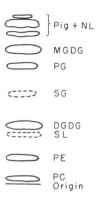

Fig. 11-2. Diagram of a one-dimensional thin-layer chromatograph of *Chlamydomonas* lipid extract separated on Silica Gel G impregnated with $(NH_4)_2SO_4$ in acetone:benzene:water (91:30:8 v/v). Abbreviations as in Fig. 11-1.

c. Sulfolipid spray. Sulfoquinovosyl diglyceride (SL) turns bright pink immediately after spraying. Other lipids turn blue or green.

Preparation: Cresyl fast violet (0.05%) is dissolved in 1% acetic acid and heated for 15 min at 60 C in a water bath.

d. 50% Sulfuric acid. Caution. Spray very lightly. After spraying, heat in a 100 C oven. All lipids will char and turn brown on prolonged heating. Sterols will turn mauve or purple.

All the major lipids may be identified using these sprays and with reference to Figs. 11-1 and 11-2. The identity of some lipids may be confirmed by analysis of their fatty acids. Phosphatidylglycerol, for example, is the only lipid to contain significant quantities of *trans*-Δ^3-hexadecenoic acid (16:1; ratio:number of carbons to number of unsaturated bonds).

D. Transesterification

The sprays recommended in the previous section are destructive and cannot be utilized when fatty acid analysis is required. A nondestructive spray may be used on similar plates run in one or two dimensions.

A solution of 0.05% dichlorofluorescein in methanol is sprayed lightly and evenly on the developed plate. The plate is then examined under UV light and the visible bands carefully outlined with a mounted needle. If the bands are not visible, the plate should be resprayed and examined until they fluoresce under UV light.

The outlined bands are carefully scraped from the plate. The scrapings are collected in culture tubes (16 × 125 mm) with Teflon-lined caps, and 3–5 ml dry 1.5 N methanolic HCl is added. The tubes are flushed with N_2 and the caps secured tightly. Transesterification is carried out overnight in an oven at 80 C.

On cooling, the fatty acid methyl esters (FAMEs) are extracted from the methanolic HCl three times with 2 ml hexane. The hexane extracts are combined in small 1- to 2-dram screw-cap vials and dried under N_2 at 40 C. A known amount of internal standard (10–25 μg) of methyl pentadecanoate (15:0) or methyl heptadecanoate (17:0) is added to the vial and the contents dried again. The FAMEs are redissolved in 50–100 μl of hexane for gas-liquid chromatography (GLC).

E. Gas-liquid chromatography of fatty acid methyl esters

1. Separation and identification of FAMEs. Algae contain a wide variety of different fatty acids, and no GLC stationary phase will adequately separate all the FAMEs found. For a complete analysis, it is often necessary to use more than one stationary phase. The final selection of the stationary phase(s) will depend on the fatty acid composition of the particular species being examined.

The very polar phases (EGSS-X and Silar 10C) provide good separation of FAMEs with the same carbon number but differing degrees of unsaturation, such as the 18-carbon series stearate (18:0), oleate (18:1), linoleate (18:2), and linolenate (18:3). They are also able to separate α- and γ-linolenate. However, the more unsaturated 16-carbon FAMEs, such as hexadecatrienoate (16:3) and hexadecatetraenoate (16:4), and the 20-carbon FAMEs, such as arichidate (20:0) and eicosenoate (20:1), are also eluted from the column at similar retention times, causing serious overlapping of peaks. A slightly less polar packing, Reoplex 400 (polypropylene glycol adipate), has proved successful in separating 16:3, 18:0, and 18:1 in that order, although complete separation of 18:0 and 18:1 is not always obtained. The 20-carbon FAMEs are eluted after 18:3. FAMEs from algae containing 16:4 and/or 18:4 in the presence of 20:0 must be resolved on a much less polar stationary phase such as OV-225. This stationary phase, however, does not separate 16:0 and 16:1 or 18:0 and 18:1. The GLC stationary phases and conditions used in our laboratory are listed in Table 11-1.

The initial identification and analysis of FAMEs may require experimentation with all three types of stationary phases before the most suitable is selected for the particular plant material being examined. The review by Wood (1974) is useful in obtaining a preliminary idea of what fatty acids to expect in a particular group of algae.

FAMEs may be identified in GLC runs by comparing retention times with authentic standards run under the same conditions. The separation of FAMEs and the performance of a stationary phase may be readily determined from a standard mixture of methyl palmitate, stearate, oleate, linoleate, and linolenate.

Table 11-1. *Recommended GLC packings and operating conditions*

Stationary phase	Support	Column temperature (C)
10% Silar-10C (Apolar-10C)	100/120 Gas-Chrom Q	190
15% EGSS-X	100/120 Gas-Chrom P	190
15% Reoplex 400	Chromosorb W HP	190
3% OV-225	100/120 Gas-Chrom Q	160

Note: Glass or stainless-steel columns, 4–6 ft in length, $\frac{1}{8}$ or $\frac{1}{4}$ in. i.d.

2. Quantitative estimation of FAMEs. The area of a peak obtained from a flame ionization detector is usually proportional to the weight of FAME injected. Therefore, if a known quantity of one FAME is added to a sample as an internal standard, the area of the peak of that FAME provides a standard from which the other FAMEs may be estimated. The quantity of each FAME in a sample may, therefore, be calculated as follows: $W_x = (A_x \times W_s)/A_s$, where W_x is the weight of the FAME to be determined, A_x is the area of that FAME, and W_s and A_s are the weight and area of the peak of the internal standard, respectively.

The internal standard chosen is usually an odd-carbon-numbered FAME. Pentadecanoate and heptadecanoate are the most commonly used, as these fatty acids are rarely found in algal lipids. Preweighed standards may be obtained from several GLC supply houses (Applied Sciences Laboratories, Chromatographic Specialties). The fatty acid compositions of lipids are often expressed as percent by weight of each fatty acid. The total amount of a lipid may be estimated from the total weight of fatty acid present.

III. Discussion

It is important during analysis of lipids, especially those containing polyunsaturated fatty acids, to prevent oxidation. This may be done by adding antioxidants, such as butylated hydroxytoluene, to lipid extracts and eliminating O_2 wherever possible by flushing and drying with N_2. Oxidation occurs very rapidly at elevated temperatures, and it is important, especially during transesterification, to flush tubes or vials thoroughly with N_2 before sealing. Techniques that require the heating of lipids or lipid extracts should be avoided where possible.

The techniques outlined here are those that have proved successful in our laboratory. Many different methods are described in the literature, but space limitations prevent their inclusion here.

IV. Acknowledgments

I gratefully acknowledge the assistance of Dr. M. Khan, Miss Nora Lem, and Mr. S. Leung in the preparation of this manuscript.

V. References

Bligh, E. G., and Dyer, W. J. 1959. A rapid method of total lipid extraction and purification. *Can. J. Biochem. Physiol.* 37, 911–17.

Folch, J., Lees, M., and Sloane-Stanley, G. 1957. A simple method for the isolation and purification of total lipids from animal tissue. *J. Biol. Chem.* 226, 497–509.

Holm-Hansen, O., Coombs, J., Volcani, B. E., and Williams, P. M. 1967. Quantitative micro-determination of lipid carbon in microorganisms. *Anal. Biochem.* 19, 561–8.

Kates, M. 1970. Plant phospholipids and glycolipids. *Adv. Lipid Res.* 8, 225–64.

Mazliak, P. 1973. Lipid metabolism in plants. *Annu. Rev. Plant Physiol.* 24, 287–310.

Pohl, P., Glasl, H., and Wagner, H. 1970. Zur Analytik pflanzlicher Glyko- und Phospholipoide und ihrer Fettsäuren. I. Eine neue Dünnschichtchromatographische Methode zur Trennung pflanzlicher Lipoide und quantitativen Bestimmung ihrer Fettsäure-Zusammensetzung. *J. Chromatogr.* 49, 488–92.

Stahl, E. (ed.). 1969. *Thin-Layer Chromatography: A Laboratory Handbook,* 2nd ed. Springer-Verlag, Berlin. 1041 pp.

Vaskovsky, V. E., and Kostetsky, E. Y. 1968. Modified spray for the detection of phospholipids on thin-layer chromatograms. *J. Lipid Res.* 9, 396.

Williams, J. P., and Merrilees, P. A. 1970. The removal of water and non-lipid contaminants from lipid extracts. *Lipids* 5, 367–70.

Wood, B. J. B. 1974. Fatty acids and saponifiable lipids. In Stewart, W. D. P. (ed.). *Algal Physiology and Biochemistry,* pp. 236–65. Blackwell, Oxford.

12: Carrageenans and agars*

J. S. CRAIGIE AND C. LEIGH

Atlantic Regional Laboratory, National Research Council of Canada, Halifax, Nova Scotia B3H 3Z1, Canada

CONTENTS

I	**Introduction**	*page*	**110**
II	**Equipment, materials, and test organisms**		**111**
III	**Methods for carrageenans**		**112**
	A. Extraction and fractionation		112
	B. Alkali modification		113
	C. Analytical procedures		114
IV	**Representative data for carrageenans**		**120**
V	**Discussion**		**121**
	A. Carrageenan analysis		121
	B. Specialized methods		122
VI	**Separation of nuclear phases**		**123**
	A. Procedure		123
VII	**Methods for agar**		**124**
	A. Extraction and fractionation		124
	B. Analytical procedures		126
VIII	**Representative data for agar**		**128**
IX	**Discussion**		**128**
X	**Acknowledgments**		**129**
XI	**References**		**129**

* Issued as NRCC No. 16228.

I. Introduction

Carrageenan is the generic term for a family of commercially important galactan sulfates extractable with hot water from certain Rhodophyceae, especially members of the Gigartinaceae, Hypneaceae, and Solieriaceae. These polysaccharides are of high molecular weight (10^5 –10^6 daltons) and are heavily sulfated (20–50% as $-OSO_3Na$). Characteristically they occur in the cell walls and intercellular matrix and make up a high proportion of the dry weight of the alga. The backbone of the polymer is formed of alternating α-1,3- and β-1,4-linked D-galactopyranose units (Fig. 12-1), which differ in the degree and sites of sulfate esterification (Percival and McDowell 1967). A major variation occurs in κ- and ι-carrageenan, where 3,6-anhydro-D-galactose units formally replace the α-1,3-D-galactose units of the basic polymer. There are considerable differences in the substitution patterns from the ideal forms, and this masks the repeating units (Rees 1972). Within the Gigartinaceae, a distinctive distribution of carrageenans is found, as λ-carrageenan occurs in the diploid, sporophytic generation, and κ-carrageenan is produced by the haploid gametophytes.

Agar is the name given to another family of galactans from red algae. Certain species (principally from the Gelidiales, the Gigartinales, and occasionally from other orders) afford, on extraction with boiling water, polysaccharides that usually gel upon cooling to room temperature. Such preparations frequently exhibit a wide hysteresis (38–85 C) between their gelling and melting temperatures. Chemical examination has established that the repeating unit of agarose is the disaccharide, agarobiose (4-O-β-D-galactopyranosyl-3,6-anhydro-L-galactose). The extensive investigations described by Araki (1966) demonstrate the presence of at least two components, agarose and agaropectin. More recent studies reported by Duckworth et al. (1971) and Izumi (1971) show a complex situation in which a family of related polysaccharides was separated from *Gracilaria* spp. These ranged from the neutral polysaccharide, agarose (Fig. 12-1), through "pyruvated agarose," an agarose containing 4,6-O-(1-carboxyethylidene-D-galactose, to a sulfated galactan of the agar type. The relative

[110]

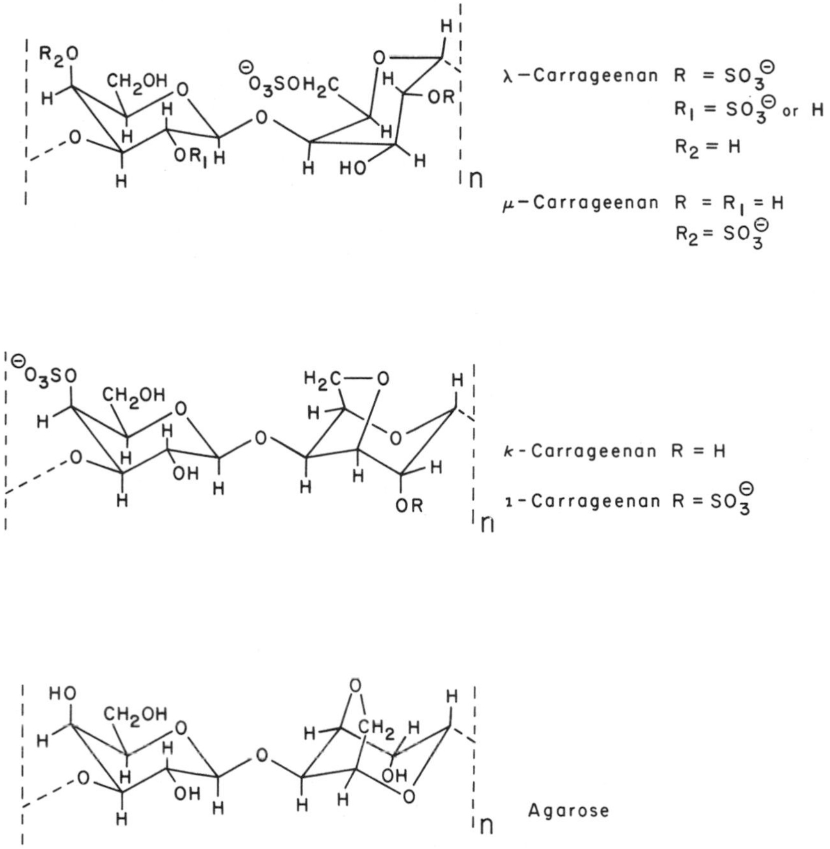

Fig. 12.1 Structure of the repeating disaccharide units in the idealized polymers representing λ- μ-, κ-, and ι-carrageenans and agarose.

proportions vary with the species examined. As in carrageenans, modification of the basic structure may mask the repeating units (Rees 1972). The structure of agaropectin was not determined, and both Duckworth et al. and Izumi suggest that the ionic polysaccharides be termed "charged agarose," and "sulfated galactan."

We describe laboratory procedures for extracting and fractionating these polysaccharides and provide appropriate simple methods for their analysis.

II. Equipment, materials, and test organisms

All chemicals were of the best quality available and were used directly unless otherwise noted. Algal specimens were collected in the field or were cultured in tanks. Careful attention to the sorting and identification of the algal material cannot be overemphasized. Voucher specimens labeled with dates and sites of collection must be set aside for future examination, especially when working with unfamiliar or exotic species.

The methods described for carrageenans were developed using *Chondrus crispus* and have been successfully applied to a number of other carrageenan-producing algae. The principal experimental alga for preparing agar was *Gracilaria* sp., but the methods have given good results when extended to other agarophytes.

III. Methods for carrageenans

A. Extraction and fractionation

1. Sort and clean the freshly collected algae and separate haploid from diploid plants where possible. The younger, growing portions of thalli are usually more free of adhering contaminants. Keep samples refrigerated until used.

2. Place samples with a blotted fresh weight of up to 20 g in the 150-ml steel chamber of a Dangoumou ball mill (Prolabo) together with the 32-mm steel ball and freeze with liquid N_2 for 3–5 min. When the excess liquid N_2 has vaporized, assemble the chamber and pulverize the sample by operating the mill for 2–3 min.

3. Transfer the frozen powder into a beaker containing 100 ml of acetone, stir, and finally allow to settle. Remove the green supernatant by decantation or vacuum filtration. Repeat this step and subsequently reextract the powder with boiling 80% ethanol, absolute ethanol, and finally at room temperature with diethyl ether. The algal powder is not sticky and may be filtered on glass fiber filter paper (Whatman GF/C) at each of these stages. Briefly air-dry the sample to remove excess ether before desiccating it overnight in vacuo over P_2O_5.

4. Add about 300 mg of the dried powder to 150 ml of 0.5 M $NaHCO_3$ in a 400-ml beaker and stir vigorously while maintaining the temperature at approximately 90 C on an oil bath. Extraction takes place over 1–2 h.

5. Pour the hot, viscous solution into a stainless-steel pressure filter (Sartorius, type SM 162 45) fitted with a 1- to 2-cm-thick pad of glass-wool resting on a membrane filter, 12 μm pore size (Sartorius No. 125 00). It is possible to use the same filter for as many as six identical samples (if the carrageenans are to be combined). Residues collected on the glass-wool and filter should be reextracted as above for 30 min.

6. Refine the warm solution by filtering through a 3.0-μm or even a

1.2-μm membrane filter. Glass-wool is not usually required at this stage. The filtrate should be essentially colorless and free from turbidity. Colorimetric tests may be carried out directly on this preparation.

7. Precipitate the carrageenans by the slow addition, with stirring, of 20 ml of 2% aqueous cetyltrimethylammonium bromide (Cetavlon). Allow the bulky, white, somewhat gummy precipitate to settle and collect it by low-speed centrifugation. Wash the precipitate twice with distilled water and then repeatedly on the centrifuge (up to seven times) with a near-saturated solution of sodium acetate in 95% ethanol. Careful stirring at this stage is essential to break up all clumps, and warming the suspensions speeds up the exchange process. At least three final washings with warm 95% ethanol are required to remove the sodium acetate. Finally, wash the sample with diethyl ether and dry it overnight in vacuo over P_2O_5. Record the weight of total carrageenan.

8. Redissolve the carrageenans in water at about 0.1% concentration by warming to 70 C with vigorous stirring, then pressure filter (1.2 μm) into centrifuge buckets, and add 0.1 volume of 3 M KCl to each. Mix the solutions well and cool them for at least 1 h in crushed ice before compacting any gel at 30 000 $\times$ g for 30 min in a refrigerated centrifuge. The insoluble carrageenan should be resuspended in 0.3 M KCl and recentrifuged. Remove the KCl from the precipitate by washing with warm 80% ethanol until the washings give a negative $AgNO_3$ test for Cl^-. If the soluble carrageenans are desired, recover them as in step 7 above, or by precipitation with 2.5 volumes of 2-propanol followed by thorough washing with 80% alcohol to remove the KCl. Wash the products with absolute ethanol and diethyl ether and dry as before. Alternatively, salts may be removed by dialysis in Visking tubing and the carrageenans recovered by lyophilization.

B. Alkali modification

In principle, the treatment of carrageenans or agars by alkali will eliminate the primary sulfate ester at C-6 of the 4-linked galactopyranose unit giving rise to 3,6-anhydrogalactose. This reaction is a valuable step in characterizing such polysaccharides (Rees 1963). A sulfate ester at C-3 will also produce 3,6-anhydrogalactose when treated with alkali.

Procedure. Place 100 mg of carrageenan into a 125-ml Erlenmeyer flask, add 20 ml distilled water and a stirring bar. Dissolve the sample and add 20 mg of $NaBH_4$; mix well and allow to stand overnight. Add a further 60 mg of $NaBH_4$ and 10 ml of 3 M NaOH. Mix and heat at 80 C for 7 h. Pour the reaction mixture into dialysis tubing and dialyze against running water overnight or until no alkali remains. Add 1 ml of 10% NaCl and filter through Whatman GF/C paper. Recover the polysaccharide by precipitation in 3 volumes of ethanol, wash with 80% ethanol until free of Cl⁻, dehydrate with absolute ethanol, rinse with diethyl ether, and dry in vacuo as above. Approximately 70–85% of the weight of original sample is usually recovered. The product may also be recovered by lyophilization immediately after dialysis.

C. Analytical procedures

1. Infrared analysis of films. This continues to be the definitive method for identifying the type of carrageenan. The technique may be used semiquantitatively and shows directly (Fig. 12-2) the presence of ester sulfate (1240 cm⁻¹) and 3,6-anhydrogalactose (928–940 cm⁻¹) and indicates the position of esterification (e.g., 820 cm⁻¹, primary hydroxyl, 6-sulfate; 830 cm⁻¹, equatorial secondary, 2-sulfate; 840–850 cm⁻¹, axial secondary, 4-sulfate). The absorption band near 805 cm⁻¹ is attributed to 3,6-anhydro-D-galactose-2-sulfate.

Weigh approximately 3 mg of carrageenan and transfer it onto a 25-mm diameter × 5-mm thick AgCl window (Harshaw). Add 8–10 drops of boiling distilled water and dissolve the polysaccharide by stirring with a small Teflon spatula. The sample will be viscous and should be spread uniformly to within 1 mm of the edge of the disk. Evaporate the water in a dark, vented box by warming it to 40–45 C with a heat lamp. Leave the glassy film on the disk and record the spectrum from 700 to 2000 cm⁻¹ with a standard infrared spectrometer such as the Perkin-Elmer 237 or 521.

The usefulness of the AgCl windows may be greatly prolonged by avoiding unnecessary exposure to light or abrasion during preparation of the film or cleaning of the disks. Films may be removed readily with water and a gentle rubbing with the fingertips coated with a mild, nonabrasive hand soap. After rinsing, the disks may be dried with acetone and stored in darkness.

2. Colorimetric measurement of 3,6-anhydrogalactose. The most reliable and widely used method for anhydrogalactose determination is based on the resorcinol reaction as described by Yaphe and Arsenault (1965). The color yield of 3,6-anhydrogalactose is 92% of that of an equimolar concentration of D-fructose, and the latter is recommended as the reference standard because of its ready availability. Other hexoses commonly encountered in polysaccharides give but little interference (e.g., galactose, 1%; glucose, 2%; and mannose, 2.5%). Pentoses such as ribose or xylose give about 5% of the color yield of fructose.

   a. Stock solutions

   i. Resorcinol: 150 mg (1.36 mmol) in 100 ml of glass-distilled water. Store in a dark bottle in refrigerator and make fresh weekly.

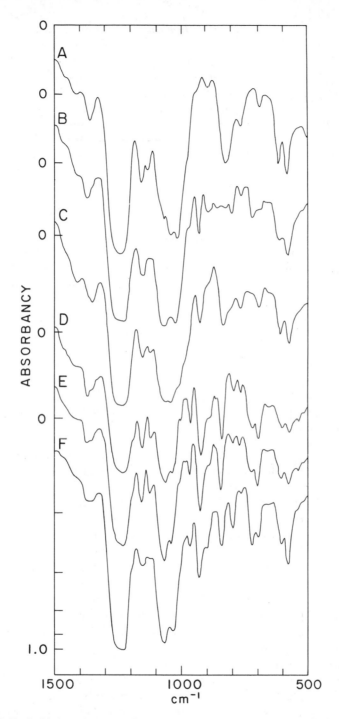

Fig. 12-2. Typical infrared spectra of carrageenans, *A*, λ-carrageenan; *B*, alkali-modified λ-carrageenan; *C*, 0.3 M KCl-soluble carrageenans; *D*, as for *C* but alkali-modified; *E*, κ-carrageenan; *F*, ι-carrageenan. Spectrum *F* is from *Eucheuma spinosum;* all others are from *Chondrus crispus* preparations.

ii. 1,1-Diethoxyethane or acetal: 82 mg (695 μmol) in 10.0 ml of distilled water. Keeps refrigerated in dark bottle up to 3 weeks. Dilute 1 ml to 25 ml in water prior to mixing the colorimetric reagent.

iii. D-fructose: 27.0 mg (150 μmol) in 50.0 ml of benzoic acid– saturated distilled water. Store under refrigeration.

b. Working solutions

i. Resorcinol-acetal reagent: Add 100 ml of conc. HCl to 9.0 ml of resorcinol stock solution and add 1.0 ml of the diluted acetal stock so- lution. Prepare freshly daily.

ii. Fructose standards: Dilute 3.0 ml of stock solution to 100.0 ml with distilled water.

c. Procedure. Place 2.0-ml samples containing an accurately known quantity (ca. 100 μg) of total hexose (III.C.3) in 20 $\times$ 150-mm culture tubes with Teflon-lined screw caps. Cool in ice and add 10.0 ml of cold resorcinol-acetal reagent to each. Mix thoroughly with a tube buzzer and warm the loosely capped tubes to 20 C for 4 min and then heat at 80 C for 10 min. Cool the tubes in an ice bath for 1.5 min and read their absorbance at 555 nm within 15 min. Avoid exposing the colored solutions to strong light.

The quantity of 3,6-anhydrogalactose is determined by reference to the standard curve for fructose and multiplying the value obtained by 1.087. Beer's law applies up to 0.25 μmol per 2 ml of sample analyzed, and at this concentration fructose should give an absorbance of 0.560 –0.580 when measured in a 1-cm cuvette.

3. Hexose analysis. Methods are based on either the direct reaction of the polysaccharide solution with a suitable reagent or the hydrolysis of the polymer followed by the separation and determination of the individual sugars released. Unfortunately, no universally applicable hydrolysis procedure is available for polysaccharides, and incomplete recovery of the constituent monomers is usual.

The colorimetric procedure of Dubois et al. (1956) employs a phe- nol-H_2SO_4 reagent (Chp. 10) that is only slightly affected by salts. The standards containing up to 120 μg of galactose per milliliter should be made up in 0.5 M NaCl if salt is present in the unknown solution. The interpretation of results is straightforward for polysaccharides con- taining a single type of sugar (e.g., λ-carrageenan). Anhydrogalactose in other carrageenans reacts strongly and thus interferes seriously with the measurement of galactose.

To examine the constituent sugars, it is necessary to hydrolyze the sample completely. This is conveniently done by weighing a sample (10 mg) into a Pyrex test tube (10 $\times$ 75 mm), moistening it with a few drops of ethanol, and adding 1 ml of 1.5 N H_2SO_4. Drop a pea-sized piece of dry ice into the tube and allow it to evaporate. When the tube is filled with CO_2 draw out the end and seal it in a hot flame. After hydrolysis for 4 h at 100 C, cool the tube and open it with a sharp file and a bead of molten glass. Neutralize the sample by pouring it into 2 ml of Dowex 2 × 8 ion-exchange resin in the carbonate form. Stir periodically until the sample is neutral when tested with pH paper. Remove the resin by filtration, rinse it with water, and concentrate the filtrate to dryness in vacuo on a rotary evaporator at 40 C. Redissolve the sugars in water for conventional paper chromatographic analysis (Wilson 1959) using ethyl acctate:pyridine:water (8:2:1 v/v/v) as the developing solvent for 25–30 h.

Reliable gas chromatographic methods are now commonly applied to the determination of sugar mixtures from hydrolysates. We routinely use the procedure for preparing and separating alditol acetates described by Sloneker (1972). The column, however, is replaced by one 3.18 mm × 1.8 m packed with 1% ECNSS-M on 80- to 100-mesh Gas-Chrom Q. The temperature is programmed from 160 to 210 at 2 C min^{-1} with a N$_2$ carrier gas flow rate of 20 ml min^{-1}. The use of a suitable internal standard such as ribitol or *myo*-inositol added before hydrolysis simplifies the analysis by eliminating the necessity for quantitative transfer of the sample at each step.

4. Sulfate analysis. Methods for determining the sulfate in polysaccharides depend on achieving complete hydrolysis of the esters followed by quantitative precipitation of the sulfate liberated. We find the submicromethod of Jones and Letham (1954) to be both convenient and reliable when applied to algal polysaccharides.

a. Glassware. All glassware must be cleaned with HNO$_3$ and thoroughly rinsed with tap and distilled water and oven-dried. Sulfate is a ubiquitous contaminant!

b. Reagents

i. 4-Amino-4'-chlorobiphenyl hydrochloride (Burdick and Jackson), 95.0 mg (396 μmol) in 50 ml of 0.10 N HCl. *Caution:* This reagent is a potential carcinogen. Use extreme care when weighing, transferring, and disposing of it. It is advisable to wear surgical gloves during these manipulations. A simple synthesis for this reagent by reduction of the corresponding nitro compound is described by Belcher et al. (1953).

ii. 20.0 ml 0.100 N H$_2$SO$_4$ (standardized solution) diluted to 500 ml with distilled water.

iii. 2% recrystallized Cetavlon in distilled water.

iv. 1 N HCl.

c. Procedure. Place approximately 4-mg samples of carrageenan in tared 10 × 75-mm test tubes and dry in vacuo at 60 C overnight over P$_2$O$_5$. Be certain the sample is introduced into the bottom of the tube. This can be achieved by lining the inside of the tube with a small cylinder of weighing paper or aluminum foil before introducing the polysaccharide. Handle the tubes with forceps or surgical gloves and reweigh to ± 0.05 mg. Moisten the sample with a few drops of 95% ethanol and add 1.0 ml of 1 N HCl. Seal the tubes with a hot flame and hydrolyze for 2 h at 105 C. Open the tubes and quantitatively transfer the contents to 50.0-ml volumetric flasks and make to volume with glass-distilled water. Place duplicate 0.50-ml samples in 3-ml centrifuge tubes and add 1 microdrop of 2% Cetavlon and 0.50 ml of aminochlorobiphenyl reagent. Cover the tubes with Parafilm and invert to mix the contents. The blank should be clear and colorless. After 2 h centrifuge the tubes and make 0.30 ml of each supernatant to 25.0 ml with 0.1 N HCl. Read the samples at 254 nm in 1.0-cm quartz cells against 0.1 N HCl. The blank should read 1.00–1.05 absorbance units. Prepare fresh reference sulfate solutions from standardized 0.100 N H_2SO_4 by diluting 2.0 ml of the acid to 50.0 ml with distilled water. This contains 2.0 μmol H_2SO_4 per milliliter (192 μg of SO_4^{2-} per milliliter). Place subsamples of 0.10–0.50 ml in 3-ml centrifuge tubes, add distilled water to make 0.50 ml volume, and carry out the analysis as described above. The absorbance values measure the amount of reagent remaining and are inversely proportional to the SO_4^{2-} present in the samples.

5. Viscosity. Reproducible measurements on dilute solutions may be made with Ostwald pipettes or Cannon-Fenske viscometers. These provide a means of measuring internal friction of liquids and require the use of dilute solutions. All glassware must be acid-cleaned, thoroughly rinsed, dried, and protected from dust. Solutions must be prepared with deaerated distilled water and filtered through fine-porosity glass filters or membranes (0.45 μm) before measurements are attempted. Good analytical technique is essential.

a. Ostwald pipettes. Select pipettes with comparable flow rates for water near 100 ± 5 sec and mark these consecutively with a diamond pencil. They must be free of chips, scratches, etc., and thoroughly cleaned and dried by rinsing with redistilled acetone followed by a jet of filtered air.

b. Viscometer constant. Prepare 1 l of filtered, distilled water, deaerated by boiling or by suction, and equilibrate a portion at 25 ± 0.1 C. Clip the Ostwald pipettes to a rack and immerse them to approximately 2 cm above the upper graduation mark. Ascertain that the pipettes are absolutely vertical by the use of a spirit level. Using a long-tipped serological pipette, add 3.0 ml of the distilled water to the wide arm of the viscometer without introducing air bubbles. Allow to equilibrate for 5 min and, with a soft rubber bulb, gently draw the liquid up the capillary arm to a level just above the upper graduation mark. Time the flow of liquid between the two graduation marks to the nearest 0.1 sec with a stop watch. Average five such determinations and calculate the viscometer constant, k, where $k = \eta/td$, and η = absolute viscosity of water = 0.8937 cps at 25 C; t = seconds; d = density of water = 0.99708 g ml^{-1} at 25 C. Repeat this for each viscometer.

c. Density of solutions. Select a clean 10-ml volumetric flask and accurately determine its tare. Fill it exactly to mark at 25 C using deacrated, distilled water. Be careful to remove any excess droplets from the upper neck with lint-free paper wipes. Determine the volume of the flask from $V = M/D$, where V = milliliters, M = grams, and D = 0.99708 g ml^{-1} at 25 C.

Measure the density, D_0, of the 0.10 M NaCl solvent at 25 C using this calibrated flask and the relationship $D_0 = M/V$. Be certain that the flask is clean and dry before use.

d. Absolute viscosity of solvent. Prepare 500 ml of 0.10 M NaCl using deaerated, distilled water and filter through a membrane (0.45 μm). Determine the absolute viscosity, η_0, of this solvent from $\eta_0 = ktd_0$ for each viscometer following the procedure outlined in b. The solvent density, d_0, is obtained as in c, above.

e. Preparation of carrageenan solutions. Use 0.10 M NaCl as the solvent throughout. Accurately weigh 75 mg of κ-carrageenan into a 50-ml volumetric flask. Add a magnetic stirring bar and approximately 40 ml of NaCl solution. Stir vigorously for 30–45 min at room temperature and observe carefully to be certain that no undissolved carrageenan remains. Retrieve the stirring bar and rinse it into the flask with solvent. Allow the solution to equilibrate to 25 C and bring it to volume. Prepare three dilutions from this stock using volumetric pipettes and glassware as follows: 15.0–25.0, 15.0–50.0, and 3.0–25.0 ml. The solutions contain 0.15, 0.09, and 0.045, and 0.018% of carrageenan in 0.10 M NaCl. For λ-carrageenan use one-half of these concentrations. All solutions should be inspected carefully for optical clarity and should be filtered if any suspended matter is seen.

f. Density of carrageenan solutions. Measure the density of each solution in the 10-ml volumetric flask as described in c, above.

g. Viscosity of carrageenan solutions. Deaerate each of the carrageenan solutions by briefly applying a vacuum to each flask. Carefully place 3.0 ml of solution into a viscometer, equlibrate 5 min at 25 ± 0.1 C and proceed as in b, above.

i. Calculate the absolute viscosity, η, for each solution from $\eta = ktd$, where k = viscometer constant, t = seconds, and d = density of the solution.

ii. Compute the specific viscosity, η_{sp}, from $\eta_{sp} = (\eta - \eta_0)/\eta_0$, where η_0 = absolute viscosity of the solvent (see d, above).

iii. The reduced specific viscosity is η_{sp}/C, where C = concentration of test solution in g/100 ml. Plot each value of η_{sp}/C vs. C on linear graph paper and draw a straight line through the data using a least-squares fit. Determine the intrinsic viscosity $[\eta]$ by extrapolating the line to zero concentration. Correct this value for moisture content in the orginal carrageenan sample by multiplying $[\eta]$ by 100/percentage of dry matter in the sample. The intrinsic viscosity is the arbitrary number used to characterize the solution and has the dimensions cubic centimeter per gram.

For very dilute solutions it is sufficiently accurate for most purposes to determine the absolute viscosity, η, from $\eta = (t/t_0)\,\eta_0$, where t and t_0 are the flow times in seconds of the solution and solvent, respectively.

Typical intrinsic viscosities recorded under these conditions for native λ-, κ-, and ι-carrageenans are 20–22, 8–11, and 6–8 cm^3 g^{-1}, respectively.

6. *Moisture content.* Algal polysaccharides are very hygroscopic, and considerable care must be exercised in handling samples for analysis. Preparations dried as described above may contain 8–15% of moisture, and this value should be ascertained before reporting analytical data. An accurately weighed sample (25–50 mg) should be dried overnight at 60 C in vacuo over a Petri dish containing P_2O_5. Release the vacuum slowly through a drying tube containing anhydrous silica gel. When the sample has reached a constant weight, the moisture content may be calculated.

IV. Representative data for carrageenans

The total carrageenan content of *C. crispus* from Nova Scotia ranges from about 50% to 70% of the alcohol-extracted dry weight of the frond. The lower values were encountered in tetrasporophytes, whereas values for the gametophytes generally lie between 60% and 66%. Extracts of the diploid tetrasporophyte are composed of λ-carrageenan and contain no gelling fraction. Alkali treatment of this extract produces an alkali-modified λ-carrageenan that possesses a characteristic infrared spectrum (Fig. 12-2), thus confirming the identification.

Fractionation of extracts from haploid plants with 0.3 M KCl yields a gelling fraction (κ-carrageenan) accounting for 75–85% of the total carrageenan. The remaining KCl-soluble fraction is a mixture that, on alkali modification, affords a κ-carrageenan as the principal product. The distinctive characteristics of these polysaccharides are clearly shown in the infrared spectra of Fig. 12-2.

Detailed analyses of typical preparations of the carrageenans in *C. crispus* from our area are given in McCandless et al. (1973).

V. Discussion

A. Carrageenan analysis

The procedures described have been successfully applied in analyses of carrageenans in *C. crispus* and related algae, including representatives of most genera of the Gigartinaceae as well as *Petrocelis* spp., *Eucheuma* spp., and *Hypnea musciformis*. The following remarks may prove helpful in adapting the procedures to new situations.

The cleaned alga should be reduced to a fine powder (III.A.2) to facilitate the solubilization of the polysaccharides. If air-dried or freeze-dried material is used, passage through the 40-mesh screen of a Wiley mill will suffice.

Pigments and compounds of low molecular weight should be removed by sequential extraction with acetone, hot 80% ethanol, ethanol, and diethyl ether. For fresh material these extractions are necessary to remove water. The resulting powder is dried in a small Petri dish in vacuo over P_2O_5 at room temperature. This gives a convenient estimate of extracted dry weight, and capped vials of dry powder may be readily stored in closed vessels containing silica gel or Drierite.

Factors to bear in mind during the actual extraction step are that the high viscosity of carrageenan solutions sets a maximum working concentration of 1–2%, that extremes of pH must be avoided, and that temperature should be kept below 100 C, in order to minimize degradation of the carrageenan. The 3,6-anhydrogalactose $\alpha(1 \rightarrow 3)$ galactopyranose linkage is particularly sensitive to acid cleavage, whereas treatment with alkali increases the 3,6-anhydrogalactose content through elimination of the 6-sulfate with concomitant ring closure (Rees 1963).

Direct alcohol precipitation of carrageenan is not advisable because of the insolubility of $NaHCO_3$ and the likelihood of coprecipitating contaminating macromolecules (e.g., starch). A quaternary base is more selective, as it combines with the ester sulfate of the carrageenans, giving an insoluble precipitate. To recover the carrageenans from this complex, it must be washed repeatedly and carefully with near-saturated solutions of sodium acetate in ethanol. This is followed by further alcohol washing to remove the salt. The use of cetylpyridinium chloride instead of Cetavlon is advantageous in that removal of the latter may be followed by the decrease in ultraviolet absorption (260 nm) of the preparation. Carrageenans cannot be efficiently precipitated from aqueous solution by alcohol unless the ionic strength is at least 0.1 M. It is helpful to increase the ionic strength with salts such as NaCl or sodium acetate, as they have moderate solubilities in 80% alcohol.

Once isolated, the carrageenan can be fractionated either by leaching (stirring for 3–5 h at room temperature) with 0.15–0.30 M KCl to leave behind the insoluble (usually κ- or ι-) carrageenan (Stancioff and Stanley 1969) or by adding KCl to a hot solution of carrageenans (III.A.8).

The isolated carrageenans should be analyzed for 3,6-anhydrogalactose, for galactose, and for sulfate content. Infrared spectra should be recorded, if at all possible, to ascertain the type of carrageenan isolated. If the carrageenan is modified by alkaline treatment under reducing conditions and reexamined, it will provide additional confirmation of the type. In no case should the identification of carrageenans be attempted solely on the basis of their solubilities in KCl solutions.

Chromatograms of carrageenan hydrolysates show principally galactose. Traces of xylose and one or two other carbohydrates are sometimes observed and are generally attributed to contaminating polysaccharides.

B. Specialized methods

Space limitations preclude any detailed discussion of the variety of methods used to detect or characterize carrageenans. The following summary provides some insight into interesting and potentially useful techniques that have been developed.

1. Enzymatic. An enzyme isolated from a marine bacterium *Pseudomonas carrageenovora* has been used to establish the presence of κ-carrageenan in various red algae. For example, κ-carrageenase degrades furcellaran, the lightly sulfated kappa type of carrageenan from *Furcellaria fastigiata,* and does not attack that of *Agardhiella tenera* (a more highly sulfated iota type). As pointed out by Weigl and Yaphe (1966), the enzyme is specific for the alternate β-1,4-linkages in κ-carrageenan. It does not attack those attached to galactose-6-sulfate, galactose disulfate, or 3,6-anhydrogalactose-2-sulfate; thus the enzyme does not degrade ι-, μ-, or λ-carrageenan. Another enzyme, also from a marine bacterium, specifically hydrolyzes ι-carrageenan (Yaphe, personal communication).

2. Immunochemical. An immunochemical technique has been developed that shows promise in the classification and even the analysis of carrageenans (Hosford and McCandless 1975). When a single type of carrageenan is injected into the blood stream of a suitable animal, specific antibodies are produced. κ-Carrageenan does not precipitate the antiserum to λ-carrageenan, and conversely.

3. Polyanionic functions. The carrageenans can be precipitated selectively by certain ions and by cationic detergents such as cetyltrimethylammonium bromide (Cetavlon) or cetylpyridinium chloride. This method has been used both in the purification of carrageenans and in their gravimetric analysis. The same principle has been used with dyes such as alcian blue (Gangolli et al. 1973), acridine orange (Cundall et al. 1973), and *o*-tolidine (Graham 1972).

4. Nuclear magnetic resonance spectroscopy. Potentially the most powerful tool for analysis of extracts may be nuclear magnetic resonance (NMR) spectroscopy as applied to agars by Izumi (1973), who intends to extend his study to carrageenans. Achieving sufficient concentrations may be difficult, but the technique would ideally give direct molar ratios of galactose and 3,6-anhydrogalactose, and perhaps the degree and sites of sulfate esterification. The region of the proton spectrum between τ 5.2 and 5.8 contains numerous signals, and the resolution of these may prove complicated.

5. Gel strength. Among the properties of carrageenans most highly regarded in commercial applications are those relating to gel strength and viscosity. The industrial methods of determining these properties involve expensive equipment and require much more sample than is usually recovered from small-scale laboratory preparations.

Goring (1956) describes a reliable, inexpensive gelometer for use with samples containing approximately 50–100 mg of carrageenan or agar. The precautions to be observed in forming reproducible gels are discussed. The observed gel strength is a function of the area of the plunger, and Goring recommends one 6 mm in diameter advancing at a rate of 0.36 mm sec^{-1}. The diameter of the plunger must be given when reporting gel strength measurements. The preparation of carrageenan gels with powdered milk is described by Pernas et al. (1967).

VI. Separation of nuclear phases

Our observation that tetrasporophytes of *C. crispus* contain λ-carrageenan, whereas gametophytes contain κ-carrageenan (McCandless et al. 1973), has now been extended to include representatives of all genera in the Gigartinaceae. This permits us to separate vegetative plants rapidly into diploid or haploid phases according to the type of carrageenan present. The procedure does not apply to carrageenophytes from other families.

A. Procedure

Excise 3- to 15-mg pieces of air-dried *C. crispus* and place them separately in 18 × 150-mm test tubes. With a little practice this quantity can be estimated visually and large numbers of plants can be processed easily. Add 5 ml of distilled water to each from an automatic syringe or pipette and autoclave the samples at 121 C for 45 min. Shake the racks of hot tubes vigorously by hand to disintegrate the tissue. Add an additional 5 ml of water and mix. Using Pasteur pipettes, place subsamples of 1 drop into clean tubes and add 2 ml of fresh resorcinol-acetal reagent (III.C.2) that has been diluted by the addition of 25 ml distilled water. Heat the tubes for 2.5 min in a boiling-water bath. Samples from diploid algae remain virtually colorless, whereas those from haploid specimens contain anhydrogalactose and exhibit clear pink to red colors. If the weight of algal sample greatly exceeds 15 mg, traces of pink color may be encountered in tubes from diploid specimens.

VII. Methods for agar

A. Extraction and fractionation

The collection, sorting, identification, and preliminary extraction of the algae follow the procedures outlined for carrageenophytes, except that larger samples of 20–100 g fresh weight should be collected. We prefer to pulverize the fresh algae in a ball mill in the frozen state. Air-dried or lyophilized plants may be ground to pass a 40-mesh screen of a Wiley mill. The powdered tissue should be solvent-extracted and dried as described in III.A.1–3.

Partially purified agarose may be prepared by the use of a quaternary ammonium base (Hjertén 1962) or by various ion-exchange procedures (Zabin 1969; Duckworth and Yaphe 1971a; Hjertén 1971; Izumi 1971; Johansson and Hjertén 1974). The latter method gives the maximum informtion on the types of polysaccharides in the agar.

1. Extraction of crude agar. Weigh 1.0 g of dried algal powder and place in a 400-ml beaker and hydrate this for at least 3 h with 75 ml of distilled water or of 0.1 M phosphate buffer at pH 6.3. Cover the beaker with aluminum foil, make a small vent hole, and autoclave for 3 h at 121 C. Filter under pressure through glass-wool and a membrane filter (12 μm). Keep solutions and filter apparatus above 60 C. The pressure filter may be wrapped with an electrical heating tape connected through a Variac. Add 5–10 g of Celite and refine the boiling filtrates sequentially through filters (3.0 and 1.2 μm). A combined magnetic stirrer and hot plate is essential. Pour the solutions into shallow trays or Petri dishes and allow them to gel at room temperature. Cut the gel into 2-cm strips with a razor blade and freeze. Thaw the gel at room temperature and filter off the free liquid with suction through Whatman No. 54 paper covered with nylon mesh. Wash the residue carefully with three 100-ml portions of distilled water followed by 80% and 95% ethanol and diethyl ether. Dry overnight in vacuo over a dish containing P_2O_5. If the extract does not gel, reduce the volume by half on a rotary evaporator at 60 C and pour the solution into 3 volumes of 95% ethanol. Collect the precipitate and wash it thoroughly with warm 80% ethanol until the solution gives a negative $AgNO_3$ test for chloride. Complete the dehydration with ethanol and diethyl ether, and dry as above. Agar may also be recovered by lyophilization following dialysis to remove salts.

In some cases where information on the native agar is not required, the agar may be modified directly by triturating the algal powder (200 mg) with 1 M NaOH (20 ml), adding 75 mg $NaBH_4$, and warming this to 95 C for 60 min. Cool the resulting slurry and thoroughly leach the gel with water until no alkali remains. The removal of alkali may be assisted by washing with 0.1 M acetic acid followed by complete removal of the acid by washing with water. The pH of the aqueous slurry should be measured and adjusted with 1 N NaOH or HCl, as required, to between pH 6.3 and 7.5 before proceeding with the extraction of the alkali-modified agar.

2. Fractionation of crude agar with DEAE-Sephadex. Wash DEAE-Sephadex A-50 (Cl⁻) on a Buchner funnel sequentially with 0.5 N aqueous solutions of HCl, NaOH, HCl, and finally with distilled water until no acidity remains. Resuspend the Sephadex in distilled water to give a slurry of approximately 1.0 g/100 ml.

Dissolve 25.0 mg of agar in hot distilled water, cool, and make to 25.0 ml volume. Prepare duplicate glass-stoppered 250-ml Erlenmeyer flasks containing 87.5 ml of distilled water and 2.5 ml of the Sephadex suspension. Add 10.0 ml of agar solution to each and stir vigorously for 1 h at 60–65 C, then transfer duplicate 5.0-ml samples into centrifuge tubes. Sediment the Sephadex and, with a plastic-tipped plunger-type pipette, carefully withdraw 1.0-ml samples for colorimetric analysis (III.C.3). Compare the absorbancy with that of the original agar solution correspondingly diluted and calculate the percentage of agarose in the sample.

Separate the Sephadex from the remaining 90 ml of agar solution by vacuum filtration on glass fiber filters (Whatman GF/C). Rinse the flask and filter with 3 × 50 ml of boiling distilled water and carefully return the filter and Sephadex to the 250-ml flask. Add 90 ml of 0.5 M NaCl and stir vigorously for 1 h at 60–65 C. Centrifuge a subsample and determine the carbohydrate content colorimetrically. This represents the charged agarose fraction, and its absorbance must be compared with that of the stock agar solution diluted 1:10 with 0.5 M NaCl and reacted under the same conditions.

For highly charged agars, repeat the Sephadex extraction step with 80 ml of 2.5 M NaCl. Measure the total carbohydrate as above. Finally extract the Sephadex with 28 ml of 5 M NaCl stirred for 1 h at 60–65 C. Centrifuge an aliquot and dilute the supernatant 2.0 ml to 5.0 ml with distilled water prior to determining the amount of carbohydrate released.

3. Isolation of partially purified agarose with a quaternary base. Dissolve 250 mg of crude agar in 100 ml of boiling distilled water. Add 25 mg of λ-carrageenan and when dissolved add 10 ml of 2% Cetavlon (or cetylpyridinium chloride) to the hot (80–100 C) solution. Then add Celite (2 g) and pressure filter through a membrane (0.8 μm). Pour the filtrate into a Petri dish to gel. Freeze and thaw the gel and recover the partially purified agarose as described in VII.A.1.

B. Analytical procedures

The methods described for the analysis of 3,6-anhydrogalactose, hexose, and sulfate in carrageenans apply equally to polysaccharides of the agar type. NMR techniques have been successfully applied to agar (Izumi 1973). Infrared spectra can be prepared, but appear to be of limited use in characterizing agar. Several techniques having special application to agar have been refined in Professor W. Yaphe's laboratory as follows.

1. Enzymatic hydrolysis: agarase. A reliable method for distinguishing agar from other polysaccharides is based on the specificity of agarase isolated from *Pseudomonas atlantica* or obtained from a commercial source (Calbiochem). The qualitative assay depends on the ability of undegraded agar to form a purplish-black color with I_2-KI reagent.

   a. Reagents

   i. I_2-KI. Dissolve 0.5 g iodine (1.97 mmol) and 1.0 g KI (6.02 mmol) in 50 ml distilled water. Store in darkness.

   ii. Naphthoresorcinol. Dissolve 100 mg naphthoresorcinol (0.63 mmol) in 50 ml ethanol. Just before use add 1 volume of this to 2 volumes of a mixture containing 100 ml conc. H_2SO_4 in 375 ml of 95% ethanol.

   b. Procedure

   i. Dissolve 10 mg of agar in 1 ml of boiling phosphate buffer (0.01 M at pH 7.0). Place in a water bath set between 42 and 50 C. The enzyme attacks agar only in the sol state.

   ii. Add 0.05 ml of agarase solution prewarmed to 42–50 C and incubate 60 min.

   iii. At 15-min intervals test 1 drop of hydrolysate with 2 drops of I_2-KI in a spot plate.

   iv. If no color develops, continue the incubation overnight at 25 C with additional enzyme (0.05 ml). If color persists after 60 min repeat steps ii–iv.

v. Prepare chromatograms of the hydrolysate on thin layers of cellulose (MN-300) using galactose as a reference sugar. Develop the plates 10 cm in *n*-butanol:ethanol:water (3:2:2) or *n*-butanol:acetic acid:water (4:1:2) by volume. Spray the air-dried plates with naphthoresorcinol reagent, develop color for 10 min at room temperature, and observe under ultraviolet light (254 nm).

c. Results. The appearance of oligosaccharides indicates that a polysaccharide of the agar family was hydrolyzed. A detailed interpretation of the various neutral, pyruvated, methylated, and sulfated oligosaccharides is given in Duckworth and Yaphe (1970b, 1971b) and Duckworth and Turvey (1969).

2. Melting temperatures. Prepare 5 ml of a 1.5% agar solution in boiling distilled water. Introduce 1-ml samples into 10 × 75-mm test tubes in triplicate. Cool overnight at 2–4 C. Place a 5-mm glass bead on top of each gel and equilibrate the tubes in a well-stirred water bath at 30 C for 10 min. Commence heating the water bath at approximately 1 C min^{-1} and record the temperature at which the beads sink.

3. Gelling temperature. The sol-to-gel transition is strongly temperature-dependent and involves complex stereochemical changes in the agarose molecules. These result in a rapid increase in viscosity that can be measured using Ostwald pipettes (III.C.5). Where larger samples are available, prepare 1.5% gels and record the temperature at which a permanent deformation of the meniscus occurs when the thermometer is withdrawn.

These methods measure the dynamic gelling temperature, which should be clearly distinguished from the isothermal gelling temperature. The latter is the temperature at which the agar will remain in the sol state indefinitely and is usually 10–15 C higher than the dynamic gelling temperature. A spectrophotometric method for determining the isothermal gelling temperature of agar is described by Ng Ying Kin and Yaphe (1972) in which they record the absorbance changes at 610 nm as an iodine-agarose complex is formed during cooling of the agar sol.

4. Viscosity. The general procedure (III.C.5) may be applied to solutions of agar in 0.1 M NaCl except that the temperature should be maintained at 50.0 ± 0.1 C.

5. Pyruvic acid. A sensitive spectrophotometric method based on lactic dehydrogenase was introduced by Duckworth and Yaphe (1970a) to measure pyruvic acid in agar hydrolysates. We have experienced difficulty in obtaining a stable absorbance value at the end of the reaction, and thus an absolute value for the pyruvate content of agar could not be calculated. The same method applied to λ-carrageenan from *C. crispus,* however, showed no reaction and gave steady absorbance values.

Izumi (1973) measured the pyruvate content in agar by NMR spectroscopy and found good agreement with values obtained using dinitrophenylhydrazine.

VIII. Representative data for agar

Native agar from cultivated *Gracilaria* sp. averaged 29% of the alcohol-extracted dry weight, and the recovery was similar whether extracted in distilled water or in 0.1 M phosphate buffer between pH 6.3 and 7.5. Hydration of the dry algal powder for at least 3 h before autoclaving was essential for reproducible recoveries. Alkali treatment (III.B.1) of the agar gave a modified product in 76% yield. Chromatographic analysis of the hydrolysate revealed principally galactose and 6-O-methylgalactose together with small amounts of glucose and an unknown sugar ($R_{galactose}$2.07 in ethyl acetate: pyridine:H_2O, 8:2:1 v/v/v).

The Cetavlon method for the partial purification of agarose was tested with Difco Bacto-Agar. The soluble fraction constituted 57.2–63.6% of the agar and was considerably more than the value for neutral agarose as determined for the same agar by Sephadex fractionation. This product contains agarose and some charged agarose and thus is only partially purified. The same procedure applied to agar from *Gracilaria* sp. gave apparent agarose contents of 22% and 42% for the native and alkali-modified agars, respectively.

Difco Bacto-Agar fractionated by the batchwide Sephadex procedure (VII.A.2) gave carbohydrate yields of 32.7% in the water fraction and 47.0% in the 0.5 M NaCl extract. Yields for native *Gracilaria* agar were 3.6, 10.8, 7.7, and 3.9% in the distilled water, 0.5, 2.5, and 5.0 M NaCl fractions, respectively. Comparable data for alkali-modified *Gracilaria* agar were 7.9, 11.2, 6.8, and 4.7%.

IX. Discussion

The identification of agar in an extract is best confirmed by the use of agarase. This step provides direct evidence for the presence of an agar, and the oligosaccharides released indicate the occurrence of methylated, pyruvated, and sulfated galactose units. The proportions of agarose and charged agarose in both native and alkali-modified agars are important diagnostic features. A quaternary base such as Cetavlon precipitates polysaccharides bearing several ionic functions leaving a mixture of agarose and related polymers with a low charge density in solution. Ion-exchange methods give virtually complete separation of agarose and ionic polysaccharides. The method described (VII.A.2) is simple, rapid, easily quantified, requires only small samples, and avoids many of the undesirable features associated with column chromatography using DEAE-Sephadex A-50. Agarose should contain less than 0.03% sulfate and no detectable pyruvate (Yaphe and Duckworth 1972).

Tests with Difco Bacto-Agar gave recoveries of approximately 80% using distilled water and 0.5 M NaCl. Nearly one-third of the agar was agarose, a value somewhat greater than the 25.5% reported by Duckworth and Yaphe (1971a), whose overall recoveries, however, were less than 60%. The *G. foliifera* examined by Duckworth et al. (1971) is the alga presently termed *Gracilaria* sp. The agarose content is very low and virtually identical to the value they reported. Alkali modification approximately doubles the agarose content of the preparation. Only 26–31% of the *Gracilaria* agar could be recovered from the Sephadex. An extensive description of the agar isolated from several species of *Gracilaria* is provided in Duckworth et al. (1971) and Kim (1970).

X. Acknowledgments

We deeply appreciate the thoughtful and constructive criticism provided by Professors E. L. McCandless and W. Yaphe. Professor Yaphe kindly made available the enzymatic procedure outlined in VII.B.1. Permission to use published material was generously granted by Academic Press, Professors G. P. Arsenault, A. S. Jones, and W. Yaphe. We thank Mr. Y. Doshi for helpful comments and for checking the procedures on agar. Dr. T. Edelstein supplied the *Gracilaria* sp. from cultures. Messrs. T. Moore and R. Gordon provided valuable laboratory assistance. The figures were prepared by Messrs. D. Johnson and W. Crosby.

XI. References

Araki, C. 1966. Some recent studies on the polysaccharides of agarophytes. *Proc. Int. Seaweed Symp.* 5, 3–17.

Belcher, R., Nutten, A. J., and Stephen, W. I. 1953. Substituted benzidines and related compounds as reagents in analytical chemistry. XII. Reagents for the precipitation of sulphate. *J. Chem. Soc.* 1334–7.

Cundall, R. B., Phillips, G. O., and Rowlands, D. P. 1973. A spectrofluorometric procedure for the assay of carrageenan. *Analyst.* 98, 847–62.

Dubois, M., Gilles, K. A., Hamilton, J. K., Rebers, P. A., and Smith, F. 1956. Colorimetric method for the determination of sugars and related substances. *Anal. Chem.* 28, 350–6.

Duckworth, M., Hong, K. C., and Yaphe, W. 1971. The agar polysaccharides of *Gracilaria* species. *Carbohydr. Res.* 18, 1–9.

Duckworth, M., and Turvey, J. R. 1969. The action of a bacterial agarase on agarose, porphyran and alkali-treated porphyran. *Biochem. J.* 113, 687–92.

Duckworth, M., and Yaphe, W. 1970a. Definitive assay for pyruvic acid in agar and other algal polysaccharides. *Chem. Ind.* 747–8.

Duckworth, M., and Yaphe, W. 1970b. Thin-layer chromatographic analysis of enzymic hydrolysates of agar. *J. Chromatogr.* 49, 482–7.

Duckworth, M., and Yaphe, W. 1971a. The structure of agar. I. Fractionation of a complex mixture of polysaccharides. *Carbohydr. Res.* 16, 189–97.

Duckworth, M., and Yaphe, W. 1971b. The structure of agar. II. The use of a bacterial agarase to elucidate structural features of the charged polysaccharides in agar. *Carbohydr. Res.* 16, 435–45.

Gangolli, S. D., Wright, M. G., and Grasso, P. 1973. Identification of carrageenans in mammalian tissues: An analytical and histochemical study. *Histochem. J.* 5, 37–48.

Goring, D. A. I. 1956. A semimicro gelometer. *Can. J. Technol.* 34, 53–9.

Graham, H. D. 1972. *o*-Tolidine and sodium hypochlorite for the determination of carrageenan and other ester sulfates. *J. Dairy Sci.* 55, 1675–82.

Hjertén, S. 1962. A new method for preparation of agarose for gel electrophoresis. *Biochim. Biophys. Acta* 62, 445–9.

Hjertén, S. 1971. Some new methods for the preparation of agarose. *J. Chromatogr.* 61, 73–80.

Hosford, S. P. C., and McCandless, E. L. 1975. Immunochemistry of carrageenans from gametophytes and sporophytes of certain red algae. *Can. J. Bot.* 53, 2835–41.

Izumi, K. 1971. Chemical heterogeneity of the agar from *Gelidium amansii*. *Carbohydr. Res.* 17, 227–30.

Izumi, K. 1973. Structural analysis of agar-type polysaccharides by NMR spectroscopy. *Biochim. Biophys. Acta* 320, 311–7.

Johansson, B. G., and Hjertén, S. 1974. Electrophoresis, crossed immunoelectrophoresis, and isoelectric focussing in agarose gels with reduced electroendosmotic flow. *Anal. Biochem.* 59, 200–13.

Jones, A. S., and Letham, D. S. 1954. A submicro method for the estimation of sulphur. *Chem. Ind.* 662–3.

Kim, D. H. 1970. Economically important seaweeds in Chile. I. *Gracilaria. Bot. Mar.* 13, 140–62.

McCandless, E. L., Craigie, J. S., and Walter, J. A. 1973. Carrageenans in the gametophytic and sporophytic stages of *Chondrus crispus*. *Planta* 112, 201–12.

Ng Ying Kin, N. M. K., and Yaphe, W. 1972. Properties of agar: Parameters affecting gel-formation and the agarose-iodine reaction. *Carbohydr. Res.* 25, 379–85.

Percival, E., and McDowell, R. H. 1967. *Chemistry and Enzymology of Marine Algal Polysaccharides*. Academic Press, London. 219 pp.

Pernas, A. J., Smidsrød, O., Larsen, B., and Haug, A. 1967. Chemical heterogeneity of carrageenans as shown by fractional precipitation with potassium chloride. *Acta Chem. Scand.* 21, 98–110.

Rees, D. A. 1963. The carrageenan system of polysaccharides. I. The relation between the κ- and λ-components. *J. Chem. Soc.* 1821–32.

Rees, D. A. 1972. Shapely polysaccharides. *Biochem. J.* 126, 257–73.

Sloneker, J. H. 1972. Gas-liquid chromatography of alditol acetates. *Methods Carbohydr. Chem.* 6, 20–4.

Stancioff, D. J., and Stanley, N. F. 1969. Infrared and chemical studies on algal polysaccharides. *Proc. Int. Seaweed Symp.* 6, 595–609.

Weigl, J., and Yaphe, W. 1966. The enzymic hydrolysis of carrageenan by *Pseudomonas carrageenovora:* Purification of a κ-carrageenase. *Can. J. Microbiol.* 12, 939–47.

Wilson, C. M. 1959. Quantitative determination of sugars on paper chromatograms. *Anal. Chem.* 31, 1199–201.

Yaphe, W., and Arsenault, G. P. 1965. Improved resorcinol reagent for the determination of fructose, and of 3,6-anhydrogalactose in polysaccharides. *Anal. Biochem.* 13, 143–8.

Yaphe, W., and Duckworth, M. 1972. The relationship between structures and biological properties of agars. *Proc. Int. Seaweed Symp.* 7, 15–22.

Zabin, B. A. 1969. *Isolation of Agarose from Agar.* U.S. Patent No. 3,423,396. 2 pp.

13: β-1,3-Glucans in diatoms and brown seaweeds

SVERRE MYKLESTAD

Institute of Marine Biochemistry,
University of Trondheim, N-7034 Trondheim-NTH, Norway

CONTENTS

I **Introduction**	*page*	**134**
A. Diatoms		134
B. Brown seaweeds		135
II **Materials and test organisms**		**135**
A. Diatoms		135
B. Brown seaweeds		136
III **Methods**		**136**
A. Diatoms		136
B. Brown seaweeds		137
IV **Sample data**		**138**
V **Discussion**		**138**
A. β-1,3-Glucan from diatoms		138
B. Laminaran		139
VI **Acknowledgment**		**140**
VII **References**		**140**

I. Introduction

A. Diatoms

The reserve product of diatoms was reported as early as 1930 and prepared in crystalline form from freshwater diatoms by von Stosch (1951). Investigation of the chemical composition and structure of this material revealed that it contained 99.5% glucose mainly linked by β-1,3-linkages with the presence of a few 1,6-linkages (Beattie et al. 1961). The name chrysolaminarin was suggested for this polysaccharide by Quillet (1955). A similar structure was reported for the reserve polysaccharide from *Ochromonas malhamensis* (Meeuse 1962). Figure 13-1 shows a β-1,3-linked glucan.

A water-soluble glucan of the laminaran type was extracted in a yield of about 14% of the dry matter from the marine diatom *Phaeodactylum tricornutum* grown under bacteria-free conditions. The glucan was shown to be β-1,3-linked and resembled closely the reserve polysaccharide of freshwater diatoms (Ford and Percival 1965).

Handa (1969) and Handa and Tominaga (1969) found the same type of glucan in a culture of the diatom *Skeletonema costatum* and from a natural population consisting mainly of *Biddulphia* sp. and *Coscinodiscus* sp., respectively. Recently a β-1,3-glucan was also found in the diatom *Chaetoceros affinis* var. *willei* (Myklestad and Haug 1972) and in eight other species of marine diatoms. A rapid production of β-1,3-glucan was observed to coincide with depletion of nitrate from the medium in dilution/batch cultures (Myklestad and Haug 1972; Myklestad 1974).

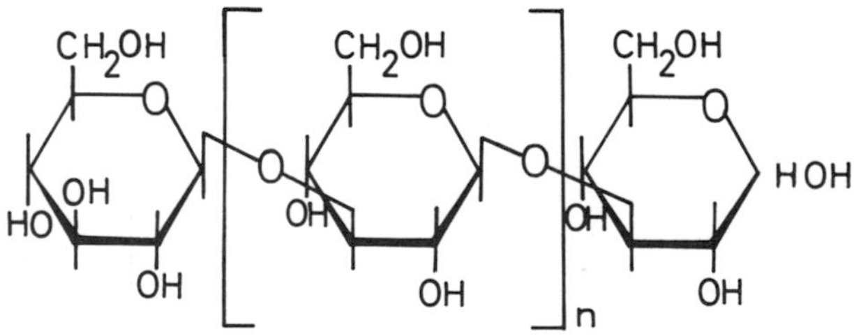

Fig. 13-1. Structure of a β-1,3-glucan.

The change in ratio of protein to carbohydrate, mainly caused by the accumulation of glucan, was shown to be a sensitive and convenient parameter for characterizing the physiological state of plankton in culture and in natural populations (Myklestad and Haug 1972; Haug et al. 1973).

Data from metabolic studies together with structural information suggest that β-1,3-glucan is a photosynthetic reserve in diatoms (Craigie 1974).

B. Brown seaweeds

Laminaran was described as early as 1885 and has been shown to be widely distributed in brown algae, where it occurs together with mannitol. The content of laminaran amounts to more than 35% of dry matter in fronds of some Laminariaceae in the autumn and shows a clear seasonal variation, with a minimum in the spring (May-June) and an increase during summer to a maximum content in autumn (September-October) (Haug and Jensen 1954). The amount in the Fucales does not exceed 7% (Black 1949).

Chemical investigation of laminaran revealed that this glucan is essentially linear and β-1,3-linked (Barry 1939; Bächli and Percival 1952), but with small amounts (ca. 2%) of mannitol as a constituent. The mannitol was found as a terminal residue of about half of the laminaran chains (Peat et al. 1958). Average degree of polymerization (DP) from about 15 to 60 was found. Laminaran isolated from *Laminaria hyperborea* precipitates easily from cold water, whereas laminaran from *Laminaria digitata* is more soluble and requires the addition of ethanol to help precipitation. The two forms of laminaran are referred to as "insoluble" and "soluble," respectively. Other chemical and physical properties are very similar (Percival and McDowell 1967). Both forms of laminaran contain small amounts of β-1,6-interchain linkages. The major chemical difference appeared to be that the degree of branching was significantly higher for the soluble form of the polysaccharide (Fleming et al. 1966).

The disappearance of the laminaran during the winter when photosynthesis is very low and its hydrolysis to glucose by enzymes present in the algae suggests that laminaran is a food reserve material.

II. Materials and test organisms

A. Diatoms

Data are given for the following species (clone): *Chaetoceros affinis* var. *willei* (Gran) Hustedt (Ch-1), *Chaetoceros debilis* Cleve (Ch-2), *Chaetoceros curvisetus* Cleve (Ch-24), *Chaetoceros socialis* Lauder (Ch-26), *Thal-*

assiosira gravida Cleve (Th-11), *Skeletonema costatum* (Grev.) Cleve (Skel-5), *Thalassiosira fluviatilis* Hustedt (Th-31C).

All species were isolated from the Trondheimsfjord, except *T. fluviatilis,* which was kindly supplied by Dr. E. Paasche, Institute of Marine Biology and Limnology, University of Oslo, Norway. The clones are maintained in the culture collection, University of Trondheim.

B. Brown seaweeds

Algae were collected at Reine, Lofoten, Norway, and samples of *Ascophyllum nodosum* (L.) Le Jol., *Laminaria digitata* (Huds.) Lamour., and *Laminaria hyperborea* (Gunn.) Fosl. were dried, milled, and analyzed.

III. Methods

A. Diatoms

Material for analysis of diatoms from cultures may be obtained in two different ways: by direct filtering of samples or harvesting by centrifugation and subsequent freeze-drying of the sample. From 10^6 to 10^8 cells will be a convenient amount, depending on cell size and physiological state of the organism.

The sample is collected (e.g., 10 ml of a medium-dense culture) on a 25-mm glass fiber filter (Whatman GF/C) and the filter transferred directly to a 25-ml glass vial. Alternatively, the cells are collected on a 5-μm nylon screen (Nytal, Switzerland) supported by a GF/C filter. The cells are then transferred to the vial by flushing with ethanol from a Pasteur pipette. The ethanol is evaporated at 60–70 C prior to extraction with acid. The glucan is extracted by 10 ml of 0.05 M H_2SO_4 for 1 h at 20 C in the glass vial, which is placed in a stirring machine.

After extraction, the mixture is filtered, the extraction repeated with 6 ml 0.05 M H_2SO_4, and the combined filtrates from the two extractions are collected for analysis (Myklestad and Haug 1972). The phenol–sulfuric acid method (Dubois et al. 1956; see Chp. 10) is used. In this method 2 ml of the sample containing 5–100 μg of carbohydrate is transferred to a test tube; 0.5 ml 3% aqueous phenol and 5 ml conc. H_2SO_4 are added from a dispenser. The mixture is immediately stirred with a glass rod. After 30 min the test tube is cooled in running water and the absorbance read in a 1.0-cm cell at 485 nm in a spectrophotometer against a distilled-water blank. The determination is run in duplicate or triplicate. The amount of reducing sugar is calculated with glucose as a standard: $A \times 100 \times V = \mu$g glucose in the sample, where A is absorbance in the spectrophotometer and V is volume (in milliliters) of the combined extracts. Amount of glucan $= \mu g$ glucose $\times (162/180)$.

When freeze-dried samples are used, 5–10 mg is weighed and transferred to the glass vial. The extraction and analysis are carried out as above. When dry matter and ash content are determined, the content of β-1,3-glucan can be calculated as percent of organic dry matter.

B. Brown seaweeds

Samples of brown seaweeds for determination of laminaran are obtained by collecting 20–50 specimens of the alga, drying at low temperature (30–50 C) in a through circulation bed dryer, and grinding the dry alga in a mill furnished with a 60-mesh screen. The samples can be stored for years in airtight boxes.

For analysis (Haug and Jensen 1954) 200 mg of dry, milled alga is transferred to 11-mm Pyrex tubes of 250-mm length fused in one end. Then 8 ml of 0.5 M H_2SO_4 is added and mixed with the meal. The tubes are sealed and the hydrolysis carried out for 5 h in a boiling-water bath. After the tubes are cooled to room temperature, dilution with distilled water to 100 ml is carried out.

1. Fermentation method. The content of reducing sugars in the hydrolysate is determined by the phenol–sulfuric acid method (see above) or by using an alkaline copper reagent (Somogyi 1952). The amount of glucose is then determined by fermentation.

Fresh bakers' yeast, *Saccharomyces cerevisiae* (1.5 g), is washed with 25 ml of distilled water and centrifuged to remove impurities. The yeast suspension is then mixed with 25 ml of diluted, neutralized hydrolysate and placed in an agitating incubator at 30 C. After 1 h the suspension is centrifuged to remove the yeast, and the amount of reducing sugars is determined as above. The content of reducing sugars in the hydrolysate calculated as glucose, minus the content of reducing sugars in fermented hydrolysate also calculated as glucose, is equivalent to the content of glucose in the hydrolysate. When the water content in the algal meal is known, the percentage of glucose in dry matter can be calculated.

The content of laminaran as percent of dry matter will equal percent glucose times 162/180.

2. Glucose oxidase method. Alternatively, glucose can be determined by an enzymatic method (Lloyd and Whelan 1969) based on the oxidation of glucose to gluconic acid catalyzed by the enzyme glucose oxidase. Reagents are prepared as follows.

  a. TRIS-phosphate-glycerol buffer. Dissolve 36.3 g TRIS and 50 g

$NaH_2PO_4 \cdot H_2O$ in water. Add 400 ml glycerol and water to 1000 ml. Adjust to pH 7.0 by addition of solid $NaH_2PO_4 \cdot H_2O$.

b. Glucose oxidase reagent. Dissolve the following in 100 ml of TRIS-phosphate-glycerol buffer: glucose oxidase (Sigma), 3 mg; *o*-dianisidine dihydrochloride (Sigma), 10 mg. Dissolve each component separately prior to mixing.

Transfer 1 ml of the hydrolysate (or a dilution) containing 0–50 μg glucose to a test tube and add 2 ml of glucose oxidase reagent. After mixing, incubate the tube at 37 C for 30 minutes. Then add 4 ml 5 M HCl, mix the solution well, and read the absorbance at 525 nm. A linear calibration curve is obtained for glucose solutions.

IV. Sample data

Data for the β-1,3-glucan content of marine diatoms grown in batch cultures are shown in Table 13-1. The zero-time cells represent exponentially growing cultures; the 2-day and 6-day samples represent slowly growing and stationary cultures, respectively. The increase in β-1,3-glucan between days 2 and 6 takes place with little change in cell density and in response to nitrogen deficiency.

The laminaran content of fronds of some brown algal species are shown in Table 13-2. Values obtained by the glucose oxidase methods are compared with earlier data obtained by the fermentation method (Haug and Jensen 1954).

V. Discussion

A. β-1,3-Glucan from diatoms

It has been shown by gas-liquid chromatography of the hydrolysates of the acid extracts from a number of different species of diatoms that glucose was the predominating sugar and that the amount of glucose left in the remaining material was small (Myklestad et al. 1972; Myklestad 1974). Therefore, the amount of carbohydrate present in the acid extract is used as an estimate of glucan. The method is thus not specific for glucan, as other reducing or potentially reducing sugars in the extract will give systematically high estimations.

The assumption is made that all the estimated glucose originates from true polymeric glucose (glucan), and the factor 162/180 was used in the calculation of the results. Unpublished work in this institute (Myklestad and Larsen) has shown that the ratio of monomeric glucose to glucan in the acid extract can vary markedly according to culture conditions and physiological state of the cells. Because of the close biochemical connection between monomeric glucose and glucan

Table 13-1. *Content of β-1,3-glucan (mg l^{-1}) in batch cultures of marine diatoms of different ages*

Species	Age (days)		
	0	2	12
Chaetoceros affinis	2.6	5.9	35.4
Chaetoceros debilis	2.1	6.1	16.1
Chaetoceros curvisetus	3.1	6.3	13.3
Chaetoceros socialis	3.1	5.1	10.8
Thalassiosira gravida	2.1	4.1	18.5
Skeletonema costatum	2.1	5.9	54.8
Thalassiosira fluviatilis	5.1	18.5	89.6

Source: Myklestad (1974).

Table 13-2. *Content of laminaran as percent of dry matter in some brown algae collected at Reine, Lofoten, Norway*

Species	Date	Fermentation method[a]	Glucose oxidase method
Laminaria hyperborea			
Frond	Sept. 1950	23.8	22.0
Frond	Oct. 1950	23.6	24.2
Old fronds	Feb. 6, 1951	4.8	4.1
Old fronds	May 3, 1951	1.9	1.2
New fronds	May 3, 1951	0.6	0.3
Ascophyllum nodosum			
	Oct. 2, 1950	3.8	1.9
	Feb. 5, 1951	1.6	0.1
Laminaria digitata			
Fronds	Oct. 3, 1951	17.6	17.6
Fronds	Feb. 26, 1952	2.1	0.5

[a] Data from Haug and Jensen (1954).

in the cells, no attempt has been made to estimate the amount of true glucan in this routine method.

B. Laminaran

No exact comparison of the two methods presented is attempted. Table 13-1 shows, however, that the agreement seems to be good for values above about 5% laminaran. At lower contents the glucose oxidase method gives significantly lower values.

It has been pointed out (Jensen 1956) that hydrolysates of some marine algae contain other sugars in addition to glucose, some of which may be fermented by a commercial bakers' yeast. Powell and Meeuse (1964) also mentioned that the fermentation method is unreliable below a content of 2% laminaran. The present results support these conclusions, as the glucose oxidase method is very specific and should give more realistic values in a mixed hydrolysate.

VI. Acknowledgment

The author is grateful to Mrs Solveig Hestmann for excellent assistance with chemical analysis.

VII. References

Bächli, P., and Percival, E. G. V. 1952. The synthesis of laminaribiose (3-β-D-glucosyl-D-glucose) and proof of its identity with laminaribiose isolated from laminarin. *J. Chem. Soc.* 1243–6.

Barry, V. C. 1939. The constitution of laminarin: Isolation of 2:4:6-trimethylglucopyranose. *Sci. Proc. R. Dublin Soc.* 22, 59–67.

Beattie, A., Hirst, E. L., and Percival, E. 1961. Studies on the metabolism of the Chrysophyceae: Comparative structural investigations on leucosin (chrysolaminarin) separated from diatoms and laminarin from the brown algae. *Biochem. J.* 79, 531–7.

Black, W. A. P. 1949. Seasonal variation in chemical composition of some of the littoral seaweeds common to Scotland. II. *Fucus serratus, Fucus vesiculosus, Fucus spiralis* and *Pelvetia canaliculata. J. Soc. Chem. Ind. (London)* 68, 183–9.

Craigie, J. S. 1974. Storage products. In Stewart, W. D. P. (ed.). *Algal Physiology and Biochemistry*, pp. 206–35. Blackwell, Oxford.

Dubois, M., Gilles, K. A., Hamilton, J. K., Rebers, P. A., and Smith, F. 1956. Colorimetric method for determination of sugars and related substances. *Anal. Chem.* 28, 350–6.

Fleming, M., Hirst, E., and Manners, D. J. 1966. The constitution of laminarin. VI. The fine structure of soluble laminarin. *Proc. Int. Seaweed Symp.* 5, 255–60.

Ford, C. W., and Percival, E. 1965. The carbohydrates of *Phaeodactylum tricornutum*. 1. Preliminary examination of the organism and characterisation of low molecular weight material and of a glucan. *J. Chem. Soc.* 7035–41.

Handa, N., 1969. Carbohydrate metabolism in the marine diatom *Skeletonema costatum. Mar. Biol.* 4, 208–14.

Handa, N., and Tominaga, H. 1969. A detailed analysis of carbohydrates in marine particulate matter. *Mar. Biol.* 2, 228–35.

Haug, A., and Jensen, A. 1954. Seasonal variations in the chemical composition of *Alaria esculenta, Laminaria saccharina, Laminaria hyperborea*, and *Laminaria digitata* from northern Norway. *Nor. Inst. Tang-Tareforsk. Rep.* 4, 1–14.

Haug, A., Myklestad, S., and Sakshaug, E. 1973. Studies on the phytoplankton ecology of the Trondheimsfjord. I. The chemical composition of phytoplankton population. *J. Exp. Mar. Biol. Ecol.* 11, 15–26.

Jensen, A. 1956. Component sugars of some common brown algae. *Nor. Inst. Tang-Tareforsk. Rep.* 9, 1–9.

Lloyd, J. B., and Whelan, W. J. 1969. An improved method for enzymic determination of glucose in the presence of maltose. *Anal. Biochem.* 30, 467–70.

Meeuse, B. J. D. 1962. Storage products. In Lewin, R. A. (ed.). *Physiology and Biochemistry of Algae*, pp. 289–313. Academic Press, New York.

Myklestad, S. 1974. Production of carbohydrates by marine planktonic diatoms. I. Comparison of nine different species in culture. *J. Exp. Mar. Biol. Ecol.* 15, 261–74.

Myklestad, S., and Haug, A. 1972. Production of carbohydrates by the marine diatom *Chaetoceros affinis* var. *willei* (Gran) Hustedt. I. Effect of the concentration of nutrients in the culture medium. *J. Exp. Mar. Biol. Ecol.* 9, 125–36.

Myklestad, S., Haug, A., and Larsen, B. 1972. Production of carbohydrates by the marine diatom *Chaetoceros affinis* var. *willei* (Gran) Hustedt. II. Preliminary investigation of the extracellular polysaccharide. *J. Exp. Mar. Biol. Ecol.* 9, 137–44.

Peat, S., Whelan, W. J., and Lawley, H. G. 1958. The structure of laminarin. II. The minor structural features. *J. Chem. Soc.* 729–37.

Percival, E., and McDowell, R. H. 1967. *Chemistry and Enzymology of Marine Algal Polysaccharides*. Academic Press, London. 219 pp.

Powell, J. H., and Meeuse, B. J. D. 1964. Laminarin in some Phaeophyta of the Pacific Coast. *Econ. Bot.* 18, 164–6.

Quillet, M. 1955. Sur la nature chimique de la leucosine, polysaccharide de réserve caractéristique des Chrysophycées, extraite d'*Hydrurus foetidus*. *C.R. Acad. Sci.* 240, 1001–3.

Somogyi, M. 1952. Notes on sugar determination. *J. Biol. Chem.* 195, 19–23.

von Stosch, H. A. 1951. Ueber das Leukosin, den Reservestoff der Chrysophyten. *Naturwissenschaften* 38, 192–3.

14: Alginic acid

BJØRN LARSEN

Institute of Marine Biochemistry,
University of Trondheim, N-7034 Trondheim-NTH, Norway

CONTENTS

I Introduction   *page* 144
II Materials   145
III Methods   145
   A. Preextraction   145
   B. Analysis   146
IV Sample data   148
V Discussion   148
VI References   149

I. Introduction

All brown algae investigated contain alginic acid as one of their major constituents and the only one so far utilized commercially. Alginic acid is a polyuronide (i.e., a polysaccharide made up only of uronic acids). In contrast to cellulose, which is a homopolymer because it contains only one type of sugar unit, alginic acid is a heteropolymer containing two different building units, D-mannuronic and L-glucuronic acids. These units are joined by 1,4-glycosidic linkages into an unbranched chain.

D-mannuronic acid   L-guluronic acid

As is true of other carboxylic acids, the uronic acids dissociate in water, and alginic acid is consequently a polyelectrolyte. The name "alginic acid" thus applies only to, and is used here only for, the polymer in its acid form. The carboxylic acid groups, by neutralization with the appropriate base, form salts with the metal or other ions. The salt form of the polymer is called an "alginate" (e.g., sodium alginate). Salts of the alkali metal ions are soluble in water; alginic acid itself and its salts with divalent and trivalent metal ions are generally insoluble.

Alginate is the major structural polysaccharide in the brown algae, and this function of the molecule is based on its ability to form very strong gels with divalent cations. In seawater, the dominating cations are Ca^{2+} and Mg^{2+}. Among other factors, the firmness of the Ca-Mg-alginate gel in the plant depends on the composition of the alginate molecule (i.e., the relative ratio of mannuronic to guluronic acid). This ratio varies considerably from species to species (Haug 1964).

Difficulties in constructing a selective and accurate analytical method for alginate stem from these two factors: the insolubility, and the variation in composition. Gravimetric and colorimetric methods

[144]

all require the complete extraction of the alginate from the alga – a multistep, time-consuming procedure – and are subject to interference from other extractable components or variation in the composition of the alginate. The method recommended is, therefore, a direct determination of the polyuronide content by acid decarboxylation in a modification introduced by Anderson (1959).

When treated with strong mineral acid at elevated temperatures, all uronic acids undergo decarboxylation, thus liberating CO_2. The method is based on a determination of the CO_2 liberated by precipitation of $BaCO_3$ from a standardized $Ba(OH)_2$ solution and back-titration of excess $Ba(OH)_2$.

II. Materials

Samples of fresh algae may be used provided the tissue is disintegrated into small particles. It is, however, recommended that the material be dried prior to analysis, as a more reliable estimate of the alginate content can be expected from the dried material (water content less than 25%). The material is then ground to pass a sieve not coarser than 60-mesh or finer than 150-mesh. The dry matter is determined by heating an accurately weighed sample in an oven at 105 C for 4 h.

III. Methods

The following reagents are required: (1) 0.2 N HCl. Dilute 17.2 ml conc. HCl to 1 l with distilled water. (2) 19% (w/w) HCl. Mix 100 ml conc. HCl (37%, analytical grade) with 113 ml distilled water. (3) 0.05 N $Ba(OH)_2$. Weigh 8.0 g of $Ba(OH)_2 \cdot 8H_2O$ and dissolve in approximately 600 ml distilled water that has previously been boiled for 30 min to remove CO_2 and cooled to room temperature. Warming the solution to 50–60 C will be advantageous. Filter the solution directly into a 1-l volumetric flask (use a Schleicher and Schüll blue-ribbon filter or corresponding) and dilute to the mark with boiled distilled water. Keep the flask tightly stoppered to exclude atmospheric CO_2. (4) 0.05 N HCl. This is the primary standard solution, and it is recommended that a commercial volumetric reagent for 0.1 N HCl be diluted in a 2-l volumetric flask. (5) Phenolphthalein solution. Dissolve 10 mg phenolphthalein in 10 ml of 96% ethanol and add 10 ml distilled water. (6) Nitrogen gas. CO_2-free N_2 is used as a carrier gas.

A. Preextraction

A sample of the dried, milled material (150–200 mg) is weighed and transferred to a 100-ml conical flask. Add 20 ml of 0.2 N HCl and

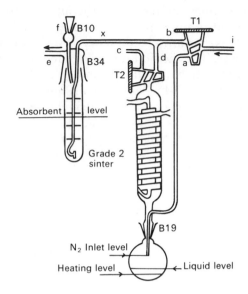

Fig. 14-1. Apparatus used for the decarboxylation analysis.

extract overnight on a slow-moving shaker. The extract is filtered through paper (ca. 6 cm diam.), and the extracted particles are transferred quantitatively to the reaction flask (Fig. 14-1) by means of 35 ml of 19% HCl. The preextraction step removes all inorganic carbonates and a substantial amount of nonalginate uronic acid-containing polysaccharides present in many brown algae.

B. Analysis

The apparatus used for the analysis is shown to scale in Fig. 14-1 and is readily constructed by a competent glassblower. A detailed description of the apparatus is given by Anderson (1959).

The reaction flask is fitted to the B19 socket and secured by springs. A small amount of silicone grease should be applied to the joint. A heating mantle, made to fit a 50-ml flask, is placed under the reaction flask. N_2 carrier gas is passed through tube i, at 15 ml min^{-1} with taps T1 and T2 in positions to let the gas flow through tubes a and c, respectively. After 15 min all CO_2 present in the condenser and flask will be expelled. Maximum current is applied to the heating mantle to bring the HCl to boiling, and at the same time tap T2 is turned, allowing the gas to flush the absorption trap during the heating period. When boiling starts, 15 ml of 0.05 N Ba(OH)$_2$ solution is introduced with a pipette through socket f and washed down with sufficient boiled distilled water to raise the liquid level almost to the third baffle.

The heat is reduced to the lowest level that will permit continuous boiling, and the reaction is allowed to proceed for 2.5 h. Tap T1 is then turned to let the carrier gas pass into the absorption tube through tube b, the heater is disconnected, and tap T2 is turned to release pressure through tube c. The reaction flask may now be exchanged for a new one. The stopper in socket f is removed, 2–4 drops of phenolphthalein solution is introduced and washed down with a few milliliters of CO_2-free water. The tip of the burette holding the standard 0.05 N HCl is introduced through socket f. In our experience, it is advantageous to equip the tip of the burette with 10–15 cm of polyethylene tubing (i.d. 0.5–0.8 mm), thereby introducing the acid into the middle of the stirred solution and minimizing the risk of overtitration. Acid is introduced from the burette until the solution changes from red to colorless. Toward the endpoint, the titration should be carried out slowly. The endpoint is reached when the solution stays colorless for 1 min. The volume, t_1, of titrant consumed is recorded. The absorption tube may now be disconnected from the apparatus, the sinter and baffles rinsed with boiled distilled water, and a new absorption tube introduced. During these last operations, tap T1 is turned to flush the reaction flask and condenser.

A blank titration of the $Ba(OH)_2$ solution should be made at regular intervals, at least once a week, using 15 ml of 0.05 N $Ba(OH)_2$ solution titrated either in a conical flask under N_2 or preferably in the apparatus using the titration procedure described above. The volume, t_2, consumed is recorded. The amount of CO_2 released during the reaction is then proportional to $t_2 - t_1$.

The acid decarboxylation of uronic acids is a fairly slow process. The yield of CO_2 depends primarily on the temperature, and the reaction time and is not necessarily stoichiometric. The apparatus should, therefore, be calibrated by weighing a suitable amount of a pure uronic acid (e.g., galacturonic acid) and carrying out the analysis procedure. With a reaction time of 2.5 h, Anderson (1959) obtained a yield of 98% of the theoretical. In our hands the yield is 93% and the analytical figures consequently corrected by dividing by 0.93. The alginic acid content may then be calculated according to the formula:

$$c = \frac{(t_2 - t_1) \times 0.05 \times 176 \times 100}{a \times 0.93} \tag{1}$$

where a = weight of sample in milligrams; 176 = molecular weight of an anhydrohexuronic acid unit; 0.05 = normality of the primary standard; and c = percent of anhydrouronic acid (i.e., percent alginic acid). The content should be based on dry weight as determined in II, above.

Table 14-1. *Range of alginic acid content in some brown algae*

Species	Alginic acid g/100 g dry matter
Fucales	
Ascophyllum nodosum (L.) Le Jol	22–30
Fucus serratus L.	20–29
Fucus vesiculosus L.	16–21
Fucus spiralis L.	17–20
Pelvetia canaliculata (L.) Dcne. et Thur.	22–25
Halidrys siliquosa (L.) Lyngb.	16–29
Himanthalia elongata (L.) S. F. Gray	22–29
Laminariales	
Laminaria digitata (Huds.) Lamour.	
Fronds	25–44
Stipes	35–47
Laminaria hyperborea (Gunn.) Fosl.	
Fronds	17–33
Stipes	25–38
Laminaria saccharina (L.) Lamour.	
Fronds	17–22
Stipes	25–31
Alaria esculenta (L.) Grev.	20–42

IV. Sample data

Table 14-1 gives the range in alginic acid content for some species of Fucales and Laminariales common to the Norwegian coast (Haug 1964). Variations in the alginic acid content with season, habitat, and age of tissue account for the range of values observed within one species.

V. Discussion

The method will respond to all components evolving CO_2 upon acid treatment. Two sources of error are important in algae: inorganic carbonates and nonalginate uronic acid-containing polysaccharides. The interference from carbonates is easily eliminated by the preextraction step described above. For samples contaminated with calcareous animals or algae, it is particularly important to allow this extraction to go on for a fairly long period.

The interference from uronic acids in polysaccharides other than alginate is more severe. All Fucaceae as well as a number of other species of Fucales appear to contain a family of polysaccharides consisting mainly of fucose, xylose, and glucuronic acid, but of varying composition (Haug 1964; Percival and McDowell 1967; Medcalf and Larsen 1977). Substantial quantities of these polysaccharides are removed by the acid preextraction, but even repeated extractions with acid fail to remove all of them. The analytical figures obtained by the decarboxylation method are, therefore, a very good estimation of the alginic acid content in the Laminariales, but overestimate in most Fucales. The error is probably within 1–3% absolute in *Ascophyllum nodosum.*

The major advantage of the method is the relatively short handling time per sample and the fact that variations in the ratio of mannuronic to guluronic acid of the alginic acid do not influence the results. The major disadvantage is the small number of analyses that can be performed during a day.

When run as a routine method the reproducibility is very good. Care should, however, be taken to clean the glass sinter regularly, as a buildup of the back-pressure owing to partial clogging increases the temperature and leads to higher yields of CO_2 than those expected from the calibration.

VI. References

Anderson, D. M. W. 1959. Studies on materials containing uronic acid. I. An apparatus for routine semimicro estimations of uronic acid content. *Talanta* 2, 73–8.

Haug, A. 1964. Composition and properties of alginates. *Nor. Inst. Tang-Tareforsk. Rep.* 30, 1–123.

Medcalf, D. G., and Larsen, B. 1977. Fucose containing polysaccharides in the brown algae *Ascophyllum nodosum* and *Fucus vesiculosus. Carbohydr. Res.* (in press).

Percival, E., and McDowell, R. H. 1967. *Chemistry and Enzymology of Marine Algal Polysaccharides.* Academic Press, London. 219 pp.

15: Fucoidan

BJØRN LARSEN

Institute of Marine Biochemistry,
University of Trondheim, N-7034 Trondheim-NTH, Norway

CONTENTS

I	Introduction	*page* 152
II	Materials	153
III	Methods	153
	A. Extraction	153
	B. Fractionation	153
	C. Colorimetric determination of fucose	154
IV	Sample data	155
V	Discussion	155
VI	References	156

I. Introduction

The name "fucoidan" was introduced by Kylin in 1913 for the mucilaginous polysaccharide exuded by *Fucus* spp. Structural investigations of this material extracted from *Fucus vesiculosus* demonstrated a heavily sulfated polysaccharide consisting mainly of L-fucose but with persistent amounts of other sugars, mainly xylose and galactose (Conchie and Percival 1950; O'Neill 1954; Côté 1959). The main structural feature was that of a polymer of α-L-fucose-4-sulfate, and this has since been accepted as the structure of fucoidan.

The occurrence of fucose as a polymeric constituent is, however, not restricted to water or acid extracts of the algae (Larsen et al. 1970), which are the classic sources of fucoidan. Subsequent extractions with chelating agents, like EDTA, alkali, or alkali at elevated temperatures will remove additional amounts of fucose.

Later investigations have demonstrated that fucose is a component of polysaccharides containing xylose and uronic acid, and possibly smaller amounts of other sugars (Percival and McDowell 1967; Larsen et al. 1970) and that these polysaccharides probably constitute large, complex molecules in the algae (Medcalf and Larsen 1977). One of the participants in the complex formation is a polysaccharide rich in fucose and sulfate, which on repeated fractionation, approaches the composition of polymeric L-fucose-4-sulfate (Medcalf and Larsen 1977). Some acid hydrolysis is required to release this polysaccharide from the complex.

The above considerations indicate clearly that the analytical procedure to be used will depend on the definition of fucoidan. As our knowledge of the chemical structure of the various fucose-containing components of the complex is rather restricted, it does not seem appropriate to introduce a definition of fucoidan different from the one traditionally used. It should be pointed out, however, that a polymeric molecule of this particular composition may not exist in the algae. The component released from the complex is a fairly close approximation to such a structure, and for the present purpose this will be considered as fucoidan. To release fucoidan from the complex and separate it from the other fucose-containing polysaccharides, the method must include a partial acid hydrolysis and a fractionation.

As already pointed out, fucose is released from the algae in all extracts as the extraction conditions become increasingly severe. How much of this fucose is contributed by a fucoidan-type molecule is unknown, and, in keeping with the above, fucoidan will be considered as extractable with acid. According to Black et al. (1952), the amount of fucoidan extracted depends on acid strength and temperature, and the extraction conditions chosen approach those reported by these authors but with a much lower alga/solvent ratio recommended. Extraction at elevated temperature leads to a concomitant degradation of the complex molecules into separate components.

II. Materials

In our laboratory the method is usually carried out on dried, milled seaweed samples. Fresh algae may also be used provided the tissue is disintegrated (e.g., by wet milling) prior to extraction.

III. Methods

The following chemicals will be required: (1) 0.2 N HCl. Dilute 17.2 ml conc. HCl to 1 l. (2) 0.02 N HCl. Dilute 100 ml of 0.2 N HCl to 1 l. (3) 0.1 M $MgCl_2$. Dissolve 20.33 g $MgCl_2 \cdot 6$ H_2O in 1 l of distilled water. (4) 1 N NaOH. Dissolve 4 g NaOH in 100 ml of distilled water. (5) 3% Phenol solution. Dissolve 3 g of analytical grade phenol in 100 ml of distilled water. (6) Dilute H_2SO_4. Mix conc. H_2SO_4 with distilled water (6:1 v/v) and cool to room temperature. (7) Cysteine solution. Dissolve 300 mg analytical grade cysteine·HCl in 10 ml of distilled water. Store in an amber bottle, well stoppered and protected from light.

A. Extraction

A 1-g sample of the seaweed meal is accurately weighed into a 100-ml conical flask. Add 25 ml 0.2 N HCl and heat the suspension to 70 C in a thermostated water bath for 1 h. The suspension should be stirred (ca. 200 rpm) or shaken during the extraction. Cool the flask in tap water, allow the particles to settle, and pour off the liquid through a filter into a 200-ml conical flask. Subsequent extracts from the same sample are filtered into the same flask. Transfer particles retained on the filter back into the extraction flask with 25 ml 0.02 N HCl. Heat the flask at 70 C for 1 h, cool in tap water, and refilter. The combined extracts contain the acid-soluble, fucose-containing polysaccharides and may be used directly in the colorimetric procedure for fucose.

B. Fractionation

To obtain the best possible fractionation, it is essential to work under standardized conditions. Therefore, it is necessary to remove excess

ions introduced and to bring the total carbohydrate content to a reference level chosen to be 1% as assayed by the phenol–sulfuric acid
method (Dubois et al. 1956; Chp. 10).

The combined extracts are neutralized to pH 6–7 with 1 N NaOH
(ca. 4 ml). The solution is transferred to a dialysis tubing, $\frac{5}{8}$ in. diameter, and the flask rinsed with distilled water into the dialysis tube. The
dialysis tubing is closed and placed in a 2-l beaker with distilled water.
Slow stirring of the water is advantageous. The outside water should
be changed once, and the total period of dialysis should be approximately 24 h. The dialysis sack is opened and the contents poured into
a 100-ml measuring cylinder. Rinse the tubing twice with a small
amount of distilled water into the same cylinder. Mix the contents of
the cylinder well and record the volume. Into a test tube pipette 0.5
ml of this solution, add 1.5 ml water followed by 0.5 ml of 3% phenol
solution. Mix well and add from a pipette 5.0 ml conc. H_2SO_4. Mix the
contents of the tube with a glass rod, allow the tube to stand for 30
min, cool in tap water, and read the absorbance at 548 nm against a
blank prepared as described above but with distilled water instead of
sample. Record the absorbance. If higher than 0.8, dilute 1.0-ml sample with 1.0 ml distilled water and repeat the analysis. If this is done,
the result of the calculation below is multiplied by 2. The total amount
of carbohydrate, A, in milligrams is then calculated from the formula:
$A = a \times 0.4 \times V$, where a is the absorbance, 0.4 is a conversion factor, and V the volume of the dialysate in milliliters.

The solution is evaporated under reduced pressure to approximately 15 ml. The solution is then transferred from the evaporator
flask to a 250-ml beaker with sufficient water to make the final solution 1% with respect to total carbohydrate. To this solution add an
equal volume of 0.1 M $MgCl_2$ and precipitate by adding ethanol to a
concentration of 50%. Allow the solution to stand for 30 min, transfer
to centrifuge tubes and centrifuge for 10 min at 12 000–15 000 × g.
Carefully pour off the liquid into an evaporating flask and remove the
ethanol under reduced pressure. The precipitate is composed of
uronic acid–containing polysaccharides and may be dissolved and analyzed directly. The aqueous solution is transferred quantitatively
from the evaporating flask to a measuring cylinder and the volume
recorded.

C. Colorimetric determination of fucose

The colorimetric method for fucose determination was introduced by
Dische (1955). In most cases it is necessary to dilute the solution obtained from the fractionation before analysis.

Pipette 1.0 ml of sample (diluted if required) into a test tube
(16 × 160 mm) placed in an ice-water bath. Carefully add 4.5 ml di-

Table 15-1. *Fucose content of brown algae (mg g^{-1})*

	Ascophyllum nodosum	Fucus vesiculosus
Total fucose extracted	74.9	108.3
Nondialyzable fucose	58.1	78.6
Insoluble MgCl$_2$/ethanol	12.2	9.7
Soluble MgCl$_2$/ethanol	40.0	59.5

lute H$_2$SO$_4$ and allow the mixture to cool for 1 min. Mix with a glass rod and place the tube in a boiling-water bath for exactly 10 min. Cool in tap water to room temperature and add 0.1 ml cysteine solution. Mix with a glass rod and leave the tube at room temperature for 30 min. Read the absorbance at 396 and 427 nm. The absorbance due to fucose is the difference between the reading obtained at 396 nm and that at 427 nm. It is strongly recommended that the dilution of the sample be adjusted so that the difference between these readings falls in the range 0.300–0.500. The total fucose, F, in milligrams is calculated from the formula $F = b \times 0.06 \times V$, where b is the difference between the absorbancy readings, 0.06 is a factor converting absorbance to amount of fucose in milligrams per milliliter, and V is the total volume in milliliters.

In the formula, b must be the difference in absorbance if 1.0 ml of the fucoidan solution were pipetted. If dilution has been carried out, b must be calculated by multiplying the actual difference by the dilution factor before the fucose content is calculated.

To obtain the approximate amount of fucoidan, the fucose content must be multiplied by a factor of 2 (see V, below).

IV. Sample data

Table 15-1 gives the analytical data obtained for *Ascophyllum nodosum* and *Fucus vesiculosus* from the Trondheimsfjord area. For other species data are rather scarce, but those obtained by Myklestad and Haug (1977) indicate that elevated fucose contents are associated with algae growing higher up in the intertidal zone.

V. Discussion

It must be emphasized again that the method described is based on a definition of fucoidan as an acid-soluble sulfated fucan. The family of polysaccharides containing uronic acids and other neutral sugars in addition to fucose (Percival and McDowell 1967; Medcalf and Larsen 1977) should, therefore, not be considered as fucoidan and must be excluded. The major part, if not all, of the fucoidan appears to be linked to other polymers in the algae and thus does not exist as a separate, free entity. This fucoidan scarcely fulfills the requirement that fucoidan be a homopolymer of L-fucose, but after acid degradation and fractionation it approaches this composition. In our experience galactose appears to be present in all fucoidan preparations of this type. The factor of 2 used to convert the total fucose to weight of fucoidan is, therefore, imprecise, and the figures obtained are only estimates.

Our knowledge thus far is restricted to a few members of the Fucales, and care should be taken when using the method on species other than *Ascophyllum nodosum* or *Fucus* spp. In *Laminaria* spp. the fucose content is usually much lower. In these species the fractionation step should be unnecessary, and fucoidan may be determined by a direct fucose assay of the combined acid extract.

VI. References

Black, W. A. P., Dewar, E. T., and Woodward, F. N. 1952. Manufacture of algal chemicals. IV. Laboratory-scale isolation of fucoidin from brown marine algae. *J. Sci. Food Agric.* 3, 122–9.

Conchie, J., and Percival, E. G. V. 1950. Fucoidin. II. The hydrolysis of a methylated fucoidin prepared from *Fucus vesiculosus. J. Chem. Soc.* 827–32.

Côté, R. H. 1959. Disaccharides from fucoidin. *J. Chem. Soc.* 2248–54.

Dische, Z. 1955. New color reactions for determination of sugars in polysaccharides. *Methods Biochem. Anal.* 2, 313–58.

Dubois, M., Gilles, K. A., Hamilton, J. K., Rebers, P. A., and Smith, F. 1956. Colorimetric method for determination of sugars and related substances. *Anal. Chem.* 28, 350–6.

Larsen, B., Haug, A., and Painter, T. 1970. Sulphated polysaccharides in brown algae. III. The native state of fucoidan in *Ascophyllum nodosum* and *Fucus vesiculosus. Acta Chem. Scand.* 24, 3339–52.

Medcalf, D. G., and Larsen, B. 1977. Fucose containing polysaccharides in the brown algae *Ascophyllum nodosum* and *Fucus vesiculosus. Carbohydr. Res.* (in press).

Myklestad, S., and Haug, A. 1977. The content of polyanionic groups and cation binding in some brown algae. *Proc. Int. Seaweed Symp.* 8 (in press).

O'Neill, A. N. 1954. Degradative studies on fucoidin. *J. Am. Chem. Soc.* 76, 5074–6.

Percival, E., and McDowell, R. H. 1967. *Chemistry and Enzymology of Marine Algal Polysaccharides.* Academic Press, London, 219 pp.

16: Phenolic compounds in brown and red algae*

MARK A. RAGAN† AND J. S. CRAIGIE

Atlantic Regional Laboratory, National Research Council of Canada, Halifax, Nova Scotia B3H 3Z1, Canada

CONTENTS

I	Introduction	*page* **158**
II	Materials and test organisms	**158**
	A. Materials	158
	B. Algae	160
III	Methods and sample data	**160**
	A. General methods for analysis of phenols	160
	B. Specialized methods and sample data: polyphloroglucinols in *Fucus vesiculosus*	162
	C. Specialized methods and sample data: bromophenols in *Polysiphonia lanosa*	167
IV	Discussion	**177**
	A. Polyphloroglucinols in brown algae	177
	B. Bromophenols in red algae	177
	C. Phenolic constituents of other algae	177
V	Acknowledgments	**178**
VI	References	**178**

* Issued as NRCC No. 16229.

† Present address: Institute of Marine Biochemistry, University of Trondheim, N-7034 Trondheim-NTH, Norway

[157]

I. Introduction

The phenolic constituents of relatively few algae have been analyzed in detail. Those that have received the most attention fall into two broad groups: the polyphloroglucinols of brown algae and the bromophenols of red algae. Separate methods were employed for examining these groups of phenols. We describe the results obtained and discuss limitations on their interpretation.

Brown algal polyphloroglucinols (consisting of 1,3,5-trihydroxybenzenoid structural units). Polyphloroglucinols are present in most Phaeophyceae. A variety of these compounds (I–IX, Fig. 16-1) has been reported, and in addition to the compounds shown, more complex polyphloroglucinols are present in some taxa (Glombitza 1974; Ragan and Craigie 1976; Ragan 1976). No evidence for naturally occurring phenolic glycosides or esters has been presented to date.

Red algal bromophenols (phenols containing covalent bromine). Although red algal bromophenols occur predominantly as sulfate esters (Weinstein et al. 1975), the difficulties associated with isolation and separation of sulfated phenols have led most researchers to investigate only the free phenols formed after hydrolysis of the ester sulfate (III.C.1). Approximately 20 bromophenols have been reported from red algae (Fenical 1975; Chevolot-Magueur et al. 1976; Landymore 1976).

II. Materials and test organisms

A. Materials

All reagents used were of analytical (AR) grade. Paper chromatography was done on 46 × 57-cm sheets of Whatman No. 1 paper. The polyamide powder used was Merck (No. 7435) or Woelm, thin-layer chromatography (TLC) grade; the former was more suitable for preparation of thin-layer chromatograms. Microcrystalline cellulose (Avicel TG-101) was purchased from Charles Tennant & Co. Silica gel powder was obtained from Baker (7GF) and Mallinckrodt (TLC-7GF); the commercial silica gel plates (20 × 20 cm) were Eastman

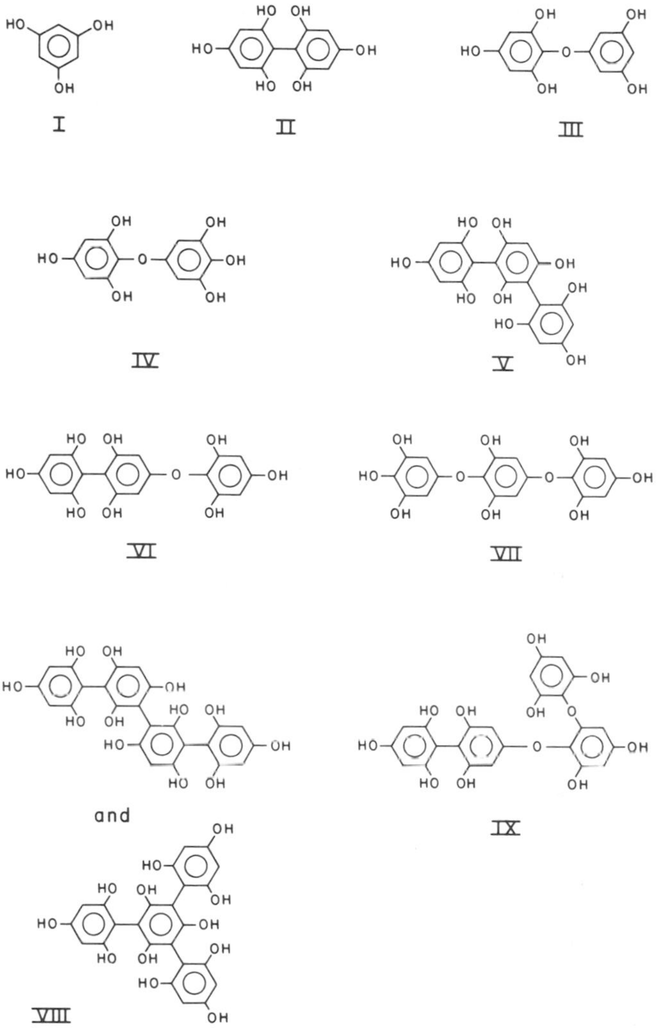

Fig. 16-1. Structures and nomenclature of brown algal polyphloroglucinols. *I*, phlor-oglucinol; *II*, 2,2'4,4',6,6'-biphenylhexol; *III*, 2,3',4,5',6-pentahydroxybiphenyl ether; *IV*, 2,3',4,4',4',6-hexahydroxybiphenyl ether; *V*, 1,5-di(2,4,6-trihydroxyphenyl)-2,4,6-trihydroxybenzene; *VI*, 2,2',4,6,6'-pentahydroxy-4'-(2,4,6-trihydroxyphenoxy) biphenyl; *VII*, 1,3-dihydroxy-2-(3,4,5-trihydroxyphenoxy)-5-(2,4,6-trihydroxyphen-oxy)benzene; *VIII*, phloroglucinol-based quaterphenyls; *IX*, 2,2',4,6,6'-pentahydroxy-4'-[2-(2,4,6-trihydroxyphenoxy)-4,6-dihydroxyphenoxy]biphenyl.

(No. 13181) and Merck (F-254). Fast Bordeaux Salt BD was pur-chased from Esbe Laboratory Supplies.

Gas chromatography was conducted using a Hewlett-Packard model 5730A instrument. Stainless-steel columns, 0.32 cm ($\frac{1}{8}$ in.) in diameter and of the indicated lengths, were packed with commercially available materials: SE-30 (3% or 10% w/w on 60- to 80-mesh Gas-Chrom Q) and *neo*pentyl glycol sebacate (0.5% w/w on 80- to 100-mesh Chromosorb G). Nitrogen (helium for combined gas chromatography and mass spectrometry, GC-MS) was the carrier gas (flow rate 20 ml min^{-1}), and flame ionization detection was employed. Trimethylsilyl (TMS) derivatives were formed with Tri-Sil (Pierce Chemical Co.).

Liquid-liquid extraction was carried out using a modified Gaebler extraction apparatus (Griffin and George Ltd.). Ultraviolet spectra were recorded on a Pye Unicam model SP 8005 spectrophotometer. A Hitachi Perkin-Elmer model 139 spectrophotometer was utilized for colorimetric tests. GC-MS was performed using a Hewlett-Packard model 5750 gas chromatograph interfaced with a Du Pont 21-491 mass spectrometer; further operating conditions are listed in the appropriate tables.

B. Algae

Fucus vesiculosus (L.) and *Polysiphonia lanosa* (L.) Tandy were examined. In each case the algae were refrigerated and were shielded from direct sunlight during transportation to the laboratory, where they were immediately extracted. Storage of either alga, but particularly of *F. vesiculosus* (or other brown algae), for longer than 24 h before use is strongly discouraged. The advantages of freeze-drying (e.g., ease of sample storage) must be balanced against the increased opportunity for formation of artifacts during this process. Under no circumstances should algae (especially brown algae) be air- or oven-dried prior to extraction of phenols.

Accurate identification of algal samples is of paramount importance, and voucher specimens should be retained for future reference.

III. Methods and sample data

A. General methods for analysis of phenols

1. Extraction of phenols. Procedures for the extraction of algal phenols must maximize the recovery of phenols while minimizing both phenol degradation and extraction of interfering compounds. This may be accomplished by stirring the milled alga in either aqueous alcohol (e.g., 80% v/v ethanol) or aqueous acid (e.g., 0.2 N H_2SO_4). Phenols are extracted much more efficiently from finely milled samples than from whole plants; with *F. vesiculosus*, milling produced a sixfold improvement in the recovery of "total phenols." In either case, the water content of the algal tissue must be included in determining the final water content of alcoholic extracts.

2. Derivatization and stability of phenols. Many phenols, especially those with two or more phenolic hydroxyl groups on a single benzene ring (polyphenols), are easily oxidized to brown derivatives. This process is retarded, although not prevented, by minimizing exposure of the phenols to oxygen, prolonged heating, direct illumination, and alkaline pH. Because silica gel TLC plates provide a large, active surface that may catalyze some oxidations, TLC operations on silica gel should be conducted as expeditiously as possible, with minimum exposure of the plates to air or strong light. Phenols must also be protected from exposure to heavy metal ions.

The stability of phenols can be increased by converting the phenolic hydroxyl groups to acetyl, methyl, benzoyl, or TMS derivatives. Such modification may give crystalline derivatives, especially of the lower molecular weight phenols. These derivatives exhibit excellent solubility in organic solvents such as chloroform, acetone, and ethyl acetate, and are generally insoluble in water (TMS ethers are decomposed by water). This step is usually necessary for analysis of phenols by gas chromatography, and TMS ethers are the most useful derivative for mass spectrometry. An important consequence of derivative formation is the concomitant reduction of reactivity with most of the detection reagents listed below (III.A.3).

Algal phenols of unknown structure should be converted to at least two different derivatives to prevent naturally occurring derivatives from being overlooked. Acetates of phenols are readily formed by heating the anhydrous phenol in acetic anhydride with a small amount of freshly fused sodium acetate in a sealed glass tube at 105 C for 1–2 h. They are moderately stable, and the acetyl groups can be removed quantitatively by mild heating of the derivative in anhydrous methanol containing a trace of sodium (or potassium) carbonate. Methylation is conducted by dissolving the phenol in ethereal diazomethane (Vogel 1959) at or below room temperature for up to 24 h if necessary. (*Caution:* Diazomethane is extremely toxic and explosive. Ground-glass apparatus must be avoided, and sharp edges of glassware should be fire-polished. Use an efficient fume hood at all times.) TMS ethers are usually prepared with commercial reagents such as Tri-Sil or BSTFA (Pierce).

3. Detection of phenols on chromatograms. The simplest method for detecting phenols on paper or thin-layer chromatograms is to expose the chromatograms to air and light for several days; phenols will give grayish to brown colors, with *o*-dihydroxyphenols discoloring especially rapidly. Ultraviolet light (especially below 300 nm) is absorbed by most phenols, and some phenols give a characteristic fluorescence under these conditions. The ultraviolet absorption is frequently enhanced or altered by ammonia vapor or by a light spraying with aqueous or ethanolic solutions of borate or .aluminum salts, or diazo-coupling reagents.

The following reagents may be sprayed onto chromatograms using commercially available aerosol spraying apparatus. (*Caution:* Use an efficient fume hood.)

a. $FeCl_3$ (0.5–2% w/v in water) reacts with phenols to give blue, gray, or yellowish colors. Much more sensitive is the ferric chloride–potassium ferricyanide reagent (equal volumes of 0.5% w/v $FeCl_3$ and 0.5% w/v $K_3Fe(CN)_6$ combined just before spraying the chromatogram), which yields blue, or occasionally green or yellow, colors with reducing compounds, especially phenols.

b. Diazo-coupling reagents such as Fast Bordeaux Salt BD (0.01% w/v in water; spray lightly, note color, then overspray with saturated aqueous Na_2CO_3) or diazotized *p*-nitroaniline (5 volumes of 0.5% w/v *p*-nitroaniline in 2 N aqueous HCl are added to a mixture of 1 volume of 5% w/v $NaNO_2$ and 20 volumes of 20% w/v aqueous sodium acetate; overspray with 15% w/v aqueous Na_2CO_3) yield a wide range of colors with phenols. Polyphenols tend to give more muted colors than do simple phenols. For spraying polyamide plates, the reagents should be prepared in 80% v/v ethanol, and ammonia solution (10% v/v conc. NH_4OH in 95% ethanol) should be used as the second spray reagent.

c. Ehrlich's reagent (0.5% w/v dimethylaminobenzaldehyde in 30:1 v/v 95% ethanol and conc. HCl) reacts at room temperature (or after mild heating) with many phenols, especially phloroglucinol.

d. Lindt's reagent (1% w/v vanillin in 9:1 v/v 95% ethanol and conc. HCl) reacts with phloroglucinol derivatives at room temperature to give a red or reddish-orange color and reacts with a limited number of other compounds after heating.

e. Sodium molybdate (2% w/v in water) reacts with *o*-dihydroxyphenols.

f. Beilstein's flame test for halogens (appearance of a bluish green colored flame upon burning the solid compound in a spatula-shaped piece of flame-oxidized copper wire) can be applied to compounds eluted from chromatograms. Halogenated solvents (e.g., chloroform) should be avoided, or removed by evaporation, before this test is carried out.

B. Specialized methods and sample data: polyphloroglucinols in Fucus vesiculosus

1. Extraction of polyphloroglucinols. Fresh *F. vesiculosus* (50 g) was frozen by immersion in liquid nitrogen. While frozen, it was milled to a fine powder in a Dangoumou ball mill (Prolabo). The resulting powder was placed in a 1-l Erlenmeyer flask, covered with methanol, flushed exhaustively with nitrogen, and shaken slowly on a reciprocating shaker overnight at 3 C in the dark. The extracted powder was removed by filtration, and the methanol was evaporated to dryness in vacuo at less than 40 C. The dried residue was immediately resuspended in distilled water, the flask flushed with nitrogen and stoppered, and the suspension shaken thoroughly. After the aqueous phase was separated by centrifugation or filtration, the pigmented residues were once again washed with water and then discarded. The combined aqueous phases were evaporated in vacuo at less than 40 C, and the solids were immediately resuspended in about 25 ml of methanol, cooled to − 15 C for several hours, and filtered while still cold through Whatman No. 50 paper. The filter cake (including mannitol and salts) was washed twice with small volumes of cold methanol, and the combined filtrates were evaporated under a jet of nitrogen to about 2 ml in preparation for column chromatography.

2. Chromatography and identification of phenols. A 2 × 10-cm column of microcrystalline cellulose (Avicel TG-101) was prepared by placing a disk of Whatman No. 5 filter paper over the sintered glass base and packing the dry cellulose tightly. The concentrated methanolic filtrate (III.B.1) was allowed to absorb evenly into the uppermost 0.5–1.0 cm of the Avicel column, after which the column was eluted first with acetone (ca. 100 ml, until the eluate was colorless), then with methanol (ca. 400 ml, until the eluate was colorless) at 3 C. These fractions were collected in round-bottomed flasks and evaporated in vacuo at less than 40 C. The solids were immediately redissolved in methanol, cooled to − 15 C, filtered through an ultrafine-porosity sintered glass funnel, and stored in darkness at − 15 C under a nitrogen atmosphere.

Polyphloroglucinols in the acetone eluate were examined by several methods:

a. Silica gel chromatography on commercial (Eastman No. 13181 or Merck F-254) or homemade (from Baker 7GF or Mallinckrodt TLC-7GF) thin-layer plates, using chloroform : methanol (7 : 3 v/v) or chloroform : methanol : water (12 : 6 : 1 v/v/v) solvent systems, resolved at least six compounds that reacted with the vanillin-HCl reagent (Table 16-1).

b. Paper chromatography (with 2% v/v formic acid in water, or water saturated with diethyl ether, as the solvent system) resolved at least three polyphloroglucinols (Table 16-1). The formic acid–water system gave more reproducible relative migration rates (R_f values) than did the ether-water system, but was not used for preparative work because commercial formic acid may contain formaldehyde,

Table 16-1. *Chromatographic data (TLC, PC, GC) and GC-MS on Fucus vesiculosus polyphloroglucinols*

Compound	A:[a] $R_{phloroglucinol}$	B:[a] $R_{phloroglucinol}$	C:[a] $R_{phloroglucinol}$	D:[a] $R_{4-hydroxybenzaldehyde}$	E:[a] $R_{gallic\ acid-TMS}$	F:[a] $R_{phloroglucinol-TMS}$	GC/MS:[b] monoisotopic M^+
Phloroglucinol (I)	1.00	1.00	1.00	0.03	0.67	1.00	m/e 342
Dimer (II)	0.63	0.34	1.24[c]	0.00	1.35	2.55	m/e 682
"Trimer" (VI)	0.47	0.23	1.53[c]	0.00	1.91	4.35	m/e 950
"Tetramer" (IX)	0.35	—	—	—	2.73–2.87[d]	—	m/e 1218
"Pentamers"	0.24	—	—	—	—	—	—
"Hexamers"	0.16	—	—	—	—	—	—

[a] A. Silica gel (Eastman No. 13181) developed with chloroform:methanol:water (12:6:1 v/v/v). B. Silica gel (Eastman No. 13181) developed with toluene:ethyl acetate:acetic acid (5:4:1 v/v/v). C. Paper (Whatman No. 1) developed with 2% (v/v) formic acid in water. D. Paper (Whatman No. 1) developed with the upper phase of 6:7:3 (v/v/v) benzene:acetic acid:water. E. Gas chromatography on 10% (w/w) SE-30; conditions are given in II.A and III.B.2.d. Values listed are for TMS ethers on seasoned columns. F. Gas chromatography on 0.5% (w/w) *neo*pentyl glycol sebacate; conditions are given in II.A and III.B.2.d. Values listed are for TMS ethers.

[b] GC-MS performed on 10% (w/w) SE-30 (for conditions, see II.A and III.B.2.d). The mass spectrometer was operated at accelerating voltage, 82 eV; source temperature, 360; scan range, m/e 18–1600 at 10 sec/decade; internal standards, perfluorokerosene (PFK) and tris-(perfluoroheptyl)-s-triazine (PCR-8).

[c] Chromatographic behavior is variable, depending on the presence or absence of other polyphloroglucinols, lipophilic materials, and salts. The values given are taken from chromatograms of mixtures of polyphloroglucinols.

[d] Two components, incompletely resolved, with m/e 1218 for both, possibly representing isomeric "tetramers."

which can cross-link phenols. With either solvent system, the R_f values vary, depending on the presence or absence of other polyphlorogluocinols, the presence of lipophilic materials, and other factors. Consequently, identifications made by co-chromatography are much more reliable than those based on R_f values.

c. Column chromatography on polyamide (Merck No. 7435 or Woelm TLC grade) developed with 3:1 v/v methanol:water is useful for small-scale preparative work. The eluate was monitored continuously at 280 nm with a continuous-flow UV monitor (e.g., the ISCO UA-2). Phloroglucinol was eluted first under these conditions, followed in seriatim by several oligomers. Because the polyamide strongly adsorbed oligomers containing five or more phloroglucinol units, these higher oligomers cannot easily be purified by this method.

d. Alternatively, the acetone-eluted fraction was trimethylsilylated for analysis by gas chromatography. After the sample had been thoroughly dried in vacuo, excess Tri-Sil was added and the tube was capped; the mixture was heated to about 60 C for 2–3 min, then was allowed to stand at room temperature for 10–15 min to ensure complete reaction. The sample was then clarified by centrifugation, transferred (if desired) to a small, dry vial, and concentrated under a jet of dry nitrogen. Resolution of the polyphloroglucinol–TMS ethers was achieved on the following GC columns.

i. Silicone gum columns. 10% (w/w) SE-30, in 0.32 × 183-cm columns with temperature programming from 100 to 280 C at 4 C min^{-1} (temperature held at 280 C thereafter), produced the best separations (Fig. 16-2) and was used for GC-MS of the polyphloroglucinols (Table

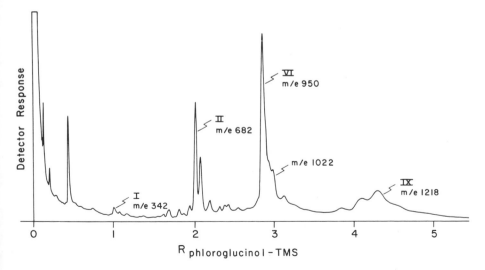

Fig. 16-2. Gas chromatographic separation of polyphloroglucinols from *Fucus vesiculosus*. For operating conditions see II.A and III.B.2.d.

16-1). Columns (305 cm) with 3% (w/w) SE-30 and identical temperature programming were adequate for some purposes, but did not provide the resolution achieved with 10% SE-30. The liquid phase OV-1 is similar to SE-30 and is sometimes preferred for GC-MS because it gives minimal background interference or "column bleed."

ii. Polyester column. 0.5% (w/w) *neo*pentyl glycol sebacate, in 0.32×183-cm columns with temperature programming from 100 to 240 C at 4 C min^{-1} (temperature held at 240 C thereafter), gave excellent resolution of compounds I, II, and VI, but did not allow recovery of the "tetramer" (IX).

Preliminary identification of the major components was accomplished by chromatographic comparison with known reference compounds (Table 16-2). More definitive identification was made by GC-MS (Table 16-1). The results demonstrate the virtually quantitative removal of mannitol from this fraction by the steps described above.

The methanol eluate from the Avicel column contained more highly polymerized polyphloroglucinols. Structural analysis of these polymers requires sophisticated techniques, including mass spectrometry, proton magnetic resonance spectrometry, and pulse Fourier transform carbon-13 nuclear magnetic resonance spectrometry (Ragan 1976). Methods have not been developed for the facile resolution of oligomeric polyphloroglucinols containing five or more phloroglucinol units, and the problem is further complicated by the probable occurrence of structural isomers of these oligomers.

3. Quantitative determination of polyphloroglucinols
    a. After isolation and purification. Of the variety of techniques available, perhaps the most readily applicable is the Folin-Denis colorimetric method for "total phenols" (Swain and Hillis 1959; Horwitz 1970), useful in the range from 0.5 to 8.0 μg of anhydrous phloroglucinol equivalents per milliliter. Other simple methods include measurement of UV absorption at an appropriate wavelength, usually near 280 nm; colorimetry using the vanillin–sulfuric acid reagent (Swain and Hillis 1959; Ragan 1976), useful in the range from 0.2 to 4.0 μg of anhydrous phloroglucinol per milliliter; and colorimetry with dyes such as Brentamine Fast Red 2G (Sieburth and Jensen 1969). The analytical results from each of these methods may be expressed as anhydrous phloroglucinol equivalents per gram of dry weight.
    b. In crude extracts and exudates. There is no universally applicable technique for determining absolute concentrations of phenolic substances in crude extracts or exudates from brown algae. For instance, the Folin-Denis reagent reacts with ascorbic acid and reducing

sugars as well as phenols, and the extinction of the colored product is greater with phloroglucinol than with the high-molecular-weight polyphloroglucinols. The vanillin-H_2SO_4 reagent detects 1,3- or 1,3,5-hydroxylated phenols, often producing colored carbanions with absorption maxima differing significantly from that produced with phloroglucinol (λ_{max} = 486 nm). Moreover, the colored product formed with phloroglucinol has a much greater extinction than that formed with high-molecular-weight polyphloroglucinols (Ragan 1976). Serious interference from nonphenolic UV-absorbing materials and poorly characterized solvent effects upon the extinction coefficients complicate the UV determination of phenols in crude extracts. The determination of polyphenols by potentiometric titration (Haug and Larsen 1958) depends on the assumed equivalent weight of the reducing species.

It is known that most of the polyphloroglucinols in *F. vesiculosus* exist in a highly polymerized form susceptible to quantitative removal by polyamide (Table 16-3 and Ragan 1976). Phloroglucinol itself is almost completely adsorbed by excess polyamide, whereas most ascorbic acid remains in solution. Thus measuring the "total phenols" by the Folin-Denis method before and after polyamide treatment provides an estimation of the concentration of extractable polyphloroglucinols (as milligrams of anhydrous phloroglucinol units per gram dry weight of algal tissue). Other quantitative methods can likewise be applied both before and after polyamide treatment. However, data obtained for brown algae are likely to be comparable only among samples from closely related species assayed by the same method. Quantitative data obtained using different methods (Table 16-3) will not be directly comparable for the reasons listed above.

C. Specialized methods and sample data: bromophenols in Polysiphonia lanosa

1. Extraction of bromophenols. Fresh *P. lanosa* (10 g) was extracted for 2 h with boiling 85% v/v ethanol (60 ml) under reflux. The extract, containing the sulfated phenols, was filtered while warm through glasswool, then was evaporated in vacuo until about 10 ml of aqueous extract remained. A 10% v/v aliquot of this solution was removed and frozen for later examination. The remainder was acidified to pH 0.5 with 2 N aqueous HCl, and sulfate esters were quantitatively hydrolyzed by heating on a steam bath for 15 min. The acidified solution was then continuously extracted with peroxide-free diethyl ether for at least 4 h. The ether extract was dried overnight at − 15 C with anhydrous Na_2SO_4, the latter was removed by vacuum filtration at − 15 C, and the filtrate was concentrated in vacuo for chromatographic examination.

Table 16-2. Chromatographic data (TLC, PC, GC) for reference compounds (phenols, polyphenols, cinnamic acids, and other compounds of compounds of interest)

Compound	A:[a] $R_{phloroglucinol}$	B:[a] $R_{phloroglucinol}$	C:[a] $R_{phloroglucinol}$	D:[a] $R_{4-hydroxybenzaldehyde}$	E:[a] $R_{gallic\ acid-TMS}$	F:[a] $R_{phloroglucinol-TMS}$
Phenol	1.08	2.13	—	1.87	0.06	0.06
Cresols (o-, m-, and p-)	1.47–1.50	2.01–2.16	—	1.84–1.86	0.10	0.12–0.13
Resorcinol (1,3-dihydroxybenzene)	1.40	1.71	1.21	0.29	0.30	0.41
Phloroglucinol (1,3,5-trihydroxybenzene) (I)	1.00	1.00	1.00	0.03	0.67	1.00
Pyrogallol (1,2,3-trihydroxybenzene)	1.30	1.45	1.17	0.16	0.51	0.69
2,2',4,4',6,6'-Biphenylhexol (II)	0.63	0.34	1.24[b]	0.00	1.35	2.50
2,2'-Furanyl-4,4',6,6'-biphenyltetrol	—	—	—	—	1.52	—
3,5-Dimethoxyphenol	1.51	2.07	1.06	1.56	0.44	0.81
Benzaldehyde	0.96	1.95	—	1.87	0.28	0.08
4-Hydroxybenzaldehyde	1.56	1.75	1.22	1.00	0.40	0.59
Benzoic acid	1.07	2.09	1.17	1.84	0.28	0.25
Salicylic (2-hydroxybenzoic) acid	1.00	1.78	1.12	1.79	0.54	c
3-Hydroxybenzoic acid	0.55	1.64	1.17	0.86	0.59	c
4-Hydroxybenzoic acid	0.88	1.74	1.11	0.80	0.65	c
3-Hydroxybenzyl alcohol	1.42	1.42	0.96	0.47	0.50	0.58
2,4-Dihydroxybenzoic acid	0.46	1.56	0.88	0.52	0.85	c
Gentisic (2,5-dihydroxybenzoic) acid	0.51	1.39	1.04	0.44	0.81	c
Protocatechuic (3,4-dihydroxybenzoic) acid	0.41	1.26	0.91	0.18	0.84	1.49[c]

Vanillin (3-methoxy-4-hydroxybenzaldehyde)	1.14	1.82	1.15	1.68	0.55	0.97
Isovanillin (3-hydroxy-4-methoxybenzaldehyde)	1.14	1.75	1.12	1.55	0.56	1.05
Vanillic (3-methoxy-4-hydroxybenzoic) acid	0.76	1.67	0.99	1.53	0.76	1.67
Syringic (3,5-dimethoxy-4-hydroxybenzoic) acid	0.84	1.58	0.90	1.53	0.90	c
Phenylacetic acid	1.31	2.01	1.31	1.87	0.32	0.31
t-Cinnamic acid	1.19	1.91	0.99	1.95	0.55	0.92c
p-Coumaric (4-hydroxy-cinnamic) acid	0.71	1.62	1.08	0.82	0.94	c
o-Coumaric (2-hydroxy-cinnamic) acid	0.83	1.67	0.88	1.20	0.82	c
Caffeic (3,4-dihydroxy-cinnamic) acid	0.31	1.03	0.56	0.22	1.12	c
Ferulic (3-methoxy-4-hydroxycinnamic) acid	0.81	1.66	0.65	1.66	1.08	c
Sinapic (3,5-dimethoxy-4-hydroxycinnamic) acid	0.91	1.57	0.55	1.60	1.25	c
Chlorogenic acid (caffeic acid, quinyl ester)	0.19	0.09	1.00	0.00	c	c
Gallic (3,4,5-trihydroxy-benzoic) acid	0.18	0.69	0.73	0.04	1.00	1.88
Mannitol	—	—	—	—	0.96	1.76
Glycerol	—	0.22	—	—	0.28	0.23

[a] See note *a*, Table 16-1.
[b] Chromatographic behavior is variable, depending on the presence or absence of other polyphloroglucinols, lipophilic materials, and salts. The values given are taken from chromatograms of mixtures of polyphloroglucinols.
[c] Poor retention characteristics under these conditions.

Table 16-3. *Quantitative determination of Folin-Denis reactive material and of*
vanillin-H_2SO_4 reactive material in sample of Fucus vesiculosus
(data are in anhydrous phloroglucinol units)

Extraction procedure	Folin-Denis reaction		Vanillin-H_2SO_4 reaction	
	In extract (mg/g dry wt)	% adsorbed on poly-amide	In extract (mg/g dry wt)	% adsorbed on poly-amide
Fresh *F. vesiculosus* apices soaked in 85% v/v ethanol	10.2	85	2.2	98
Milled *F. vesiculosus* apices stirred in 85% v/v ethanol	59.3	98	13.7	99
Milled *F. vesiculosus* apices stirred in 0.2 N aq. H_2SO_4	48.2	96	22.9	99

2. Chromatography and identification of phenols. Bromophenols released
by acid hydrolysis were examined by the following methods:

a. TLC was performed on commercial silica get (Eastman No.
13181 or Merck F-254) plates, with toluene:ethyl acetate:acetic acid
(5:4:1 v/v/v) or chloroform:methanol (15:1 to 39:1 v/v, as required) solvent systems. Chromatography on homemade plates of
Merck No. 7435 polyamide was done with methanol:water (3:1 v/v)
as the solvent. In each case, the major phenol co-chromatographed
with 2,3-dibromo-4,5-dihydroxybenzyl alcohol (lanosol) (Table 16-4).
Phenolic components other than lanosol were present in much
smaller amounts.

b. Paper chromatography using *n*-butanol:acetic acid:water
(4:1:2 v/v/v) or 2% (v/v) formic acid in water as solvent systems revealed a major component that co-chromatographed with lanosol.
Two, or possibly three, very minor phenolic components were also
present (Table 16-4). The phenolic sulfate esters (prior to hydrolysis)
could also be chromatographed on paper; with the 2% (v/v) aqueous
formic acid solvent system, the major component (2,3-dibromo-5-hy-
droxybenzyl-1′,4-disulfate) showed an R_f = 0.84 ($R_{gallic\ acid}$ = 1.84).

c. Alternatively, the hydrolyzed phenols were trimethylsilylated
and analyzed by gas chromatography using temperature programming conditions described above (III.B.2.d). The best results were obtained with a 183-cm column of 10% (w/w) SE-30 (Fig. 16-3).
*Neo*pentyl glycol sebacate (0.5% w/w, 183 cm) produced excellent
chromatograms of lanosol–TMS ether ($R_{phloroglucinol-TMS}$ = 2.07), but

Table 16-4. *Chromatography (TLC, PC, GC) and GC-MS of Polysiphonia lanosa bromophenols*

Compound	A:[a] $R_{\text{3-hydroxybenzyl alcohol}}$	B:[a] $R_{\text{gallic acid}}$	C:[a] $R_{\text{gallic acid}}$	D:[a] $R_{\text{gallic acid}}$	E:[a] $R_{\text{gallic acid}}$	F:[a] $R_{\text{gallic acid-TMS}}$	GC/MS:[b] monoisotopic M+
2,3-Dibromo-4,5-dihydroxybenzyl alcohol	0.59	1.45	0.68	0.78	1.25	1.21	512
2,3-Dibromo-4,5-dihydroxybenzaldehyde	—	—	—	—	—	0.67	438
Unidentified, minor components[c]	0.10 0.07	0.41 0.17 0.05	—	—	—	—	—

[a] A. Silica gel (Eastman No. 13181) developed with 5% (v/v) methanol in chloroform. B. Silica gel (Eastman No. 13181) developed with toluene:ethyl acetate:acetic acid (5:4:1 v/v/v). C. Polyamide (Merck No. 7435) developed with 3:1 (v/v) methanol:water. D. Paper (Whatman No. 1) developed with 2% (v/v) formic acid in water. E. Paper (Whatman No. 1) developed with *n*-butanol:acetic acid:water (4:1:2 v/v/v). F. Gas chromatography on 10% (w/w) SE-30; conditions are given in II.A and III.B.2.d. Values listed are for TMS ethers on seasoned columns.

[b] GC-MS performed on 10% (w/w) SE-30; for conditions see II.A and III.B.2.d. The mass spectrometer was operated at accelerating voltage, 82 eV; source temperature, 320; scan range, m/e 18–800 at 10 sec/decade; internal standard, PFK. See also Pedersén et al. (1974).

[c] Accounting for less than 5% of the total bromophenols in *P. lanosa* hydrolysates.

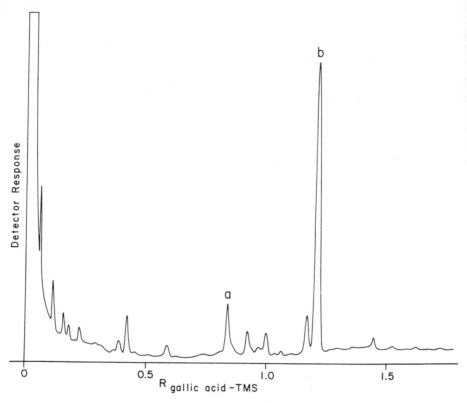

Fig. 16-3. Gas chromatographic separation of bromophenols (*a* = 2,3-dibromo-4,5-di-hydroxybenzaldehyde; *b* = 2,3-dibromo-4,5-dihydroxybenzyl alcohol) from *Polysiphonia lanosa* hydrolysate. For operating conditions see II.A and III.B.2.d.

performed poorly with most of the other bromophenols listed in Table 16-5.

Preliminary identification of the major phenol, lanosol, was accomplished by chromatographic comparison with known compounds (Table 16-5), and this was confirmed by GC-MS (Table 16-4). The identification of compounds of natural origin cannot be considered entirely conclusive unless confirmed by melting point data (including mixed melting points) of the compound and of its crystalline derivatives, or by examination with appropriate spectrometric techniques (IR, NMR, etc.).

2,3-Dibromo-4,5-dihydroxybenzaldehyde was detected as a minor component (Table 16-4) and may be an artifact of the extraction or hydrolytic procedures (Weinstein et al. 1975). The other minor phenolic constituents were not identified.

Table 16-5. Chromatographic data (TLC, PC, GC) for reference bromophenols and hydroxybenzyl alcohols

Compound	A:[a] $R_{\text{3-hydroxybenzyl alcohol}}$	B:[a] $R_{\text{gallic acid}}$	C:[a] $R_{\text{gallic acid}}$	D:[a] $R_{\text{gallic acid}}$	E:[a] $R_{\text{gallic acid}}$	F:[a] $R_{\text{gallic acid:d−TMS}}$
3-Bromo-4-hydroxy-5-methoxybenzaldehyde	1.39	2.15	0.74	1.03	1.30	0.73
3-Bromo-4-hydroxy-5-methoxybenzoic acid	0.59	2.03	0.62	0.78	1.31	0.99
2-Bromo-4-hydroxy-5-methoxybenzaldehyde	1.41	2.20	0.61	1.00	1.30	0.73
3-Bromo-4,5-dihydroxybenzaldehyde	0.61	1.72	0.51	0.99	1.27	0.83
3-Bromo-4,5-dimethoxybenzoic acid	0.40	2.16	0.40	—	1.32	0.85
3-Bromo-4,5-dimethoxybenzoic acid, methyl ester	1.43	2.38	0.84	—	1.32	0.70
3,5-Dibromo-4-hydroxybenzyl alcohol	1.30	2.17	0.77	1.14	1.31	0.96
3,5-Dibromo-4-hydroxybenzyl alcohol 1'-methyl ether	1.45	2.31	0.67	1.29	1.32	0.82
3,5-Dibromo-4-hydroxybenzoic acid	0.36	2.04	0.19	0.67	1.33	1.09
3,5-Dibromo-2-hydroxybenzoic acid	0.43	1.64	0.17	0.65	1.29	0.94
2,3-Dibromo-4-hydroxy-5-methoxybenzyl alcohol	1.29	1.76	0.64	0.70	1.32	1.13
2,3-Dibromo-4-hydroxy-5-methoxybenzaldehyde	0.91	2.25	0.21	0.00	1.29	0.99
2,3-Dibromo-4,5-dihydroxybenzyl alcohol	0.59	1.45	0.68	0.78	1.25	1.21
2,3-Dibromo-4,5-dihydroxybenzoic acid	0.50	1.90	0.10	0.16	1.37	1.30
2,3-Dibromo-4,5-dimethoxybenzyl alcohol	1.45	2.03	0.64	—	1.34	1.02

Table 16-5. (*cont.*)

Compound	A:[a] $R_{3-hydroxybenzyl\ alcohol}$	B:[a] $R_{gallic\ acid}$	C:[a] $R_{gallic\ acid}$	D:[a] $R_{gallic\ acid}$	E:[a] $R_{gallic\ acid}$	F:[a] $R_{gallic\ acid-TMS}$
2,3-Dibromo-4,5-dimethoxybenzaldehyde	1.40	2.35	0.64	—	1.32	0.85
2,3-Dibromo-4,5-dimethoxybenzoic acid	0.13	2.00	0.12	—	1.29	—
2,3-Dibromo-4-methoxy-5-hydroxybenzyl alcohol	1.13	1.89	0.57	1.05	1.34	1.06
3-Hydroxybenzyl alcohol	1.00	1.89	1.25	1.64	1.22	0.38
4-Hydroxybenzyl alcohol	0.95	1.83	1.30	1.63	1.24	0.42

[a] See note *a*, Table 16-4.

3. Quantitative determination of red algal phenols. P. lanosa contains only one major phenol (lanosol disulfate), and the quantitative determination of lanosol in hydrolysates of extracts from *P. lanosa* will provide an adequate estimate of the total phenol content. The Folin-Denis colorimetric method (III.B.3.a) can be applied to measurement of red algal phenols, with lanosol serving as the standard; the results can be expressed as milligrams of lanosol equivalents per gram dry weight.

Lanosol itself was measured as follows:

a. UV spectrophotometry. Phenolic compounds were extracted from a 2.0-g sample of *P. lanosa* and hydrolyzed as described (III.C.1). The Na_2SO_4-dried ether phase was evaporated in vacuo, and the phenols were redissolved in methanol (2.0 ml). A carefully measured aliquot of the methanolic solution (5–30 μl) was streaked onto a 5×20-cm polyamide (Merck No. 7435) TLC plate, which was then developed in methanol: water (3:1 v/v). The lanosol band (R_f ca. 0.45) was located under UV light and was transferred in methanol to a 5-ml volumetric flask using the vacuum device shown in Fig. 16-4. The solution was made to volume with methanol, and its UV absorption at 292 nm was determined. A standard curve was constructed using known quantities of lanosol (50–200 μg) that had been similarly chromatographed, recovered, diluted to 5-ml volumes, and measured at 292 nm. Typically, *P. lanosa* hydrolysates contained 20–30 mg of lanosol per gram dry weight.

b. Gas chromatography. Phenolic compounds were extracted from a 2.0-g sample of *P. lanosa* and were hydrolyzed (III.C.1). A known amount (1.00 ml) of gallic acid stock solution containing 30 μmol ml^{-1} was added, and the sample was trimethylsilylated (III.B.2.d). The Tri-Sil was evaporated to dryness under a N_2 jet, and the sample was redissolved in chloroform. A subsample was then analyzed on a 183-cm column of 10% (w/w) SE-30 with temperature programming from 160 to 270 C at 8 C min^{-1}. Under these conditions, the ratio of peak heights (lanosol–TMS ether to gallic acid–TMS ether) was directly proportional to the molar ratio (lanosol to gallic acid) present in the hydrolysate. The relationship lanosol peak height/gallic acid peak height was plotted against the mole ratio of lanosol to gallic acid on linear graph paper, and from this standard curve the amount of lanosol in the hydrolyzate was determined. Other phenols may be measured similarly or other internal standards used, if each elicits a GC detector response that is linear over the concentration range being examined and if internal standards do not chromatograph at the same position as any of the naturally occurring compounds. Lanosol concentrations calculated by this method were in close agreement with values obtained by UV spectrophotometry.

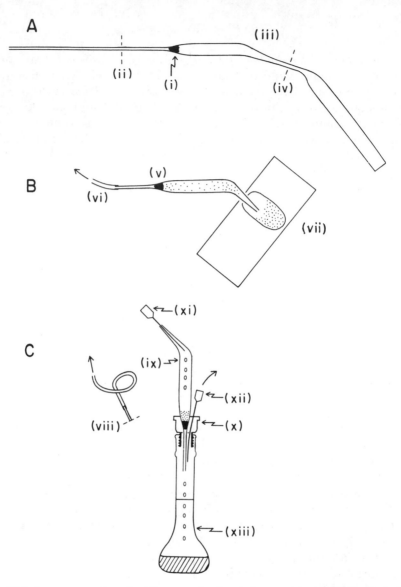

Fig. 16-4. Apparatus and method for removing bands from polyamide TLC plates. *A*. A Pasteur pipette is (i) tightly plugged with thoroughly washed, nonabsorbent cotton, (ii) cut where shown, (iii) heated and drawn as shown, then (iv) cut smoothly. *B*. The cotton is moistened slightly with methanol (v), a vacuum line (Tygon tubing) is attached (vi), and the desired band is vacuumed from the TLC plates (vii). *C*. After the Tygon vacuum line is removed (viii), the pipette device (ix) is inserted through a hole punched in (x) a No. 13 rubber Suba-Seal cap (Griffin and George Ltd.); methanol is injected through the open end (xi), and a No. 22 hypodermic needle attached to a vacuum line (xii) is inserted through the Suba-Seal cap to accelerate filtration into a 5-ml volumetric flask (xiii). The filtration apparatus is then removed from the flask, and volume is taken to 5 ml with methanol in preparation for UV spectrophotometry.

IV. Discussion

A. Polyphloroglucinols in brown algae

Methods have been presented for the analysis of polyphloroglucinols to the level of the "tetramer" (IX). Higher oligomers occur as structural isomers that so far have resisted chromatographic separation. At least in *F. vesiculosus* and *Ascophyllum nodosum* (Ragan 1976), and probably in many other Phaeophyceae, most of the polyphloroglucinols are complex, high-molecular-weight polymers for which analytical methods have not been developed.

Alternative analytical methods, based on initial acetylation of the lower molecular weight polyphloroglucinols, have been described by Glombitza et al. (1975).

Although methods exist for the quantitative analysis of individual, purified polyphloroglucinols (III.B.3.a), there is no method for the unequivocal quantification of "total phenols" in exudates or in crude extracts from brown algae. The method described (III.B.3.b) can be used to obtain valid comparative data for extracts from closely related species of Phaeophyceae, but will neither provide absolute values of "total phenols" nor serve as a means of comparing "total phenol" concentrations among taxonomically disparate brown algae (e.g., Ectocarpales, Laminariales, and Fucales).

B. Bromophenols in red algae

The principal sulfated bromophenol in *P. lanosa* is usually isolated by direct crystallization from extracts of kilogram-sized samples; however, sulfated bromophenols occurring as minor constituents are not easily recovered by this method. Therefore, mild acid hydrolysis (III.C.1) is usually used to convert the sulfate esters into free phenols, which are more easily analyzed by chromatographic methods (III.C.2). Some oxidation and/or etherification of the benzyl alcohol side chain may occur during the extraction and hydrolytic procedures employed. It is possible that small quantities of nonsulfated or oxidized bromophenols may be present naturally as biosynthetic intermediates.

C. Phenolic constituents of other algae

A variety of phenolic materials has been found among the relatively few algae investigated so far. There is evidence that some blue-green (Pedersén and DaSilva 1973) and some brown algae (Pedersén and Fries 1975) may possess bromophenols. A phenolic dihydrostilbene, lunularic acid, was reported in small quantities in representatives from six algal classes (Pryce 1972). Nakabayashi (1955) has presented evidence for the occurrence of gallotannins in *Spirogyra arcta*. Tocher and Craigie (1966) have demonstrated the occurrence of 3-hydroxy-tyramine in the green alga *Monostroma fuscum*. Landymore (1976) has demonstrated that tyrosine can be metabolized through a series of phenolic intermediates in *Isochrysis galbana* and *Navicula incerta*. The investigation of algae from the classes Phaeophyceae, Rhodophyceae, Cyanophyceae, Chlorophyceae, Bacillariophyceae, Xanthophyceae, and Chrysophyceae has yielded some 50 phenolic compounds.

V. Acknowledgments

We thank Dr. W. D. Jamieson and Mr. D. J. Embree for assistance with GC-MS, and Mr. T. Moore for technical assistance. M.A.R. gratefully acknowledges the financial support of the National Research Council of Canada.

VI. References

Chevolot-Magueur, A.-M., Cave, A., Potier, P., Teste, J., Chiaroni, A., and Riche, C. 1976. Composés bromés de *Rytiphlea tinctoria* (Rhodophyceae). *Phytochemistry* 15, 767–71.

Fenical, W. 1975. Halogenation in the Rhodophyta: A review. *J. Phycol.* 11, 245–59.

Glombitza, K.-W. 1974. Highly hydroxylated phenols from Phaeophyceae. In preprint, VIII International Seaweed Symposium, Bangor.

Glombitza, K.-W., Rauwald, H.-W., and Eckhardt, G. 1975. Fucole, Polyhydroxyoligophenyle aus *Fucus vesiculosus*. *Phytochemistry* 14, 1403–5.

Haug, A., and Larsen, B. 1958. Phenolic compounds in brown algae. I. The presence of reducing compounds in *Ascophyllum nodosum* (L.) LeJol. *Acta Chem. Scand.* 12, 650–7.

Horwitz, W. (ed.). 1970. *Official Methods of Analysis of the Association of Official Analytical Chemists*, 11th ed., p. 154. The Association, Washington, D.C.

Landymore, A. F. 1976. The degradation of phenylalanine, tyrosine, and related aromatic compounds by a marine diatom and a haptophycean alga. Ph.D. thesis, University of British Columbia, Vancouver. 158 pp.

Nakabayashi, T. 1955. Studies on tannins of *Spyrogyra arcta*. 5. On the structure of *Spyrogyra* tannin. *J. Agric. Chem. Soc. Jpn.* 29, 897–9.

Pedersén, M., and DaSilva, E. J. 1973. Simple brominated phenols in the blue-green alga *Calothrix brevissima* West. *Planta* 115, 83–6.

Pedersén, M., and Fries, L. 1975. Bromophenols in *Fucus vesiculosus*. *Z. Pflanzenphysiol.* 74, 272–4.

Pedersén, M., Saenger, P., and Fries, L. 1974. Simple brominated phenols in red algae. *Phytochemistry* 13, 2273–9.

Pryce, R. J. 1972. The occurrence of lunularic and abscisic acids in plants. *Phytochemistry* 11, 1759–61.

Ragan, M. A. 1976. Physodes and the phenolic compounds of brown algae. Ph.D. thesis, Dalhousie University, Halifax. 112 pp.

Ragan, M. A., and Craigie, J. S. 1976. Physodes and the phenolic compounds of brown algae: Isolation and characterization of phloroglucinol polymers from *Fucus vesiculosus* (L.). *Can. J. Biochem.* 54, 66–73.

Sieburth, J. McN., and Jensen, A. 1969. Studies on algal substances in the sea. II. The formation of gelbstoff (humic material) by exudates of Phaeophyta. *J. Exp. Mar. Biol. Ecol.* 3, 275–89.

Swain, T., and Hillis, W. E. 1959. The phenolic constituents of *Prunus domestica*. I. The quantitative analysis of phenolic constituents. *J. Sci. Food Agric.* 10, 63–8.

Tocher, R. D., and Craigie, J. S. 1966. Enzymes of marine algae. II. Isolation and identification of 3-hydroxytyramine as the phenolase substrate in *Monostroma fuscum. Can. J. Bot.* 44, 605–8.

Vogel, A. I. 1959. *A Text-book of Practical Organic Chemistry*, 3rd ed., pp. 969–73. Longmans, London.

Weinstein, B., Rold, T. L., Harrell, C. E., Jr., Burns, M. W., III, and Waaland, J. R. 1975. Reexamination of the bromophenols in the red alga *Rhodomela larix. Phytochemistry* 14, 2667–70.

17: Brown seaweeds: analysis of ash, fiber, iodine, and mannitol

BJØRN LARSEN

Institute of Marine Biochemistry,
University of Trondheim, N-7034 Trondheim-NTH, Norway

CONTENTS

I	Introduction	*page* 182
II	Materials	182
III	Methods	182
	A. Mannitol	182
	B. Ash	184
	C. Fiber	184
	D. Iodine	185
IV	Sample data	187
V	Discussion	187
VI	References	188

I. Introduction

In this context major components will be considered to be constituents regularly occurring at a level of 1% or more of the dry matter. Some of these components are common to all plant materials (e.g., lipids and protein) and will not be considered here. The major components most characteristic of brown algae are alginic acid, fucoidan, laminaran, mannitol, and iodine. The first three are covered elsewhere in this book (Chp. 13, 14, and 15). For many species of benthic brown algae, phenolics (Chp. 16) are no doubt among the major components (Haug and Larsen 1958).

Individual variations are quite large in the brown algae (Baardseth and Haug 1953), and the sampling is therefore very important. The sample used for analysis should represent an average of at least 30 plants. Care should also be taken to avoid an unwanted selection of special tissues (e.g., young tissue), as there may be considerable differences in chemical composition among different parts of the same plant.

II. Materials

The methods described in this chapter should be applied only on dried samples milled to pass a 60-mesh sieve.

III. Methods

A. Mannitol

1. Reagents. (a) 4 N sulfuric acid. Dilute 109 ml conc. H_2SO_4 to 1 l with distilled water. (b) 0.1 N sulfuric acid. Dilute 2.7 ml conc. H_2SO_4 to 1 l with distilled water. (c) 0.1 N periodic acid. Dissolve 11.4 g $HIO_4 \cdot 2H_2O$ in 500 ml of distilled water. Protect the solution from light. (d) 0.1 N sodium thiosulfate. Dissolve 24.82 g $Na_2S_2O_3 \cdot 5H_2O$ in 1 l of distilled water. This is the primary standard, and it is strongly recommended that a prestandardized volumetric reagent be used. (e) Potassium iodide, solid. (f) Starch indicator solution. Weigh 1 g soluble starch and dissolve by heating to 80–100 C in 100 ml of distilled water. Cool and add a few drops of toluene.

2. Method. The method recommended was first published by Cameron et al. (1948) and is based on the very rapid oxidation of sugar alcohols by periodic acid. Sugars and many polysaccharides (not laminaran) react more slowly with periodic acid, so it is imperative that the reaction be timed exactly. To obtain as accurate an estimate as possible of the mannitol content, the analysis should be carried out in two steps.

Weigh accurately an amount of sample (ca. 0.1 g) into a 100-ml conical flask. Add 5 ml 0.1 N H_2SO_4 and place in a slow-moving shaker for 30–60 min. Add 5.0 ml 0.1 N periodic acid solution, and after a reaction time of exactly 60 sec (stop watch) add 20 ml 4 N H_2SO_4, 2–3 g solid KI, and a few drops of starch solution, and immediately start titration of the free iodine with the standard sodium thiosulfate solution from a burette. When approaching the endpoint, the titration must be carried out slowly to allow iodine inside the algal particles to diffuse into the solution. The endpoint is reached when the solution stays colorless for 30 sec. Carry out a calibration titration by the same procedure, but without the seaweed sample.

To obtain comparable results between samples, it is important that excess periodic acid in the reaction mixture be the same in all experiments. The result from the first titration is, therefore, used to calculate the amount of sample that should be used to obtain a consumption of periodic acid corresponding to 7–8 ml of 0.1 N sodium thiosulfate. The amount of periodic acid consumed is the difference between the titers for the calibration and the sample titrations. If, for instance, you weighed exactly 100 mg and the consumption of periodic acid was 5.0 ml, the amount of sample required to give a titration of 7.5 ml will be:

$$x = \frac{100 \times 7.5}{5.0} = 150 \text{ mg} \tag{1}$$

This amount is weighed into a 100-ml conical flask and the analysis carried out as described above. Triplicate determinations are recommended.

The consumption of periodic acid is calculated by subtracting the sample titration (in milliliters of 0.1 N sodium thiosulfate) from the calibration titration. Thus, 1 ml 0.1 N sodium thiosulfate solution corresponds to 1.821 mg mannitol, and the percentage of mannitol, *M*, in the sample is:

$$M = \frac{1.821 \times d \times 100}{w} \tag{2}$$

where d is the consumption of periodic acid and w the weight of sample in milligrams.

B. Ash

The most important factors for obtaining reproducible results in the ash determination are a good temperature control and an adequate supply of air. Ashing over an open flame should not be used.

Weigh 1.5–2 g of the sample into a porcelain crucible, preferably a shallow, wide type. Place in a cold oven, allow the oven to heat to 400 C and hold this temperature for at least 6 h. Remove the crucible from the oven while the temperature is still above 150 C and allow the cruicible to cool to room temperature in a desiccator. Reweigh the crucible and calculate the percentage of ash in the sample.

C. Fiber

Fiber is usually considered to be the part of a sample remaining after treatment with strong acid and subsequently with strong alkali at elevated temperatures. Chemically it is an ill-defined fraction, but it will contain the fibrous cellulose present in the material. The method is based on removing all water-, acid-, and alkali-soluble materials by degradation and extraction at elevated temperatures and determining the organic part of the remainder as weight loss upon combustion. The method is essentially the one recommended by the Association of Analytical Chemists.

1. Reagents. (a) Dilute sulfuric acid. Dilute 1.25 g of conc. H_2SO_4 to 100 ml with distilled water. (b) Sodium hydroxide solution. Dissolve 2.5 g NaOH in 100 ml distilled water. (c) Asbestos powder, coarse. Any good grade of coarse asbestos powder may be used. The powder is usually contaminated with acid-soluble material, and it is recommended that the powder be washed thoroughly with H_2SO_4, then with water, and dried before use. The combustion loss should be checked in each batch and should be negligible if the powder is dried 4 h at 105 C before combustion.

2. Method. Accurately weigh approximately 1.5 g of sample and transfer quantitatively to a 500-ml conical flask equipped with a socket to fit a reflux condenser. Add 150 ml dilute H_2SO_4 and let the mixture boil under reflux for 30 min. Cool to room temperature and filter with moderate suction through a coarse sintered glass filter equipped with a piece of linen cloth cut to fit above the sintered plate. Wash the filter cake with 200 ml of distilled water. Return the particles on the filter to the conical flask with 100 ml distilled water. Add 100 ml 2.5% NaOH and a trace of an antifoaming agent and allow the mixture to reflux for 30 min. During this period prepare the Gooch sintered crucible for filtration of the particles. Use a Gooch crucible of approximately 2 cm bottom diameter and 5 cm in height. The sintered bottom should be fairly coarse. Place the crucible on a filter flask and apply gentle suction from a water aspirator. Place a layer of 2–3 mm of the acid-washed asbestos on top of the sintered plate and allow the layer to spread evenly by washing with a little distilled water.

After boiling the solution for 30 min, filter it through the crucible and wash thoroughly with boiling distilled water (at least 200 ml). Under suction allow the last of the wash water to drain off and wash once with approximately 15 ml ethanol. Remove the crucible from the filtration apparatus and dry at 105 C for at least 4 h. Allow the crucible to cool in a desiccator. Accurately weigh the crucible and record the weight, w_1. Place the crucible in an oven, heat to 450 C, and allow the combustion to continue overnight. Again cool the crucible to room temperature in a desiccator and record the weight, w_2, after combustion. The weight loss, $w_1 - w_2$, represents the fiber content of the weighed sample.

D. Iodine

Iodine is a normal constituent of marine algae and is particularly abundant in *Laminaria* spp. In this species the major part of the iodine is located in the peripheral tissue of the stipes (Larsen and Haug 1961), where it may reach concentrations up to 4.7%. No information is available about the form in which the iodine is stored.

The major problem in the analysis is to secure a total release of iodine without removing constituents that will interfere in the subsequent titration of iodine with thiosulfate. In our laboratory we prefer to destroy the organic material by fusing it with solid NaOH and then liberate iodine in the dissolved melt by adding bromine prior to titration.

1. Reagents. (a) Sodium hydroxide. Solid. (b) Oxidation mixture. Mix (dry) 23.5 g KNO_3, 33 g Na_2CO_3 (anhydrous), and 43.5 g K_2CO_3. (c) Phosphoric acid. Syrupy. (d) Bromine solution. Place approximately 10 g of Br_2 at the bottom of a glass-stoppered, 500-ml amber-colored flask. Add approximately 300 ml distilled water, shake vigorously for 2–3 min, and leave in the fume hood overnight. Before removing solution for the analysis, always shake the flask. More water may be added to the flask as required provided liquid Br_2 is present. Store in the fume hood. (e) Potassium iodide solution. Dissolve 10 g analytical grade KI in 100 ml distilled water. (f) 0.02 N sodium thiosulfate solution. Dissolve 4.964 g $Na_2S_2O_3 \cdot 5H_2O$ in 1 l of distilled water. This is the primary standard, and it is strongly recommended that a ready-made volumetric reagent be used. (g) Soluble starch. See III.A.1, above. (h) Methyl red indicator solution. Dissolve 20 mg methyl red in 60 ml of ethanol and add 40 ml water.

2. Special equipment. It is necessary to carry out the destruction of the organic material in nickel crucibles. A convenient size is about 5 cm in diameter and 6–7 cm in height. Stirring is necessary during the destruction, and a nickel spatula about 15 cm in length is recommended for this purpose.

3. Method. Weigh 6–8 g of NaOH pellets into a nickel crucible and mix with an accurately weighed sample of between 0.2 and 2 g (usually 1–1.5 g), depending on the iodine content. It is important that the sample does not contain larger particles than those passing a 60-mesh sieve, as charring may take place with concomitant loss of iodine. The crucible is placed in a suitable holder and heated very gently with a gas burner. During the first period of heating, frothing will occur, and it is necessary to stir continuously with the nickel spatula until frothing ceases. Increase the rate of heating and add, in small portions, a total of 2 g of the oxidation mixture. When all oxidation mixture is added, the crucible is heated to a red glow for a few minutes. The melt should then be white and free of visible particles. Allow the crucible to cool to room temperature and place it, together with the spatula used for stirring, in a 250-ml beaker. Add 70 ml distilled water and stir occasionally until the melt is completely dissolved. Rinse the spatula into the crucible with a little distilled water and remove it. Add 4–5 drops of methyl red solution and neutralize the solution with syrupy H_3PO_4. It is very important to obtain the correct pH in this neutralization step. If a pH meter is available it is recommended that the final pH lie between 3 and 4. If not, a buffer solution of pH 3.5 should be made to contain the same concentration of methyl red as the test solution and H_3PO_4 should be added to the test solution until the colors match.

Add Br_2 solution until a yellow color persists in the test solution, indicating an excess of Br_2, and leave the solution for 2–3 h. Excess free Br_2 must be removed by evaporation, and the solution is therefore boiled until colorless or until the smell of Br_2 is not detectable. This usually requires 15–30 min boiling. Br_2 is extremely toxic and it is very important that all the steps during oxidation and evaporation be carried out in a good fume hood. Cool the solution to room temperature and add 5 ml 10% KI. Leave the solution for 5 min, add a few drops of soluble starch solution, and titrate with 0.02 N $Na_2S_2O_3$ until the solution stays colorless for at least 30 sec. A consumption of 1 ml of the $Na_2S_2O_3$ solution corresponds to 0.423 mg iodine, and the iodine content as percent of sample is calculated according to the for-

Table 17-1. *Percentage of mannitol, ash, fiber, and iodine in brown algae from an open sea locality on the Norwegian coast*

Species	Mannitol	Ash	Fiber	Iodine
Laminaria digitata				
Frond	7.6	23.2	6.1	0.53
Stipe	5.8	28.4	8.5	0.53
Laminaria hyperborea				
Frond	8.7	21.1	6.5	0.38
Stipe	5.5	31.8	8.0	0.49
Alaria esculenta	7.8	19.8	5.3	0.03
Ascophyllum nodosum	6.5	22.4	4.3	0.07
Fucus vesiculosus	7.6	21.5	4.7	0.03
Fucus serratus	8.0	23.0	5.1	0.04
Pelvetia canaliculata	5.6	21.5	2.3	0.04
Halidrys siliquosa	13.3	18.8	4.5	0.08

mula: $I = (42.3 \times a)/w$, where a is the consumption of $Na_2S_2O_3$ solution in milliliters and w is the weight of sample in milligrams.

IV. Sample data

Table 17-1 lists the content of mannitol, ash, fiber, and iodine in a number of brown algae. Tissues from *Laminaria* spp. were separated into fronds and stipes before drying and milling; other species were dried as whole plants. The figures given refer to samples collected in January. Seasonal variations are considerable, particularly in the *Laminaria* fronds (Haug and Jensen 1954), and the figures given must not be considered as typical for all samples of these species.

V. Discussion

It is very difficult to assess the accuracy of the mannitol method. In principle, other carbohydrate constituents of the algae will be attacked by the periodic acid, thus leading to an overestimation of the mannitol content. These reactions are, however, generally slower than the oxidation occurring with the contiguous hydroxyls of the sugar alcohols, except for the reducing end of a polysaccharide, and the systematic error is probably fairly low.

A series of complicated chemical reactions will occur during the ashing of the seaweed samples. Under the conditions used, most of the cations in the sample will be found in the ash as carbonates in which the carbon atom originates from organic substances. Combustion of the sulfated polysaccharide in the sample will liberate H_2SO_4, which in turn displaces carbonate and chloride ions. It is easy to demonstrate that a considerable loss of chloride ions takes place during ashing. The ash analysis will, therefore, hardly give a correct picture of the amount of inorganic components of the algae.

The fiber method determines the amount of nonextractable organic material in the sample under the conditions specified. As very little information is available on the chemistry of this fraction, the results must be considered only as an estimate of nonextractable organic components, not necessarily of cellulose.

The method recommended for iodine is reliable and without interference from the large excess of chloride ions or from other constituents of the algae. The crucial step in the procedure is the alkaline fusion, which must be carried out very thoroughly.

VI. References

Baardseth, E., and Haug A. 1953. Individual variation of some constituents in brown algae and reliability of analytical results. *Nor. Inst. Tang-Tareforsk. Rep.* 2, 1–23.

Cameron, M. C., Ross, A. G., and Percival E. G. V. 1948. Methods for the routine estimation of mannitol, alginic acid, and combined fucose in seaweeds. *J. Soc. Chem. Ind. (London)* 67, 161–4.

Haug, A., and Jensen, A. 1954. Seasonal variations in the chemical composition of *Alaria esculenta, Laminaria saccharina, Laminaria hyperborea* and *Laminaria digitata* from northern Norway. *Nor. Inst. Tang-Tareforsk. Rep.* 4, 1–14.

Haug, A., and Larsen, B. 1958. Phenolic compounds in brown algae. I. The presence of reducing compounds in *Ascophyllum nodosum* (L.) Le Jol. *Acta Chem. Scand.* 12, 650–7.

Larsen, B., and Haug, A. 1961. The distribution of iodine and other constituents in stipe of *Laminaria hyperborea* (Gunn.) Foslie. *Bot. Mar.* 2, 250–4.

18: Quantitation of the macromolecular components of microalgae

GARY KOCHERT

Botany Department,
University of Georgia, Athens, Georgia 30602

CONTENTS

I Introduction   *page* **190**
II Methods   **190**
  A. Extraction of low-molecular-weight components   190
  B. Extraction of lipids   191
  C. Determination of lipids   191
  D. Determination of proteins and carbohydrates   192
  E. Determination of DNA   192
  F. Determination of RNA   193
III Discussion   **193**
  A. Acid-soluble components   193
  B. Lipids   194
  C. Nucleic acids   194
IV References   **194**

I. Introduction

The purpose of this chapter is to describe methods whereby the major macromolecular components of microalgal cells can be quantitatively determined. Unfortunately, there is no single procedure that will work for all organisms. The procedure described below is a good starting point. If difficulties are encountered, one should consult the voluminous literature in the field (Hutchison and Munro 1961; Ingle 1963; Munro and Fleck 1966; Dittmer and Wells 1969; Stewart 1975).

Nearly all procedures for the determination of macromolecular components involve a chemical fractionation of the cells to remove substances that would interfere with subsequent chemical determinations. A typical first step is an extraction with cold dilute acid to remove small molecules. Lipids and phospholipids are then removed by extraction with organic solvents. The acid-washed, lipid-free precipitate can be used directly for the estimation of protein and carbohydrate, but additional chemical fractionation is necessary to separate the nucleic acids from contaminants that interfere with the standard colorimetric assays. Methods for the extraction of undegraded nucleic acids are not quantitative. Extreme conditions, which lead to degradation of the nucleic acids, are needed for quantitative extraction.

II. Methods

A. Extraction of low-molecular-weight components

Collect enough cells by centrifugation to yield a packed cell volume of about 0.5 ml. Three samples need to be collected. Discard the supernatant. Place the tubes containing the cell pellets on ice and add 10 ml of ice-cold 0.2 N $HClO_4$. Thoroughly resuspend the cell pellets by vortexing or inversion. After 15 min at 4 C, centrifuge the sample in a refrigerated centrifuge and carefully remove the supernatant. (This and the subsequent supernatant should be retained if it is desired to quantitate low-molecular-weight components.) Repeat the extraction with a further 10 ml of 0.2 N $HClO_4$. The supernatants contain inorganic phosphate, nucleotides, amino acids, sugars, and a great variety

[190]

of other low-molecular-weight components. The pellets contain the macromolecular components. Methods for quantitative determination of low-molecular-weight components in $HClO_4$, alcohol, or hot-water extracts are described or referred to in Stewart (1975). See also Chp. 28 and Roberts et al. (1963) for detailed methods of extraction and analysis of such metabolites.

B. Extraction of lipids

Add 10 ml of chloroform-methanol solution (2:1 v/v) to the pellets from the $HClO_4$ extraction. Resuspend the pellets and allow the suspension to stand 5 min at room temperature. Centrifuge the samples and remove and retain the supernatants. (The pellet is poorly compacted because of the high density of the chloroform-methanol solution. If difficulty is encountered in removing the supernatant, add more methanol up to 1:1 chloroform:methanol.) Repeat the extraction with an additional 5 ml of chloroform-methanol. Combine the supernantants and retain for lipid determinations. Save the acid-extracted, lipid-free pellets for determination of nucleic acids, protein, and carbohydrate.

C. Determination of lipids

1. Reagent. Palmitic acid (United States Biochemical Corp.).

2. Solutions. (a) Lipid standard solution. Dissolve palmitic acid in chloroform to a final concentration of 1 mg ml^{-1}. (b) Dichromate solution. Dissolve 2.5 g of $K_2Cr_2O_7$ in 1 l of conc. H_2SO_4.

3. Procedure

a. Add 0.2 volumes of water to the combined chloroform-methanol extracts of the cells. Shake the solution for 5 min to mix well and centrifuge to separate the phases. Collect the organic (lower) phase, leaving behind a precipitate that forms at the interphase. Discard the aqueous (upper) phase.

b. Evaporate the chloroform-methanol solution under a stream of N_2 to a final volume of 2 ml.

c. Transfer 0.05, 0.1, 0.15, 0.20, 0.25, and 0.30 ml of the lipid standard solution to marked 5- or 10-ml screw-capped tubes. Transfer 0.1, 0.2, and 0.5 ml of the unknown lipid sample to marked tubes. Evaporate all tubes to dryness under vacuum or a stream of N_2.

d. Add 2 ml of dichromate solution to all tubes and cap with Teflon-lined caps. Place all tubes in a boiling-water bath for 45 min. Shake the tubes two or three times during the heating.

e. Cool the tubes, remove 1.0 ml of each, and dilute to 10.0 ml with water. Read the absorbance of each tube at 350 nm against an H_2O

blank. Plot a standard curve with the known lipid samples and determine the unknowns graphically or by Beer's law. Note that the assay is based on the disappearance of absorbance at 350 nm as the dichromate is reduced by increasing amounts of lipids. It is thus convenient to plot the standards as the reciprocal of absorbance against lipid concentration.

D. Determination of proteins and carbohydrates

1. Add 1 ml of 1 N NaOH to one of the acid-extracted, lipid-free pellets and heat 10 min in a boiling-water bath to dissolve the pellet.

2. Assay aliquots of the sample for protein by the dye-binding assay (Chp. 9) and for carbohydrate by the phenol–sulfuric acid assay (Chp. 10).

E. Determination of DNA

1. Reagents. Calf thymus DNA and diphenylamine (Sigma).

2. Solutions. (a) DNA standard solution. Calf thymus DNA is dissolved in 5 mM NaOH at a concentration of 0.4 mg ml^{-1}. A working solution is made every 3 weeks by mixing an aliquot with an equal volume of N $HClO_4$ and heating at 70 C for 45 min. (b) Aqueous acetaldehyde. Acetaldehyde is cooled and 1 ml is transferred in a cooled pipette into 50 ml of H_2O. This solution is stable at 4 C for several months (acetaldehyde is flammable and should be stored in the cold and thoroughly chilled before opening). (c) Diphenylamine solution. Dissolve 1.5 g of diphenylamine in 100 ml of glacial acetic acid and add 1.5 ml of conc. H_2SO_4. Stable at 4 C for up to 3 months. Just before use add 0.1 ml of aqueous acetaldehyde per 20 ml of reagent.

3. Procedure
  a. Add 5 ml of 0.5 N $HClO_4$ to another of the acid-extracted, lipid-free pellets and resuspend the sample by vortexing or inversion. Incubate the sample at 70 C for 45 min. Centrifuge and remove the supernatant.
  b. Transfer 0.1, 0.2, 0.3, 0.4, and 0.5 ml of the calf thymus DNA stock solution to marked screw-cap test tubes and adjust the volume to 2 ml with 0.5 N $HClO_4$. Include a tube containing 2 ml of 0.5 N $HClO_4$ to serve as a reagent blank.
  c. Transfer 0.5, 1.0, and 2.0 ml of the unknown DNA sample to marked screw-cap test tubes and adjust all volumes to 2 ml with 0.5 N $HClO_4$.
  d. Add 4 ml of the diphenylamine reagent to all tubes and mix well. Incubate all tubes at 30 C for 16–20 h and read absorbance at 600 nm. Plot a standard curve with the standards and determine unknown DNA concentrations.

F. Determination of RNA

1. Reagent. Purified yeast RNA (Worthington). Orcinol (Calbiochem). The orcinol is purified by dissolving in boiling benzene (caution) decolorizing with charcoal, and crystallizing after adding hexane.

2. Solutions. (a) RNA standard solution. Dissolve yeast RNA in water to a concentration of 100 μg ml^{-1}. Store frozen. (b) Cupric ion solution. Dissolve 0.15 g of $CuCl_2 \cdot 2H_2O$ in 100 ml of conc. HCl. Stable at room temperature. (c) Orcinol stock solution. Dissolve 12.5 g of recrystallized orcinol to a final volume of 25.0 ml in 95% ethanol. (d) Orcinol reagent. Mix 2 ml of orcinol stock solution with 100 ml of cupric ion solution. This solution should be freshly prepared for each determination.

3. Procedure

a. Add 2 ml of 0.3 N KOH to the remaining acid-extracted, lipid-free pellet and incubate for 18–24 h at 30 C.

b. After the hydrolysis period, cool the sample on ice and acidify the solution to pH 1.0 (pH paper) with conc. $HClO_4$. Centrifuge and carefully remove the supernatant. Wash the pellet with 1 ml of cold 0.2 N $HClO_4$. Combine the supernatants.

c. Transfer 0.2, 0.4, 0.8, 1.0, 1.5, and 2.0 ml of the RNA standard solution to marked screw-cap tubes. Transfer 0.1, 0.2, and 0.5 ml of the unknown RNA solution to marked tubes and adjust the volume of each to 2.0 ml with H_2O. Include a blank containing 2.0 ml of H_2O.

d. Add 2 ml of the orcinol reagent to each tube, seal with Teflon-lined caps, and place the tubes in a boiling-water bath. After boiling them for 35 min, cool the tubes in running water and read the absorbance of each tube at 665 nm. Plot the absorbance of the RNA standards against the RNA concentration to obtain a standard curve. Determine the concentration of the unknowns against the standard curve.

III. Discussion

A. Acid-soluble components

If it is desired to quantitate acid-soluble, low-molecular-weight components, several procedures are available. The ninhydrin reaction (Moore and Stein 1948) can be used to quantitate amino acids and peptides, but will also react with other compounds such as amino sugars. Reducing sugars can be detected by the phenol–sulfuric acid assay (Chp. 10). Nucleotides are best determined after fractionation on ion-exchange columns (Ingle 1963).

B. Lipids

The method described here is a nonspecific, general method for total lipids. It is based on the ability of lipids to reduce an acid dichromate solution (Amenta 1964). The lipid fraction of cells is a very complex mixture of compounds with diverse chemical properties, and not all react with the same intensity in this assay. Fractionation into classes is thus necessary for really accurate determinations. Detailed information on lipid characterization is covered in Chp. 11.

C. Nucleic acids

1. DNA. The most satisfactory method for DNA determinations of microalgal cells is the fluorometric method of Holm-Hansen et al. (1968). This method (Chp. 8) is very sensitive and is least subject to interference by contaminants. I have described the classic diphenyl-amine reaction here as an alternative if a fluorometer is not available. Protein in DNA samples will interfere with this assay. This is why the DNA is extracted with hot $HClO_4$; most proteins are insoluble under these conditions. Further information on this method may be found in Burton (1968).

2. RNA. RNA determinations present the greatest problem likely to be encountered in analysis of algal cells. The method described here is that of Lin and Schjeide (1969). The main problem countered will be interference by the carbohydrates, which are abundant in plant extracts. These may cause spuriously high RNA concentrations. If such a difficulty is encountered, it will be necessary to purify the nucleotide fraction before analysis. A simple ion-exchange procedure is described by Ingle (1963).

IV. References

Amenta, J. S. 1964. A rapid chemical method for quantification of lipids separated by thin-layer chromatography. *J. Lipid Res.* 5, 270–2.

Burton, K. 1968. Determination of DNA concentration with diphenylamine. *Methods Enzymol.* 12B, 163–6.

Dittmer, J. C., and Wells, M. A. 1969. Quantitative and qualitative analysis of lipids and lipid components. *Methods Enzymol.* 14, 482–530.

Holm-Hansen, O., Sutcliffe, W. H., Jr., and Sharp, J. 1968. Measurements of deoxyribonucleic acid in the ocean and its ecological significance. *Limnol. Oceanogr.* 13, 507–14.

Hutchison, W. C., and Munro, H. N. 1961. The determination of nucleic acids in biological materials: A review. *Analyst* 86, 768–813.

Ingle, J. 1963. The extraction and estimation of nucleotides and nucleic acids from plant material. *Phytochemistry* 2, 353–70.

Lin, R. I., and Schjeide, O. A. 1969. Micro-estimation of RNA by the cupric ion catalyzed orcinol reaction. *Anal. Biochem.* 27, 473–83.

Moore, S., and Stein, W. H. 1948. Photometric ninhydrin method for use in the chromatography of amino acids. *J. Biol. Chem.* 176, 367–88.

Munro, H. N., and Fleck, A. 1966. The determination of nucleic acids. *Methods Biochem. Anal.* 14, 113–75.

Roberts, R. B., Cowie, D. B., Abelson, P. H., Bolton, E. T., and Britten, R. J. 1963. *Studies of Biosynthesis in Escherichia coli. Carnegie Institution of Washington Publication* 607. Carnegie Institution, Washington, D.C. 521 pp.

Stewart, P. R. 1975. Analytical methods for yeast. *Methods Cell Biol.* 12, 111–47.

19: ATP, ADP, and AMP determinations in water samples and algal cultures

DAVID M. KARL AND
OSMUND HOLM-HANSEN

Institute of Marine Resources, A-018,
Scripps Institute of Oceanography, University of California, San Diego,
La Jolla, California 92093

CONTENTS

I	**Introduction**	*page* **198**
II	**Equipment and reagents**	**199**
	A. Equipment	199
	B. Reagents	199
III	**Methods**	**200**
	A. Sample collection	200
	B. Extraction	200
	C. Enzyme preparation	201
	D. Standard ATP solutions	201
	E. ATP determinations	201
	F. ATP, ADP, and AMP determinations	202
IV	**Calculations**	**203**
	A. ATP data	203
	B. ATP, ADP, and AMP data, and calculation of adenylate energy charge	203
V	**Capabilities**	**203**
	A. Sensitivity	203
	B. Accuracy and reproducibility	204
VI	**Limitations and troubleshooting**	**204**
	A. Internal standards	204
	B. Enzyme specificity	204
VII	**Acknowledgments**	**205**
VIII	**References**	**205**

I. Introduction

The ubiquitous nature of adenine nucleotides (ATP, ADP, and AMP) in living cells, and their fundamental importance as intermediates linking energy-yielding and energy-requiring functions of the cell, have provided an impetus for the development of analytical methods for their quantitative determination (Holm-Hansen 1970; Holm-Hansen and Paerl 1972). Among the various techniques for separating and measuring ATP are (1) precipitation of insoluble salts and assay by acid hydrolysis of the isolated material (LePage 1945); (2) chromatographic separations, including paper (Cohn and Carter 1950), thin-layer ion-exchange (Randerath 1966), and high-pressure liquid chromatography (Brown 1970); (3) radioisotopic assay procedures involving enzymatic transfer of the ^{32}P-labeled terminal phosphate (Schneider 1969); (4) hexokinase enzyme coupling assay (Strehler and Totter 1954); (5) adenylic deaminase enzyme coupling assay (Kalckar 1947); (6) firefly luciferase enzyme coupling assay (Strehler and Totter 1952).

Each method has its advantages and disadvantages, but the firefly bioluminescent reaction is by far the most specific, sensitive, and reproducible assay procedure known. This method makes use of the fact that firefly luciferin and luciferase react with ATP in the presence of Mg^{2+} and oxygen to yield one photon of light for every ATP molecule hydrolyzed (McElroy et al. 1969). This procedure for the quantitative determination of ATP has had wide application in phycological research, including the study of energy-transfer and energy-coupling mechanisms (both oxidative phosphorylation and photosynthesis), the regulation of enzyme function and biosynthesis, cell viability studies, phosphorus metabolism, and various ecological studies. This chapter describes the complete experimental procedures and equipment necessary for the measurement of ATP, ADP, and AMP, and discusses briefly some of the analytical principles and limitations of these techniques.

II. Equipment and reagents

A. Equipment

1. Filtration. (a) A side-armed filter flask (1- to 4-l capacity), provided with a Millipore or Gelman filter base and funnel for 25-mm diam. filters. (b) An adjustable vacuum source with a vacuum gauge. (c) Microfine glass fiber (Reeve-Angel, 984-H) or membrane (Millipore 0.45 μm pore size, HA) filters, 25 mm diam.

2. Boiling-water bath. To contain a test tube rack.

3. Special glassware. All glassware used for adenine nucleotide analyses should be cleaned in a hot chromic acid bath (1 h, 80 C) and rinsed thoroughly (10–20 times) with distilled water. Alternatively, glassware rinsed in clean distilled water can be baked in a muffle furnace (500 C, 2–3 h) to eliminate contamination. (a) Pyrex glass test tubes (15 × 100 mm) calibrated at 5-ml intervals. (b) Pyrex glass Erlenmeyer flasks, 50 and 250 ml. (c) Glass scintillation vials or round-bottom glass vials (18 × 45 mm). (4) Disposable glass culture tubes (12 × 75 mm) for ADP and AMP analyses.

4. Stainless-steel forceps

5. Pipettes. (a) Glass or disposable plastic pipettes, 5 and 10-ml. (b) Automatic pipettes of 50- and 200-μl volumes, with disposable tips. (c) Automatic pipette to deliver 0.1–1.0 ml, with disposable tips.

6. Centrifugation. (a) A microcentrifuge is required to concentrate enzyme suspensions. It should be capable of handling volumes between 0.1 and 0.5 ml. (b) A clinical centrifuge with swinging rotor to accommodate 6–8 of the 15 × 100-mm test tubes.

7. Incubation. A water bath is required for ADP and AMP analyses to maintain a uniform temperature (30 C) during the enzyme reaction period.

8. Instrumentation for quantitative light detection. Amino Chem-Glow Photometer (American Instrument), Du Pont 760 Luminescence Biometer (Du Pont), or ATP Photometer, Model 2000 (SAI Technology).

9. Strip chart recorder or oscilloscope (optional)

10. Automatic pipette for peak height analysis

B. Reagents

(1) Tris(hydroxymethyl)aminomethane buffer, 0.02 M, pH 7.75. (2) K_2HAsO_4 buffer, 0.1 M, pH 7.4. (3) $MgSO_4$ solution, 0.04 M. (4)

Adenosine-5'-triphosphate (ATP), disodium salt from equine muscle (Sigma). (5) Firefly lantern extract, FLE-50 (Sigma). (6) Pyruvate kinase suspension, rabbit muscle (Sigma). (7) Myokinase suspension, rabbit muscle (Sigma). (8) Potassium phosphate buffer, 75 mM, pH 7.3, containing 15 mM $MgCl_2$. (9) Potassium phosphate buffer, 75 mM, pH 7.3, containing 15 mM $MgCl_2$ and 0.5 mM phosphoenolpyruvate (trisodium salt). (10) Synthetic D-luciferin, crystalline (Sigma) (*optional*).

III. Methods

A. Sample collection

1. Subsamples from algal cultures should be gently pipetted into a clean Erlenmeyer flask and extracted immediately.
2. Environmental samples may be collected using any clean, nontoxic water sampler. Caution should be taken to reduce light and temperature shock to naturally occurring phytoplankton populations.

B. Extraction

Whenever possible, between 50 and 200 μl of the algal culture should be pipetted directly into 5 ml of boiling TRIS buffer. When direct pipetting methods are employed, it is essential also to monitor the ATP, ADP, and AMP that may be dissolved in the growth medium. Filter a portion of the culture and then directly extract between 50 and 200 μl of the filtrate. This must be done for each sampling period. If the cell density is too low, samples may be concentrated by vacuum filtration (do not exceed 25 cm Hg). Centrifugation is not recommended. Membrane filters are satisfactory (Millipore, HA), although microfine glass fiber filters (Reeve Angel, 984-H) have the advantage of a much faster flow rate. When approximately 1–2 ml of liquid remains in the filter funnel, the vacuum pump should be turned off, and the residual vacuum in the filter flask will complete the filtration operation. *Caution:* Do not suck the filter dry. Immediately after the last portion of liquid is filtered, quickly transfer the filter into a test tube containing 5 ml of boiling TRIS buffer using stainless-steel forceps. *Caution:* It is essential that the TRIS buffer is at a temperature of 100 C. Extract the filters for 3–5 min, remove the test tubes from the boiling-water bath, and allow to cool. After the extraction has been completed, restore the volume to 5 ml with distilled water and store the extracts frozen (− 20 C) for subsequent analysis. Frozen extracts are stable for up to 6 months.

C. Enzyme preparation

Each vial of lyophilized firefly lantern extract (50 mg/vial) is reconstituted by adding 5 ml of distilled water, 10 ml of 0.1 M K_3AsO_4 buffer (pH 7.4), and 10 ml of 0.04 M $MgSO_4$. The enzyme mixture is then allowed to stand for 6–8 h at 23–25 C to reduce the level of endogenous ATP. After aging, the insoluble residue is removed by vacuum filtration or centrifugation. The clear filtrate or supernatant is transferred to a clean flask, and the enzyme is ready for use.

D. Standard ATP solutions

For each enzyme preparation (1 or more vials), a standard curve must be prepared. Stock ATP solutions containing 1 μg ATP ml^{-1} (ca. 2 μM) are prepared in TRIS buffer and stored frozen (-20 C) in 5-ml aliquots. Working solutions are prepared on the day of each assay by diluting the stock solution with TRIS buffer, providing 6–7 standards ranging from 0.5 ng ATP ml^{-1} (ca. 1 nM) to 50 ng ATP ml^{-1} (ca. 100 nM). If a large number of samples is analyzed in a single day, a standard curve should be analyzed at the beginning of the assay procedure and again every 35–40 samples. If fewer than 40 samples are analyzed, a single standard curve (measured halfway through the experiment) is sufficient.

E. ATP determinations

ATP can be assayed either by measuring the peak intensity of light emission or by integrating a portion of the subsequent light decay curve. All frozen samples should be thawed in a water bath (25 C) and centrifuged for 5 min at 1800 $\times$ g prior to analysis. *Caution:* All ATP assays should be conducted in reduced illumination, and fluorescent lighting should be avoided.

1. Peak height measurement. Pipette 1 ml of the enzyme preparation into a round-bottom glass vial. Place the vial into the photometer and record the background light emission (0–6 sec). Position the vial in the photometer. Using the automatic pipette, withdraw 0.2 ml of the sample extract or ATP standard solution and inject it into the enzyme mixture. Record the peak height digital display. Blanks are determined by injecting 0.2 ml of TRIS buffer into the enzyme mixture as described above.

2. Integrated measurement. Pipette 0.5 ml of the enzyme preparation into a clean glass scintillation vial and record the background light emission. Remove the vial from the photometer and inject 0.5 ml of sample or ATP standard solution. Mix the vial thoroughly and place into the photometer. The light flux is then integrated (either electronically or manually by plotting the kinetics with a strip chart recorder and measuring the area under the curve with the use of a planimeter) over a predetermined period of the subsequent decay curve (i.e., 15–75 sec after injection of the sample). Record digital readout or calculated area. Blanks are determined by injecting 0.5 ml of TRIS buffer into the enzyme preparation as described above. *Note:* Although commercial ATP photometers (see II.A.8) are superior in terms of convenience and ease of operation, the light emitted from the firefly bioluminescent reaction can be measured using any one of a number of light-detecting instruments (Strehler 1968).

F. ATP, ADP, and AMP determinations

Each sample extract is first assayed for ATP, then after the appropriate enzymatic conversions of ADP and AMP to equivalent concentrations of ATP, the samples are reassayed. For ATP determinations, 0.2 ml of each sample extract is pipetted into a culture tube (12×75 mm) containing 50 μl of a solution of 75 mM potassium phosphate buffer (pH 7.3) and 15 mM $MgCl_2$. For ATP plus ADP determinations, 0.2 ml of each extract is pipetted into a culture tube containing 50 μl of a solution of 75 mM potassium phosphate buffer (pH 7.3), 15 mM $MgCl_2$, 0.5 mM phosphoenolpyruvate, and 20 μg pyruvate kinase. For ATP plus ADP plus AMP determinations, 0.2 ml of each sample is pipetted into a culture tube containing 50 μl of a solution of 75 mM potassium phosphate buffer (pH 7.3), 15 mM $MgCl_2$, 0.5 mM phosphoenolpyruvate, 20 μg pyruvate kinase, and 25 μg myokinase (adenylate kinase). Standard ATP solutions should be prepared at the same time as the samples in a manner identical to each of the three procedures described above. *Note:* Prior to use, the pyruvate kinase and myokinase enzyme suspensions should be centrifuged ($2000 \times g$, 3–5 min) and the pellet resuspended in an equivalent volume of 75 mM potassium phosphate buffer (pH 7.3).

All samples are then incubated at 30 C for 30 min and then placed in a boiling-water bath for 2 min to deactivate the pyruvate kinase enzyme. This procedure prevents the rapid production of ATP from ADP present in the crude enzyme preparations. When the samples cool to room temperature, they are ready for analysis by the peak height method described previously. Enzyme blanks (TRIS buffer plus enzyme solutions) should also be determined, as certain enzyme preparations contain contaminating adenine nucleotides. *Note:* The peak emission mode is essential when measuring ADP and AMP. Reliable peak height measurements require specially designed equipment for rapid and reproducible injection velocities. To our knowledge, only three instruments are commercially available for such ATP determinations (see II.A.8).

IV. Calculations

A. ATP data

Determine the net emission by subtracting the enzyme blank value from each of the sample or standard measurements. A standard curve is then prepared by plotting net peak intensities or integrated areas on the ordinate and ATP concentrations on the abscissa. From this curve, the ATP concentration in sample extracts can be calculated, and by correcting for the proportion of the sample actually assayed and the volume of medium originally extracted, the ATP ml^{-1} (or ATP cell^{-1}, ATP chl a^{-1}, etc.) of the original culture or natural water sample can be determined.

B. ATP, ADP, and AMP data, and calculation of adenylate energy charge

Determine the net light emission by subtracting the appropriate blank values from each of the sample determinations. Standard curves are prepared by plotting net peak intensities on the ordinate and ATP concentrations on the abscissa for each of the three sets of standard data. From these curves, the ATP concentration in each of the three different reaction mixtures (ATP, ATP plus ADP, and ATP plus ADP plus AMP) can be determined. ADP and AMP concentrations are calculated by the difference between these measured values. The adenylate energy charge, as defined by Atkinson and Walton (1967), is determined from the measured adenylate concentrations.

$$\text{energy charge} = \frac{[\text{ATP}] + \frac{1}{2}[\text{ADP}]}{[\text{ATP}] + [\text{ADP}] + [\text{AMP}]} \tag{1}$$

Note: In addition to the data-reduction methods described above, at least two computer-assisted techniques for processing ATP sample data have recently been published (Booth 1975; Erkenbrecher et al. 1976). Both these methods, however, require integrated light flux measurements and are, therefore, not capable of calculating ADP and AMP concentrations or adenylate energy charge ratios in their present form.

V. Capabilities

A. Sensitivity

The lower limit of ATP detectability is influenced by the amount of endogenous ATP in the enzyme preparation (enzyme blank) and by the sensitivity and thermionic emission (tube noise) of the photomultiplier tube. With a commercial ATP photometer (SAI Technology), the firefly bioluminescent reaction procedure will detect ATP concentrations as low as 0.1 ng ml⁻¹ (ca. 0.2 nM) using crude enzyme preparations (FLE-50). If synthetic D-luciferin is added to saturate the enzyme system, the sensitivity can be extended nearly 1000-fold (Karl and Holm-Hansen 1976). When low concentrations of adenine nucleotides (<1 ng ml⁻¹) are measured, extreme caution should be taken to ensure that all glassware has been properly cleaned (see II.A.3).

B. Accuracy and reproducibility

The maximum statistical variation for the extraction and assay of ATP from environmental samples and algal cell suspensions is $\pm10\%$ at the 95% confidence interval for samples containing 12 ng ATP ml⁻¹ (ca. 24 nM). For a given operator, variation in standard sample analyses is generally $\pm5\%$.

VI. Limitations and troubleshooting

A. Internal standards

The emitted light in the firefly bioluminescent reaction is influenced by the presence of interfering cations and anions, turbidity, and the color of the final cell extracts. The magnitude of this combined interference can be determined, and corrected for, through the use of an ATP internal standard. The chemical ATP standard (in TRIS buffer) can be added at any point in the extraction procedure. The concentration of the added standard should be approximately equal to the measured concentration in the extract. To determine the percent recovery of the internal standard, calculate the ATP concentration in both extracts (i.e., sample and sample plus internal standard) and use the following relationship:

$$\% \text{ recovery} = \frac{[\text{sample} + \text{internal standard}] - [\text{sample}]}{[\text{internal standard}]} \times 100 \quad (2)$$

To correct for all sources of interference, divide the final measured ATP concentration by the decimal equivalent of the percent recovery.

B. Enzyme specificity

Highly purified firefly luciferase responds only to ATP (McElroy and Green 1956), but most commercial reagents contain contaminating enzymes that will produce ATP from nucleotide triphosphate precursors in the sample extracts. The contribution of light emission from these additional reactions can be reduced by measuring the peak height of light emission rather than using the integral mode of light detection.

VII. Acknowledgments

We thank Drs. A. F. Carlucci and R. W. Eppley for their helpful comments during the preparation of this manuscript. Much of the methodology described in this chapter was developed under ERDA Contract No. EY-76-C-03-0010, PA 20. We thank Ms. Janice Stearns for typing the manuscript.

VIII. References

Atkinson, D. E., and Walton, G. M. 1967. Adenosine triphosphate conservation in metabolic regulation: Rat liver citrate cleavage enzyme. *J. Biol. Chem.* 242, 3239–41.

Booth, C. R. 1975. Instrument development and data processing for the ATP photometer. In Borun, G. A. (ed.). *ATP Methodology Seminar,* pp. 104–30. SAI Technology, San Diego, California.

Brown, P. R. 1970. The rapid separation of nucleotides in cell extracts using high-pressure liquid chromatography. *J. Chromatogr.* 52, 257–72.

Cohn, W. E., and Carter, C. E. 1950. The separation of adenosine polyphosphates by ion exchange and paper chromatography. *J. Am. Chem. Soc.* 72, 4273–5.

Erkenbrecher, C. W., Crabtree, S. J., and Stevenson, L. H. 1976. Computer-assisted analysis of adenosine triphosphate data. *Appl. Environ. Microbiol.* 32, 451–4.

Holm-Hansen, O. 1970. ATP levels in algal cells as influenced by environmental conditions. *Plant Cell Physiol.* 11, 689–700.

Holm-Hansen, O., and Paerl, H. W. 1972. Applicability of ATP determination for estimation of microbial biomass and metabolic activity. *Mem. Ist. Ital. Idrobiol. Dott. Marco de Marchi Pallanza Italy* 29 (Suppl.), 149–68.

Kalckar, H. M. 1947. Differential spectrophotometry of purine compounds by means of specific enzymes. *J. Biol. Chem.* 167, 445–59.

Karl, D. M., and Holm-Hansen, O. 1976. Effects of luciferin concentration on the quantitative assay of ATP using crude luciferase preparations. *Anal. Biochem.* 75, 100–12.

LePage, G. A. 1945. Methods for the analysis of phosphorylated intermediates. In Umbreit, W. W., Burris, R. H., and Stauffer, J. F. (eds.). *Manometric Techniques,* 1st ed., pp. 268–81. Burgess, Minneapolis.

McElroy, W. D., and Green, A. 1956. Function of adenosine triphosphate in the activation of luciferin. *Arch. Biochem. Biophys.* 64, 257–71.

McElroy, W. D., Seliger, H. H., and White, E. H. 1969. Mechanism of bioluminescence, chemiluminescence and enzyme function in the oxidation of firefly luciferin. *Photochem. Photobiol.* 10, 153–70.

Randerath, K. 1966. *Thin-Layer Chromatography.* Academic Press, New York. 285 pp.

Schneider, P. B. 1969. An enzymatic assay for adenosine-5'-triphosphate (ATP) and other nucleotide triphosphates and determination of the specific radioactivity of the terminal P. *Anal. Biochem.* 28, 76–84.

Strehler, B. L. 1968. Bioluminescence assay: Principles and practice. *Methods Biochem. Anal.* 16, 99–181.

Strehler, B. L., and Totter, J. R. 1952. Firefly luminescence in the study of energy transfer mechanisms. I. Substrate and enzyme determination. *Arch. Biochem. Biophys.* 40, 28–41.

Strehler, B. L., and Totter, J. R. 1954. Determination of ATP and related compounds. *Methods Biochem. Anal.* 1, 341–56.

Section III

Enzymes

20: Ribulose-bis-phosphate carboxylase in Chlamydomonas

S. S. BADOUR

Department of Botany,
University of Manitoba, Winnipeg, Manitoba R3T 2N2, Canada

CONTENTS

I Introduction   *page* 210
II Materials and test organism   210
     A. Equipment   210
     B. Reagents   211
     C. Test organism   211
III Method   211
     A. Preparation of solutions   211
     B. Preparation of cell-free extracts   212
     C. Enzyme assay   212
     D. Calculations   213
IV Sample data and discussion   214
     A. Possible problem areas   214
     B. Alternative techniques   215
     C. Ribulose-bis-phosphate
        oxygenase   215
V Acknowledgments   215
VI References   215

I. Introduction

Ribulose-1,5-bis-phosphate (RuBP) carboxylase (3-phospho-D-gly-cerate carboxy-lyase [dimerizing],EC 4.1.1.39) is the fundamental carboxylation enzyme in photosynthetic organisms. The enzyme catalyzes the Mg^{2+}-dependent carboxylation of RuBP to form two molecules of 3-phosphoglycerate (Paulsen and Lane 1966).

RuBP carboxylase is a very abundant protein in algal cells. Rabinowitz et al. (1975) found that this chloroplast protein accounts for ca. 9% of the total protein in *Euglena gracilis*. The interthylakoidal crystals (0.3–3 μm in diam.) observed in *Chlamydomonas segnis* (Badour et al. 1973) may contain RuBP carboxylase. Codd and Stewart (1976) have provided evidence that the polyhedral bodies (carboxysomes) of heterocystous nitrogen-fixing blue-green algae contain RuBP carboxylase.

The RuBP carboxylase activity in microalgae varies in response to changes in environmental conditions (Morris and Farrell 1971; Beardall and Morris 1975), indicating that it may play an important role in biochemical adaptations. Furthermore, the enzyme offers an excellent system for investigating the molecular mechanisms controlling the synthesis of chloroplast proteins in microalgae (Givan and Criddle 1972; Davis and Merrett 1975) and the modulation of synthetic control through phylogenetic evolution (McFadden 1973; Lord et al. 1975; Rabinowitz et al. 1975).

II. Materials and test organism

A. Equipment

In addition to the usual glassware, centrifuge tubes (15 ml), micropipettes (Lang-Levy 20 and 100 μl or the SMI Micro/pettor), and serum bottles (15–30 ml) with stoppers are required.

Other requirements include an analytical balance, pH meter, magnetic hot plate stirrer, a refrigerator with a freezer compartment, a laboratory centrifuge, and a temperature-controlled water bath. Special equipment needed includes a high-speed centrifuge such as the

RC-5 automatic superspeed refrigerated centrifuge (Sorvall, DuPont Company), equipped with the proper rotors for polypropylene centrifuge tubes (50 ml) and bottles (250–500 ml) with plastic sealing caps; French press (American Instrument Co.) or a sonicator (such as Insonator, model 1000 with a microtip standard, Ultrasonic Systems); and a liquid scintillation counter.

B. Reagents

Chemicals used are reagent grade and include the following: (1) Tris-hydroxymethyl-aminomethane (TRIS). (2) HCl, 2 M. (3) Ethylenediamine tetraacetate, disodium salt (EDTA). (4) DL-dithiothreitol (DTT). (5) Glutathione, reduced form (GSH). (6) $MgCl_2 \cdot 6H_2O$. (7) $NaHCO_3$. (8) $NaH^{14}CO_3$ (Amersham/Searle, specific activity, 60 mCi $mmol^{-1}$). (9) D-ribulose-1,5-bis-phosphate (RuBP), tetrasodium salt (Sigma). (10) Tricholoroacetic acid (TCA), 24% w/v.

C. Test organism

Chlamydomonas segnis Ettl (a culture has been deposited at UTEX) was grown autotrophically in the mineral nutrient medium of Kuhl and Lorenzen (1964), in which 10 mM NH_4Cl and 10 mM KCl instead of 10 mM KNO_3 were used. In order to maintain the pH between 6.5 and 7.0 during growth, the medium was enriched with 5 mM K_2HPO_4. The cultures were aerated with 5% vol. CO_2 in air and exposed to light (11 klux) in a growth chamber at 25 C. Further information on the morphology and growth of this organism is given in Badour et al. (1973).

III. Method

A. Preparation of solutions

1. Homogenization buffer. This contains 100 mM TRIS-HCl pH 8, 1 mM EDTA, and 1 mM dithiothreitol, and is prepared with double-distilled water as follows: (a) TRIS (3.02 g) is dissolved in 200 ml water and adjusted to pH 8 with 2 M HCl. (b) EDTA (93.0 mg) and 38.6 mg dithiothreitol are dissolved in TRIS-HCl buffer, and the solution is made to 250 ml with H_2O, stored at 2–4 C, and used for cell homogenization.

2. Substrates and cofactors. Prepare 10-ml stock solutions of 1 M TRIS-HCl buffer pH 8, 1 mM EDTA, 80 mM $MgCl_2 \cdot 6H_2O$, 50 mM GSH, 200 mM $NaH^{14}CO_3$ (0.4 μCi mol^{-1}). Prepare 5 ml of 10 mM RuBP separately. (a) Dissolve TRIS (1.214 g) in 6 ml H_2O, adjust to pH 8 with 2 M HCl, and make to 10 ml with H_2O. (b) Dissolve $NaHCO_3$

(168 mg) in 8 ml H_2O. Using a micropipette, add 40 μl of $NaH^{14}CO_3$ containing 80 μCi and make to 10 ml with H_2O. (c) Dissolve each in 10 ml H_2O: 3.72 mg EDTA, 162.7 mg. $MgCl_2 \cdot 6H_2O$, and 154 mg GSH. (d) Dissolve RuBP (19.9 mg) in 5 ml H_2O.

Each of the prepared solutions is kept in a labeled serum bottle sealed with a rubber stopper and placed in a container filled with crushed ice. These solutions are used for enzyme assay.

B. Preparation of cell-free extracts

1. Harvesting and washing of cells. (a) During the late exponential phase of growth (ca. 5×10^6 cells ml^{-1}), 1 l of the algal suspension is transferred to polypropylene centrifuge bottles (250- or 500-ml capacity), sealed with plastic caps, and centrifuged at $4000 \times g$ for 10 min. The supernatant is discarded. (b) The cells are resuspended in about 250 ml H_2O and recentrifuged for 10 min at $4000 \times g$ at 4 C.

2. Disruption of cells. (a) The pellet obtained from the previous step is resuspended in 18 ml of the homogenization buffer and is transferred to a graduated beaker (30-ml capacity) using a Pasteur pipette, made to a final volume of 20 ml with the same buffer, and placed in crushed ice. (b) The precooled cells are broken by passing this dense algal suspension through a chilled French pressure cell at 20 000 lb in.$^{-2}$ Alternatively, cell disruption is achieved by two 30-sec bursts of ultrasonic radiation (e.g., Insonator, model 1000 with a microtip at a power setting of 1.5). During sonication, the sample is surrounded by crushed ice to avoid temperature changes. The power setting and the sonication period are selected to allow cell disruption without foaming.

3. Centrifugation of cell homogenate. The cell homogenate is transferred to a 50-ml polypropylene centrifuge tube and centrifuged at 4 C for 30 min at $40\ 000 \times g$. The yellowish green supernatant is collected in a labeled serum bottle, sealed with a rubber stopper, and placed in crushed ice. This extract is the crude enzyme preparation.

4. Determination of protein. The protein content of the crude enzyme preparation is estimated (Lowry et al. 1951) using bovine serum albumin as a standard. Extracts obtained from samples disrupted by either French pressure cell or sonication (step B.2.b) contain 5.0 ± 0.5 mg ml^{-1}.

C. Enzyme assay

The RuBP-dependent incorporation of $NaH^{14}CO_3$ into an acid-stable product is determined in a reaction mixture of 1.0 ml final volume. This mixture contains 100 mM TRIS-HCl, pH 8; 0.1 mM EDTA; 8

mM $MgCl_2 \cdot 6H_2O$; 5 mM GSH; 20 mM $NaH^{14}CO_3$, specific radioactivity 0.04 μCi μmol^{-1}; 1 mM RuBP; 0.1 ml cell-free extract containing 0.5 mg of protein. The enzyme assay is carried out in duplicate with incomplete assay systems (lacking either RuBP or enzyme in the reaction mixture), which serve as controls.

A graduated pipette (0.5 ml) is used for each solution and a micropipette (100 μl) is employed for the $NaH^{14}CO_3$ solution.

1. Preincubation of enzyme. Aliquots (0.1 ml) of each of the prepared stock solutions (a, b, and c given in III.A.2) are pipetted into each of four glass centrifuge tubes (15-ml capacity). H_2O (0.3 ml) is then added to give a total volume of 0.8 ml. The tubes are placed in a water bath maintained at 30 C, and 0.1 ml of boiled cell extract is added to one tube and labeled as a control (minus enzyme). To each of the other three tubes, 0.1 ml of cold cell-free extract (III.B.3) is added. The four tubes are then left for 10 min (preincubation period) before the reaction is initiated.

2. Initiation and termination of reaction. The reaction is started by the addition of RuBP (III.A.2.d). H_2O (0.1 ml) is added to one tube, which serves as a control (minus RuBP). RuBP solution (0.1 ml) is added to each of the other tubes including the one with boiled cell extract. After 10 min incubation at 30 C, the reaction is terminated by adding 1.0 ml of 24% (w/v) trichloroacetic acid to each tube.

3. Determination of radioactivity. The tubes are placed in a fume hood and their contents evaporated to dryness by a continuous current of air. The residue is dissolved in 1.0 ml warm H_2O and centrifuged for 10 min at 5000 × g. Aliquots (0.1 ml) of each of the supernatants are pipetted into scintillation vials and mixed with 10 ml of scintillation fluid (Bray 1960), and the radioactivity is then determined. All counts are made to a predetermined accuracy of 1.5% ± 2σ. Counting efficiency, determined by the channels ratio method (Wang and Willis 1965), may vary from 70% to 80%.

D. Calculations

The resultant counts (disintegrations per minute; dpm) which have been determined in 0.1 ml of the reaction mixture, are multiplied by 10 to obtain the radioactivity in the original 1.0 ml of reaction mixture. The disintegrations per minute are then converted to micromoles of $NaHCO_3$ using a factor of ca. 1×10^{-5} (1 μmol $NaH^{14}CO_3$, specific activity 0.4 μCi, corresponds to 0.888×10^5 dpm) calculated from measuring the radioactivity of 0.1 ml of the prepared $NaH^{14}CO_3$ solution (III.A.2.b). In calculating RuBP carboxylase activity, any fixation of $NaH^{14}CO_3$ by the control samples is subtracted from that in

Table 20-1. *RuBP carboxylase activity in cell-free extracts of Chlamydomonas segnis during autotrophic growth in cultures bubbled either with air or 5% CO_2 in air*

Method	Phase of growth	RuBP carboxylase activity	
		nmol min^{-1} mg^{-1} protein	fmol min^{-1} cell^{-1}
Air culture	Exponential	50.0	4.0
	Stationary	86.5	3.5
5% CO_2 culture	Exponential	35.0	2.8
	Stationary	74.0	3.0

the presence of RuBP and enzyme. The enzyme activity is usually expressed as micro- or nanomoles of CO_2 fixed per minute per milligram of protein.

IV. Sample data and discussion

Preparations of *C. segnis* give RuBP carboxylase activities that are readily measured. Table 20-1 shows that the activity of the enzyme depends on the physiological state of the cells.

A. Possible problem areas

Extremely low enzyme activities, or the failure to detect the enzyme in extracts of photosynthetic algae, may be attributed to one or more of the following factors.

1. Size of algal sample. Insufficient algal material may have been used for the preparation of extracts. This may, however, be overcome by concentrating the enzyme protein using $(NH_4)_2SO_4$ precipitation (Paulsen and Lane 1966).

2. Composition of homogenization buffer. Loss of sulfhydryl compounds from buffers caused by long storage periods may contribute to partial inactivation of the enzyme during extraction. The enzyme requires −SH groups (Rabin and Trown 1964). Exclusion of EDTA may exert a similar effect, as EDTA chelates traces of heavy metals that oxidize −SH groups. Mg^{2+} (Lord and Brown 1975) and $NaHCO_3$ (Tabita and McFadden 1974) may be included in the buffer, as they bind to the enzyme and maintain it in an active form (Chu and Bassham 1975).

3. Drastic disruption of cells. Foaming caused by improper sonication technique may result in denatured protein.

4. Frozen cell extracts. The enzyme loses activity on repeated thawing and freezing. Immediate use of the prepared cell extract is recommended.

5. Initiation of reaction by enzyme. The use of the enzyme rather than RuBP to start the reaction results in lower activity. The enzyme requires preincubation with Mg^{2+} and $NaHCO_3$ to acquire the active form (Chu and Bassham 1975).

6. RuBP solution. This solution is unstable even at low temperature, and the use of stored solution may account for low enzyme activity.

7. Endogenous inhibitors. The presence of particular metabolites (e.g., some intermediates and related compounds of the Calvin cycle) at certain concentrations in the crude extract may inhibit the enzyme (Buchanan and Schürmann 1973).

B. Alternative techniques

1. The spectrophotometric assay for RuBP carboxylase, based on the measurement of 3-phosphoglycerate-dependent NADH oxidation by a coupled enzyme system, is described by Racker (1962). This assay is improved by addition of an ATP-regenerating system (Lilley and Walker 1974).
2. Techniques for more detailed investigations of RuBP carboxylase are described in articles edited by Wood (1975) on carboxylases.

C. Ribulose-bis-phosphate oxygenase

This enzyme catalyzes the irreversible oxidation of RuBP to 2-phosphoglycolate and 3-phosphoglycerate (Andrews et al. 1973). It is suggested that RuBP carboxylase also functions as an oxygenase (Beardall and Morris 1975; Lord and Brown 1975).

V. Acknowledgments

I thank Dr. G. G. C. Robinson for reading the manuscript and for many helpful suggestions.

This work was supported by a grant-in-aid (A6720) from the National Research Council of Canada.

VI. References

Andrews, T. J., Lorimer, G. H., and Tolbert, N. E. 1973. Ribulose diphosphate oxygenase. I. Synthesis of phosphoglycolate by fraction-1 protein of leaves. *Biochemistry* 12, 11–18.

Badour, S. S., Tan, C. K., Van Caeseele, L. A., and Isaac, P. K. 1973. Observations on the morphology, reproduction and fine structure of *Chlamydomonas segnis* from Delta Marsh, Manitoba. *Can. J. Bot.* 51, 67–72.

Beardall, J., and Morris, I. 1975. Effects of environmental factors on photosynthesis patterns in *Phaeodactylum tricornutum* (Bacillariophyceae). II. Effect of oxygen. *J. Phycol.* 11, 430–4.

Bray, G. A. 1960. A simple efficient liquid scintillator for counting aqueous solutions in a liquid scintillation counter. *Anal. Biochem.* 1, 279–85.

Buchanan, B. B., and Schürmann, P. 1973. Regulation of ribulose 1,5-diphosphate carboxylase in the photosynthetic assimilation of carbon dioxide. *J. Biol. Chem.* 248, 4956–64.

Chu, D. K., and Bassham, J. A. 1975. Regulation of ribulose 1,5-diphosphate carboxylase by substrates and other metabolites: Further evidence for several types of binding sites. *Plant Physiol.* 55, 720–6.

Codd, G. A., and Stewart, W. D. P. 1976. Polyhedral bodies and ribulose 1,5-diphosphate carboxylase of the blue-green alga *Anabaena cylindrica*. *Planta* 130, 323–6.

Davis, B., and Merrett, M. J. 1975. The glycolate pathway and photosynthetic competence in *Euglena*. *Plant Physiol.* 55, 30–4.

Givan, A. L., and Criddle, R. S. 1972. Ribulose diphosphate carboxylase from *Chlamydomonas reinhardi:* Purification, properties and its mode of synthesis in the cell. *Arch. Biochem. Biophys.* 149, 153–63.

Kuhl, A., and Lorenzen, H. 1964. Handling and culturing of *Chlorella. Methods Cell Physiol.* 1, 159–87.

Lilley, R. McC., and Walker, D. A. 1974. An improved spectrophotometric assay for ribulose diphosphate carboxylase. *Biochim. Biophys. Acta* 358, 226–9.

Lord, J. M., and Brown, R. H. 1975. Purification and some properties of *Chlorella fusca* ribulose 1,5-diphosphate carboxylase. *Plant Physiol.* 55, 360–4.

Lord, J. M., Codd, G. A., and Stewart, W. D. P. 1975. Serological comparison of ribulose-1,5-diphosphate carboxylase from *Euglena gracilis, Chlorella fusca* and several blue-green algae. *Plant Sci. Lett.* 4, 377–83.

Lowry, O. H., Rosebrough, N. J., Farr, A. L., and Randall, R. J. 1951. Protein measurement with the Folin phenol reagent. *J. Biol. Chem.* 193, 265–75.

McFadden, B. A. 1973. Autotrophic CO_2 assimilation and the evolution of ribulose diphosphate carboxylase. *Bacteriol. Rev.* 37, 289–319.

Morris, I., and Farrell, K. 1971. Photosynthetic rates, gross patterns of carbon dioxide assimilation and activities of ribulose diphosphate carboxylase in marine algae grown at different temperatures. *Physiol. Plant.* 25, 372–7.

Paulsen, J. M., and Lane, M. D. 1966. Spinach ribulose diphosphate carboxylase. I. Purification and properties of the enzyme. *Biochemistry* 5, 2350–7.

Rabin, B. R., and Trown, P. W. 1964. Inhibition of carboxydismutase by iodoacetamide. *Proc. Natl. Acad. Sci. U.S.A.* 51, 497–501.

Rabinowitz, H., Reisfeld, A., Sagher, D., and Edelman, M. 1975. Ribulose diphosphate carboxylase from autotrophic *Euglena gracilis. Plant Physiol.* 56, 345–50.

Racker, E. 1962. Ribulose diphosphate carboxylase from spinach leaves. *Methods Enzymol.* 5, 266–70.

Tabita, F. R., and McFadden, B. A. 1974. One-step isolation of microbial ribulose-1,5-diphosphate carboxylase. *Arch. Microbiol.* 99, 231–40.

Wang, C. H., and Willis, D. L. 1965. *Radiotracer Methodology in Biological Science.* Prentice-Hall, Englewood Cliffs, N.J. 382 pp.

Wood, W. A. 1975. Carbohydrate metabolism. *Methods Enzymol.* 42, 457–87.

21: Nitrate reductase in marine phytoplankton

RICHARD W. EPPLEY

Institute of Marine Resources, A-018,
University of California, San Diego, La Jolla, California 92093

CONTENTS

	I Introduction	*page* 218
II	Equipment	218
	A. Filtration	218
	B. Homogenizer	218
	C. Centrifugation	218
	D. Incubation	219
	E. Spectrophotometer	219
	F. Other glassware	219
	G. Reagents	219
III	Methods	219
	A. Sample collection	219
	B. Procedure	219
	C. Standardization	221
IV	Sample data	221
V	Capabilities	221
VI	Limitations and troubleshooting	222
VII	Acknowledgments	222
VIII	References	222

I. Introduction

The assimilatory nitrate reductase (NADH: nitrate oxidoreductase, EC 1.6.6.2) is conveniently assayed in crude enzyme preparations of homogenized diatoms by measuring the rate of formation of nitrite. The present method was developed for the marine diatom, *Ditylum brightwellii* (Eppley et al. 1969), and is based on assay methods used for leaf extracts. It works equally well for a variety of marine phytoplankton, both from cultures and natural marine assemblages, and has been widely used especially in oceanography. Some examples include Collos and Lewin (1974) and Packard and Blasco (1974) and references cited therein.

The nitrate reductase activity of marine diatoms is induced by nitrate, is repressed in vivo by ammonium at micromolar concentrations, and shows diel periodicity and activity related to irradiance, as shown in the above reports. A somewhat different procedure was developed by Lui and Roels (1972), but their report includes no information related to optimum conditions for measuring the enzyme activity. Purification of nitrate reductase from a marine diatom was reported by Amy and Garrett (1974).

II. Equipment

A. Filtration

(1) A filter flask (1- to 4-l capacity) provided with Millipore or Gelman filter base and funnel for 47-mm diam. filters. (2) A vacuum source. (3) Glass fiber filters, Whatman GF/C or Gelman type A, 4.25 cm diam. to fit the 47-mm filter base.

B. Homogenizer

(1) A 15-ml glass homogenizer tube with Teflon pestle. (2) A motor drive for the pestle (a 0.25-in. electric shop drill will do). (3) A crushed-ice slurry in each of two 250-ml beakers.

C. Centrifugation

A clinical centrifuge with swinging rotor to accomodate six to eight 15-ml tubes. A supply of borosilicate glass tubes for centrifugation.

D. Incubation

A water bath is needed for maintaining the reaction tubes at a uniform temperature during incubation. This can be as simple as a beaker of tap water (room temperature) or more elaborate if activity vs. temperature is to be measured.

E. Spectrophotometer

A grating or prism spectrophotometer set to 543 nm and provided with 1- and 10-cm absorption cells (10-ml capacity) is best. Instruments accommodating only 1- or 5-cm cells will have reduced sensitivity, but will be adequate for most purposes.

F. Other glassware

Serological pipettes, 1- and 5-ml capacity. An automatic pipette with disposable tips delivering 0.1–1.0 ml is handy.

G. Reagents

(1) Phosphate buffer, 0.2 M, pH 7.9 (refrigerate). (2) Dithiothreitol solution (Cleland's reagent), 0.01 M (store frozen). (3) Polyvinylpyrrolidone, dry powder. (4) NADH (Sigma, type III, in 1- to 5-mg preweighed vials). (5) $MgSO_4$ solution, 0.05 M. (6) KNO_3 solution, 0.1 M. (7) Ethanol, 95%. (8) Zinc acetate solution, 1.0 M. (9) Sulfanilamide solution. Dissolve 5 g of sulfanilamide in a mixture of 50 ml of conc. HCl and about 300 ml of distilled water and dilute to 500 ml with water. The solution is stable at room temperature. (10) N-(1-naphthyl)-ethylenediamine·2HCl solution. Dissolve 0.50 g in 500 ml of distilled water. Store in a dark bottle and keep in the refrigerator. Prepare fresh about monthly. (11) Standard solution of sodium nitrite. Dissolve 0.345 g of anhydrous analytical reagent grade $NaNO_2$ in 1000 ml of distilled water. Store in a dark bottle in the refrigerator with 1 ml of chloroform as a preservative. The solution is stable for several months (1 ml = 5 μmol NO_2).

III. Methods

A. Sample collection

Any clean sampling device appropriate to the circumstances will do, but the sample *must* contain a quantity of phytoplankton equivalent to 10 μg of chlorophyll *a* for successful measurements.

B. Procedure

1. Thaw the dithiothreitol solution. Chill a small volume of the 95% ethanol.

2. Prepare the homogenizing solution. Pipette 4.0 ml of the phosphate buffer and 0.4 ml of dithiothreitol solution into the homogenizing tube; with a small spatula add 5–20 mg of dry polyvinylpyrrolidone. Place the tube in an ice bath.

3. To two reaction tubes (centrifuge tubes) add 0.2 ml KNO_3 solution and 0.1 ml $MgSO_4$ solution. Set these aside in the water bath.

4. Prepare the NADH solution by adding 0.79 ml of the phosphate buffer to each 1 mg NADH in its preweighed vial. Shake the vial to dissolve the NADH and place it in an ice bath for the moment.

5. Filter the sample to dryness and immediately fold the filter two or three times with forceps and place it in the cold homogenizing solution. Homogenize the filter with diatom cells 1–2 min or 30–40 vertical traverses of the tube over the rotating pestle, all the while holding the tube in the ice bath. (Practice by homogenizing a blank filter is recommended for developing dexterity.)

6. Decant the homogenate into a cold centrifuge tube and centrifuge it 5 min at about 2000 × g (full speed on most clinical centrifuges). The homogenate should remain below room temperature in this brief procedure, but chilling the metal centrifuge tube holders before use is worthwhile. The homogenate should now appear clear with a distinct brownish hue.

7. Now pipette 1.5 ml of the homogenate into the two reaction tubes containing KNO_3 and $MgSO_4$ solutions. To one of the tubes add 0.1 ml NADH solution, and to the other (the blank) add 0.1 ml of phosphate buffer.

8. Place the tubes in a water bath at the desired temperature and incubate 30 min. (Once confidence is achieved, a time-course experiment should be run to determine if the reaction rate is indeed constant over the 30-min period.)

9. Stop the reaction by adding 0.2 ml of zinc acetate solution (do not pipette by mouth) and 6.0 ml of cold 95% ethanol. Stopper the tubes and shake them briefly. The zinc acetate precipitates pyridine nucleotides that can interfere with the subsequent nitrite determination. However, this step may not always be necessary (Liu and Hellebust 1974).

10. Centrifuge the reaction tubes 5 min at 2000 × g and decant the clear supernatant into clean test tubes.

11. Add 1.0 ml sulfanilamide solution to the supernatants and, within 5 min, 1.0 ml of N-(1-naphthyl)-ethylenediamine·2HCl solution. Stir the contents of the tubes after each addition. A pinkish-red color should now appear in the reaction mixture with NADH if nitrate reductase activity was present.

12. Decant the contents of the tubes into the spectrophotometer cells and record absorbance at 543 nm. Subtract the absorbance of the

blank (reaction mixture without NADH) from that of the complete reaction mixture.

C. Standardization

Prepare a standard curve for absorbance vs. moles of nitrite by adding different volumes of the standard nitrite solution to a series of tubes containing buffer in place of the enzyme extract. Add the zinc acetate solution and 95% ethanol and carry through the procedure as before, maintaining the same solution volumes, total = 10.1 ml. Calculate a conversion factor, F, which equals micromoles of NO_2 per absorbance unit.

The standard nitrite solution contains 5 μmol NO_2 ml^{-1}. A 1:25 dilution of this provides a solution containing 0.2 μmol NO_2 ml^{-1}. Adding 1.0 ml of this, in place of 1.0 ml of buffer, to the reaction mixture will result in an absorbance reading about 1.0 with a 1-cm cell. Adding 1.0 ml of a 1:250 dilution of the standard solution will give an absorbance value of about 1.0 using 10-cm cells.

IV. Sample data

Calculate the nitrite production rate as (absorbance − blank) × F/time of incubation. In practice, F will be about 0.2 and 0.02 μmol nitrite per absorbance unit when 1- and 10-cm spectrophotometer cells are used, respectively.

For example, assume the absorbance of the complete reaction mixture was 0.243 with a 1-cm cell and that of the mixture without NADH was 0.060, F was 0.213, and incubation time was 30 min:

$$\mu\text{mol } NO_2 \text{ formed per hour} = \frac{(0.243 - 0.060)0.213}{0.50 \text{ h}} = .078 \quad (1)$$

The nitrite production rate per milliliter of enzyme extract can be calculated, and from this the production rate per liter of cell suspension. For example, assume 1 l of cell suspension was filtered and this was then extracted in 4.4 ml of homogenizing solution and 1.5 ml of this was added to the reaction mixture:

μmol NO_2 formed per liter per hour

$$= \frac{0.078}{1.5 \text{ ml extract}} \times \frac{4.4 \text{ ml extracted}}{1.0 \text{ l filtered}} = 0.23 \quad (2)$$

V. Capabilities

The method will detect the pink nitrite-produced color, and so will the eye, equivalent to about 0.01 absorbance units. In equation 1 this will be equivalent to 4 nmol of nitrite formed per hour. The upper limit is restricted only by the linearity of the nitrite determination and the depletion of NADH during the incubation. The latter has occurred rarely with samples from very dense cultures. The precision of the method has not been reported. However, differences of 20% are probably significant. We feel the method is best regarded as semi-quantitative.

VI. Limitations and troubleshooting

The chief limitation of this method is the rather large amount of plant material required, equivalent to at least 10 μg chlorophyll a. In developing the method for *D. brightwellii* it was noted that no activity could be recovered without phosphate; hence, the use of phosphate buffer. The pH optimum was 7.9. The reduced sulfhydryl compound, dithiothreitol, had to be present during homogenization if activity was to be recovered. Glutathione (reduced) but not cysteine, could be substituted for dithiothreitol, although both gave activity with flagellates. NADH was inactive with *D. brightwellii* and FAD could not be substituted for NADH.

The conversion of nitrite formed in the reaction to ammonium via nitrite reductase, present in the extracts, was not a problem with *D. brightwellii,* as the latter enzyme was inactive with NADH.

The polyvinylpyrrolidone, added to remove phenolics, and the $MgSO_4$ were not absolute requirements, but improved recovery of activity with *D. brightwellii* extracts.

VII. Acknowledgments

I thank J. L. Coatsworth, Jane Rogers, E. H. Renger, W. G. Harrison, and T. T. Packard for assistance and moral support in developing and applying this method. Students at the University of Washington, Friday Harbor Laboratories, summer session 1972, helped in identifying the difficulties to be expected among the inexperienced. This chapter is dedicated to them.

VIII. References

Amy, N. K., and Garrett, R. H. 1974. Purification and characterization of the nitrate reductase from the diatom *Thalassiosira pseudonana. Plant Physiol.* 54, 629–37.
Collos, Y., and Lewin, J. C. 1974. Blooms of surf-zone diatoms along the coast of the Olympic Peninsula, Washington. IV. Nitrate reductase activity in natu-

ral populations and laboratory cultures of *Chaetoceros armatum* and *Asterionella socialis*. *Mar. Biol.* 25, 213–21.

Eppley, R. W., Coatsworth, J. L., and Solórzano, L. 1969. Studies of nitrate reductase in marine phytoplankton. *Limnol. Oceanogr.* 14, 194–205.

Liu, M. S., and Hellebust, J. A. 1974. Utilization of amino acids as nitrogen sources, and their effect on nitrate reductase in the marine diatom *Cyclotella cryptica*. *Can. J. Microbiol.* 20, 1119–25.

Lui, N. S. T., and Roels, O. A. 1972. Nitrogen metabolism of aquatic organisms. II. The assimilation of nitrate, nitrite and ammonia by *Biddulphia aurita*. *J. Phycol.* 8, 259–64.

Packard, T. T., and Blasco, D. 1974. Nitrate reductase activity in upwelling regions. 2. Ammonia and light dependence. *Tethys* 6, 269–80.

22: Threonine deaminase in marine microalgae

N. J. ANTIA

Pacific Environment Institute, Environment Canada,
West Vancouver, British Columbia V7V 1N6, Canada

R. S. KRIPPS

Division of Human Nutrition, School of Home Economics,
University of British Columbia, Vancouver, British Columbia V6T 1W5, Canada

CONTENTS

I Introduction   *page* 226
II Harvest and storage of algae   227
III Preparation of crude enzyme
    dispersions   228
IV Enzyme assay   228
    A. Incubations   229
    B. Keto-acid determination   229
V Sample data and discussion   230
VI References   231

I. Introduction

The stability of an enzyme in crude extracts or purified preparations is expected to depend on its *intrinsic* molecular properties and on the *extrinsic* milieu (physical and chemical) provided during long-term storage. A practical strategy for storing relatively unstable enzymes consists in leaving the enzyme "undisturbed" inside the plant cell while storing the whole cell as a freeze-dried powder under vacuum at -20 to -30 C. Each time an enzyme preparation is required for studies, this cell powder is gently warmed to room temperature under desiccation, and samples are removed for preparation of enzyme extracts. Using this strategy, we have obtained, from algal powders stored for more than 1 year, reliable and reproducible preparations of an allosteric type of enzyme, L-threonine deaminase, known to be notoriously unstable (bacteria:Desai 1968, Harding 1969; yeast: Katsunuma et al. 1971; protozoa: Raizada and Rao 1972; and higher plants: Modi and Mazumder 1966). The methodology involved is illustrated below in the case of threonine deaminase in cryptomonads, and this approach has proved equally successful with other classes of microalgae (Desai et al. 1972; Antia et al. 1975).

L-threonine deaminase (L-threonine hydrolyase [deaminating], EC 4.2.1.16) catalyzes the conversion of L-threonine to ammonia and α-ketobutyric acid. It is the first enzyme in the biosynthetic pathway from L-threonine to L-isoleucine and is also closely connected with the pathway to L-valine (Allaudeen and Ramakrishnan 1968). It is generally subject to specific feedback inhibition by L-isoleucine and sensitive activation by L-valine, and shows sigmoid substrate saturation kinetics (Antia et al. 1975).

Unlike all other threonine deaminases, the cryptomonad biosynthetic enzyme has shown an additional requirement for disulfide groups; however, this factor is an inherent part of the enzyme molecule and does not need to be supplied externally so long as the milieu is free of extraneous reductants or oxidants other than air (Antia et al. 1972a). Algal threonine deaminases have pH optima in the range 8.5

−9.5 (Desai et al. 1972; Antia and Kripps 1973) and an absolute requirement for high concentrations of K^+ or NH_4^+ (Antia et al. 1972b). There appears to be no divalent metal ion requirement for the threonine deaminases hitherto studied.

II. Harvest and storage of algae

A culture of the cryptomonad *Chroomonas salina* or *Hemiselmis virescens* is centrifuged (10 000–15 000 × g for 20 min), preferably near the temperature at which it has been grown (15–20 C), in order to avoid cell lysis from temperature shock. Batch or continuous centrifugation may be employed, depending on the culture volume and facilities available. Continuous centrifugation of 5-l cultures is readily accomplished at higher speeds (ca. 20 000 × g) in about 2 h, using a model RC-2 Sorvall centrifuge with four to eight steel tubes and the Szent-Györgyi and Blum continuous-flow attachment leading directly from the culture flask to the steel tubes receiving the algal material.

The clear colorless supernatant is carefully decanted, and the tubes containing the packed algal material are drained as completely as possible, inverting them on filter paper for a few minutes. The algal pellet is then freeze-dried directly in a standard freeze-drier (Virtis or Thermovac). The final drying is accelerated later by placing the tubes in a vacuum desiccator containing P_2O_5 (spread within two or three Petri dishes) and evacuated with a high-vacuum pump at 15–20 C for ca 24 h. If the P_2O_5 becomes wet during this process, it should be replaced. About 2–3 days are generally required for the combined harvesting-drying operation, after which the algal "crust" is scraped from the centrifuge tubes, rapidly pulverized with mortar and pestle, and transferred to a screw-capped vial. The vial containing the dry algal powder is placed immediately in a vacuum desiccator provided with a color-indicating type of desiccant (Tel-Tale activated silica gel). The desiccator is evacuated ca. 15 min with a high-vacuum pump, and finally stored indefinitely in a deep-freezer cabinet at −20 to −30 C. It is important to keep the algal powder *both dry and cold* during this long-term storage. To facilitate rapid withdrawal of aliquots without losing enzyme activity from the remaining algal powder, we use specially designed, minidesiccators (pistol type), capable of being rapidly warmed (merely by standing) to room temperature before aliquot withdrawal and rapidly evacuated and cooled to the deep-freezer temperature. This minimizes moisture condensation on the algal powder during these temperature changes. We believe that maintaining the dry state, during aliquot withdrawal as well as during long-term storage, may be critical in preserving reproducible enzyme activity.

The procedure described is equally applicable to unicellular species of other algae. However, it has been hitherto applied only to marine algae.

III. Preparation of crude enzyme dispersions

Aliquots (2–10 mg) of an algal powder are precisely weighed into tubes for direct enzyme assay. The weight of the aliquots will be governed by the total protein content as well as the level of enzyme activity present in an algal powder. We have found that 2-mg aliquots (32–54% total protein and specific enzyme activity of 8–85 units mg^{-1} protein) are adequate for cryptomonad powders (Desai et al. 1972). The tubes (polyethylene or polycarbonate is recommended) must be suitable for containing about 2 ml total of enzyme-incubation mixture and subsequent additions, as well as for fitting into the sonicator and centrifuge.

In the standard assay, 0.5-ml aliquots of 0.2 M (chilled) potassium tricine buffer (Good et al. 1966), pH 8.5, are added and the algal powder is brought into suspension by vortex mixing. Batches of tubes, capped with Parafilm, are then packed in crushed ice in the cup of a Raytheon 10-kcycle magnetostrictive (ultrasonic) oscillator set at a maximum current output of 1.1 A. External cooling of the sonicator cup is ensured by cold running tap water. The sonicator is operated for 5 min, after which the tubes are removed to a beaker of crushed ice. The sonicates so obtained are ready for enzyme assays without further treatment. Each tube holds a crude enzyme dispersion suitable for one assay.

Two important precautions are required for obtaining reproducible enzyme dispersions: (1) the sonication temperature must be as low as possible (usually 0–4 C), and (2) the buffer used for the dispersions must maintain the desired optimal pH and the minimal constant amount of K^+ ions. If TRIS is used as the buffering agent, it must be supplemented with a constant concentration of KCl (usually 0.1 M in the final enzyme-incubation mixture).

IV. Enzyme assay

The assay procedure (Friedemann 1957) is based on the spectrophotometric determination of the colored 2,4-dinitrophenylhydrazone of α-ketobutyric acid, the product of enzyme action on its substrate, L-threonine, during incubation of the latter with the crude algal dispersions.

A. Incubations

The standard incubation mixture contains 0.5 ml crude enzyme dispersion, 0.1 ml aqueous 1 mM pyridoxalphosphate, 0.2 ml aqueous 0.4 M L-threonine, and 0.2 ml distilled water, so that the final volume of the total mixture is 1 ml. The distilled-water fraction is intended to offer flexibility in enzyme characterization studies where it could be replaced by appropriate solutions of various enzyme effectors (Desai et al. 1972).

The entire incubation mixture is not made up immediately, but stepwise as follows. The distilled-water (or effector-reagent) aliquot is first added (with vortex mixing) to each enzyme-containing tube, and this partial mixture is preincubated 15 min at ambient room temperature (20–25 C). Next, the pyridoxalphosphate aliquot is added with mixing, and this partial mixture is further preincubated for 5 min at 37 C. Finally, the substrate is added with mixing, and this complete mixture is incubated at 37 C for a strictly timed period depending on the activity present in the original algal powder. We have found that 15 min incubation is adequate for 2-mg portions of cryptomonad powders (32–54% protein; specific activity, 20–50 units). Incubation periods 5 min shorter or longer are recommended, respectively, for those powders (with comparable protein) of considerably higher (70–90 units) or lower (10–15 units) specific activity.

To correct for possible interference from the various reagents used, every assay batch must include at least one blank control incubation mixture containing the same ingredients without the algal powder.

B. Keto-acid determination

Each incubation is terminated by immediate mixing with 0.5-ml aliquots of 50% aqueous trichloroacetic acid, followed by centrifugation of the precipitated protein at $3000 \times g$ for 20 min. A 1-ml aliquot of the supernatant is next treated for 5 min at room temperature with 1 ml of 1% 2,4-dinitrophenylhydrazine in 2 N HCl, after which 1 ml ethanol and 3 ml benzene are incorporated with thorough mixing. This emulsion is clarified by centrifugation at $2000 \times g$ for 5 min. A 2-ml aliquot of the organic (upper) layer is next mixed thoroughly with 3 ml of 10% aqueous Na_2CO_3, and the resulting emulsion is again clarified by centrifugation. A 2-ml aliquot of the aqueous (lower) layer is now mixed with 2 ml of 1.5 N aqueous NaOH, and the resulting absorbance (OD) at 435 nm is read on a spectrophotometer in a 1-cm cell. The reading is corrected by subtracting that of the blank control taken without algal powder. The resulting OD is used to determine the amount of keto-acid (nanomoles of keto-acid = $1000 \times$

Enzymes

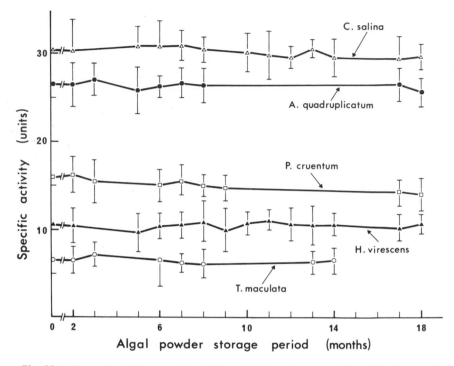

Fig. 22-1. Examples of the retention of threonine deaminase specific activity by stored algal powders of two cryptomonads (*Chroomonas salina* and *Hemiselmis virescens*), one rhodophyte (*Porphyridium cruentum*), one unicellular cyanophyte (*Agmenellum quadruplicatum*), and one prasinophycean species (*Tetraselmis maculata*). The unit of specific activity is nanomoles of keto acid produced (at pH 8.5) per minute per milligram of protein. Each point shown is the mean enzyme activity of replicate samples with standard deviations denoted by the bars (or half-bars to avoid overlapping).

OD), where the conversion factor (1000) is obtained from a rectilinear calibration curve constructed by using concentration standards of sodium α-ketobutyrate (Sigma) in place of enzyme extracts.

It is possible for an experienced worker to handle a daily batch of 12–15 assay tubes and complete the entire determination (starting from the dry algal powder) within 8 h, if the required reagents have been prepared previously.

V. Sample data and discussion

Figure 22-1 shows the stability of the enzyme from several microalgal species over a period of 16–18 months when prepared by the above methods. Each activity has changed so little that any change is masked by the standard deviation. We reckon that, using this approach, one axenic mass culture of a species should provide enough algal powder for at least 400–500 enzyme assays on 2-mg portions of powder. Depending on their nutritional status (autotrophic, heterotrophic, various N sources for growth, etc.), such mass cultures of *C. salina* and *H. virescens* have yielded quantities of dry algal powder in the range of 0.82–2.97 g per 5-l culture. Thus, the techniques described permit the complete characterization of the native threonine deaminase from the one-shot culture of an alga, without the monotonous labor and uncertainty of metabolic variations involved in repeated culturing.

Our methodology makes no effort to separate the cytoplasmic (soluble) portion from the membrane-bound "insoluble" portions associated with the intracellular organelles. In the case of threonine deaminase, preliminary tests have shown that the soluble/insoluble activity ratio, after sonication, is about 9:1 for a cryptomonad, but considerably less for other algae. Although such fractionation has not been systematically pursued for cryptomonads, we believe our strategy for the long-term preservation of the total enzyme activity in algal powders should be an asset in such studies.

A different approach, involving the preparation of acetone powders, was described for phosphatase activity in planktonic algae (Antia and Watt 1965). This approach has the advantage that a considerable amount of lipid and lipoid pigments is removed by the acetone treatment before enzyme extraction, thereby facilitating spectrophotometric enzyme assays in which such substances could interfere. It was used to a limited extent for the characterization of aldolase and enolase in cryptomonads and other planktonic algae (Antia et al. 1966; Antia 1967). This approach was particularly successful in the total extraction of phospholipase C from a marine chrysomonad (Bilinski et al. 1968), but the same method failed to extract the total phospholipase D activity of a unicellular red alga (Antia et al. 1970). We therefore recommend the algal powder approach unless it is precluded by the method for determining the enzyme.

VI. References

Allaudeen, H. S., and Ramakrishnan, T. 1968. Biosynthesis of isoleucine and valine in *Mycobacterium tuberculosis* H_{37} R_V. *Arch. Biochem. Biophys.* 125, 199–209.

Antia, N. J. 1967. Comparative studies on aldolase activity in marine planktonic algae, and their evolutionary significance. *J. Phycol.* 3, 81–5.

Antia, N. J., Bilinski, E., and Lau, Y. C. 1970. Identification and characterization of phospholipase D in a unicellular red alga (*Porphyridium cruentum*). *Can. J. Biochem.* 48, 643–8.

Antia, N. J., Kalmakoff, J., and Watt, A. 1966. Enolase activity in marine planktonic algae. *Can. J. Biochem.* 44, 449–54.

Antia, N. J., and Kripps, R. S. 1973. L-threonine deaminase in marine planktonic algae. IV. Studies of substrate specificity and inhibition from substrate analogs. Evidence for additional L-serine deaminase activity in the cryptophyte *Chroomonas salina. Arch. Mikrobiol.* 94, 29–46.

Antia, N. J., Kripps, R. S., and Desai, I. D. 1972a. L-threonine deaminase in marine planktonic algae. II. Disulfide and sulfhydryl group requirements of enzyme activity in two cryptophytes. *J. Phycol.* 8, 283–9.

Antia, N. J., Kripps, R. S., and Desai, I. D. 1972b. L-threonine deaminase in marine planktonic algae. III. Stimulation of activity by monovalent inorganic cations and diverse effects from other ions. *Arch. Mikrobiol.* 85, 341–54.

Antia, N. J., Kripps, R. S., and Desai, I. D. 1975. L-threonine deaminase in marine planktonic algae. V. Kinetic evaluation of feedback effects from structural analogs of L-isoleucine on the deaminase activity of seven unicellular species and elucidation of the allosteric basis for anomalous feedback control shown by a diatom and a cryptomonad. *J. Phycol.* 11, 60–9.

Antia, N. J., and Watt, A. 1965. Phosphatase activity in some species of marine phytoplankters. J. Fish. Res. Board Can. 22, 793–9.

Bilinski, E., Antia, N. J., and Lau, Y. C. 1968. Characterization of phospholipase C from a marine planktonic alga (*Monochrysis lutheri*). *Biochim. Biophys. Acta* 159, 496–502.

Desai, I. D. 1968. Cell preparation for the assay of threonine dehydratase in *Escherichia coli. Biochim. Biophys. Acta* 167, 467–9.

Desai, I. D., Laub, D., and Antia, N. J. 1972. Comparative characterization of L-threonine dehydratase in seven species of unicellular marine algae. *Phytochemistry* 11, 277–87.

Friedemann, T. E. 1957. Determination of α-keto acids. *Methods Enzymol.* 3, 414–8.

Good, N. E., Winget, G. D., Winter, W., Connolly, T. N., Izawa, S., and Singh, R. M. M. 1966. Hydrogen ion buffers for biological research. *Biochemistry* 5, 467–77.

Harding, W. M. 1969. Relationships between stability of threonine deaminase and its apparent kinetics. *Arch. Biochem. Biophys.* 129, 57–61.

Katsunuma, T., Elsässer, S., and Holzer, H. 1971. Purification and some properties of threonine dehydratase from yeast. *Eur. J. Biochem.* 24, 83–7.

Modi, S. R., and Mazumder, R. 1966. A biosynthetic L-threonine dehydratase from spinach. *Indian J. Biochem.* 3, 215–8.

Raizada, M. K., and Rao, V. K. M. 1972. L-threonine dehydratase activity of axenically grown *Hartmannella culbertsoni. Arch. Mikrobiol.* 84, 119–28.

23: Phosphoenolpyruvate carboxylase of blue-green algae

BRIAN COLMAN

Department of Biology,
York University, Downsview, Ontario M3J 1P3, Canada

CONTENTS

I	**Introduction**	*page* **234**
II	**Materials**	**234**
	A. Reagents	234
	B. Algae	235
III	**Methods**	**235**
	A. Preparation of cell extract	235
	B. Enzyme assay	235
IV	**Discussion**	**236**
V	**Acknowledgments**	**237**
VI	**References**	**237**

I. Introduction

Phosphoenolpyruvate (PEP) carboxylase (EC 4.1.1.31) catalyzes the irreversible carboxylation of PEP by bicarbonate to form oxalacetate. It is conveniently assayed in crude extracts of algae by measuring the incorporation of ^{14}C from $NaH^{14}CO_3$ into acid-stable products in the presence of PEP (Slack and Hatch 1967).

To assay constituent enzymes, extracts of blue-green algae can be prepared by grinding or vigorous homogenization of the cells. During this process, however, a considerable loss of enzyme activity may occur, and the cells can be more gently disrupted by osmotic lysis of their spheroplasts. Spheroplasts are prepared from the cells by removal of elements of the cell wall with lysozyme (Biggins 1967a). The spheroplasts retain their integrity when maintained in an osmoticum, but undergo lysis when the osmoticum is replaced with buffer. Relatively high levels of PEP carboxylase can be detected in blue-green algae when cell extracts are prepared by this procedure (Colman et al. 1976).

II. Materials

A. Reagents

1. Potassium phosphate buffers, 0.03 M. Make up two solutions: (a) 5.23 g K_2HPO_4 in 1000 ml of water and (b) 4.08 g KH_2PO_4 in 1000 ml of water. Potassium phosphate buffer, 0.03 M, pH 6.9, is then prepared by mixing 555 ml of solution a and 445 ml of solution b. Potassium phosphate buffer, 0.03 M, pH 8.0, is prepared by mixing 833 ml of solution a and 167 ml of solution b.

2. 0.03 M potassium phosphate–0.55 M sorbitol buffer. Dissolve 99.0 g sorbitol in 0.03 M potassium phosphate buffer, pH 6.9, and make up to 1 l with the same buffer.

3. Lysozyme, 0.06%. Lysozyme (egg white muramidase, grade I, Sigma), 0.06 g, is dissolved with stirring at room temperature in 100 ml 0.03 M potassium phosphate–0.55 M sorbitol buffer, pH 6.9.

4. PEP, 0.02 M. PEP (tricyclohexylamine salt, Sigma), 27.9 mg is dissolved in distilled water and made up to 3.0 ml.

5. Assay medium. TRIS-HCl buffer, 0.1 M, pH 8.0, containing mercaptoethanol, sodium glutamate, and $MgCl_2$. Tris-(hydroxymethyl)-aminomethane, 1.21 g; $MgCl_2 \cdot 6H_2O$, 203 mg; sodium glutamate, 168 mg; and mercaptoethanol, 140 μl, are dissolved in 90 ml of distilled water, adjusted to pH 8.0 with conc. HCl, and made up to 100 ml with distilled water. The resulting solution is kept tightly sealed and made up fresh weekly.

6. $NaH^{14}CO_3$, 0.1 M. A solution of $NaH^{14}CO_3$ is diluted with appropriate volumes of distilled water and 0.5 M $NaHCO_3$ to give a solution of about 100 μCi ml^{-1} and 100 μCi 10 μmol^{-1} HCO_3^-. This solution should be kept tightly sealed and its specific activity should be known accurately.

7. Phenylhydrazine–4 N HCl. Conc. HCl, 15 ml, is added to 30 ml distilled water, and phenylhydrazine is added until the solution is saturated.

B. Algae

Bacteria-free cultures of *Anacystis nidulans* Richt. (UTEX 1550), *Anabaena flos-aquae* (Lyngbye) Bréb. (UTEX 1444), and *Coccochloris peniocystis* Kütz (UTEX 1548) are maintained and grown as previously described (Miller et al. 1971). Cells of exponential phase cultures are harvested by centrifuging at 5000 × g for 20 min and washed with distilled water.

III. Methods

A. Preparation of cell extract

Washed algal cells (about 500–800 mg dry weight) are resuspended in 100 ml 0.06% lysozyme in 0.03 M potassium phosphate–0.55 M sorbitol buffer at pH 6.9 and incubated at 30 C for 2–3 h in the dark in a shaker bath with continuous agitation of about 100 strokes min^{-1}.

The resulting spheroplasts are centrifuged at 5000 × g for 20 min, the lysozyme solution decanted off, the spheroplasts washed twice with 100 ml 0.03 M phosphate–0.55 M sorbitol buffer pH 8.0, and then lysed by the addition of 20 ml 0.03 M phosphate buffer pH 8.0. The extract is gently stirred, centrifuged at 12 000 × g for 20 min, and the supernatant used for the assay of the enzyme.

B. Enzyme assay

The following reagents are mixed in a vial to give a total volume of 0.4 ml: 0.2 ml TRIS-HCl buffer, pH 8.0, containing $MgCl_2$, sodium glutamate, and mercaptoethanol; 0.05 ml 0.02 M PEP; 0.05 ml 0.1 M NaH^{14}CO$_3$; 0.1 ml of the algal extract, to start the reaction. The vial is tightly sealed with a serum bottle stopper and incubated at 20 C for 5 min. Assays are run in duplicate and duplicate controls are run (1) with the 0.2 M PEP replaced with an equal volume of distilled water and (2) with 0.1 ml of algal extract that has been heated at 90 C for 5 min.

The reaction is stopped by the addition of 0.5 ml of saturated phenylhydrazine–4 N HCl solution to each tube and, after thorough mixing, 100-μl aliquots are transferred to planchets, dried under an infrared lamp, and the radioactivity of the ^{14}C-labeled products determined with a gas-flow counter. Alternatively, the aliquots can be dried in scintillation vials, a suitable scintillation fluid added, and the radioactivity determined. The micromoles of HCO$_3^-$ incorporated can be calculated from the specific activity of the NaH^{14}CO$_3$ used in the experiment. Enzyme activity is expressed as micromoles of HCO$_3^-$ per minute per milligram of protein. Protein is assayed according to Lowry et al. (1951) using bovine serum albumin as the standard.

IV. Discussion

Relatively high activities of PEP carboxylase can be measured in blue-green algae when the assay is carried out immediately after spheroplast lysis. The rate of reaction is constant for 5–8 min and is proportional to protein concentration. Little activity is detected in controls without added PEP, but if such activity is detected, the activity measured in the presence of PEP should be corrected for this endogenous activity.

Osmotic lysis of spheroplasts has proved effective in producing cell-free extracts of blue-green algae with high metabolic activity. It has allowed high rates of multienzyme processes, such as photophosphorylation, to be measured (Biggins 1967b) and also high activities of single enzymes. For example, PEP carboxylase activity is 3- to 5-fold higher in extracts produced by lysis than in those produced by homogenization. Similarly, carbonic anhydrase activity is 5- to 10-fold higher and glycolate dehydrogenase 1.5- to 2-fold higher. The lysis medium can be modified to permit the separation of soluble and membrane-bound enzymes; lysis of spheroplasts in 0.3 M sorbitol or mannitol buffer prevents the solubilization of glycolate and succinic dehydrogenases, and they remain largely lamellar-bound (Grodzinski and Colman 1976). The spheroplast lysis technique should be used with caution because the cells of some blue-green algae lose considerable metabolic activity (Grodzinski and Colman 1973), or lose the activity of individual enzymes (Grodzinski and Colman 1975), during spheroplast preparation, apparently as a result of the osmotic stress required to maintain spheroplast integrity.

V. Acknowledgments

This work was supported by grants from the National Research Council of Canada.

VI. References

Biggins, J. 1967a. Preparation of metabolically active protoplasts from the blue-green alga, *Phormidium luridum. Plant Physiol.* 42, 1442–6.

Biggins, J. 1967b. Photosynthetic reactions by lysed protoplasts and particle preparations from the blue-green alga, *Phormidium luridum. Plant Physiol.* 42, 1447–56.

Colman, B., Cheng, K. H., and Ingle, R. K. 1976. The relative activities of PEP carboxylase and RuDP carboxylase in blue-green algae. *Plant Sci. Lett.* 6, 123–7.

Grodzinski, B., and Colman, B. 1973. Loss of photosynthetic activity in two blue-green algae as a result of osmotic stress. *J. Bacteriol.* 115, 456–8.

Grodzinski, B., and Colman, B. 1975. The effect of osmotic stress on the oxidation of glycolate by the blue-green alga *Anacystis nidulans. Planta* 124, 125–33.

Grodzinski, B., and Colman, B. 1976. Intracellular localization of glycolate dehydrogenase in a blue-green alga. *Plant Physiol.* 58, 199–202.

Lowry, O. H., Rosebrough, N. J., Farr, A. L., and Randall, R. J. 1951. Protein measurements with the Folin reagent. *J. Biol. Chem.* 193, 265–75.

Miller, A. G., Cheng, K. H., and Colman, B. 1971. The uptake and oxidation of glycolic acid by blue-green algae. *J. Phycol.* 7, 97–100.

Slack, C. R., and Hatch, M. D. 1967. Comparative studies on the activity of carboxylases and other enzymes in relation to the new pathway of photosynthetic carbon dioxide fixation in tropical grasses. *Biochem. J.* 103, 660–5.

24: Aldolases of multicellular marine algae

KAZUTOSI NISIZAWA

Department of Fisheries, College of Agriculture and Veterinary Medicine, Nihon University, Shimouma-3, Setagaya-ku, Tokyo, 154 Japan

CONTENTS

I Introduction   *page* 240
II Enzyme extraction   240
III Enzyme assay   241
IV Purification and properties   241
  A. *Ulva pertusa*   241
  B. *Hizikia fusiformis*   242
  C. *Porphyra suborbiculata*   242
  D. *Oscillatoria* sp.   243
V Distribution of aldolases in marine algae   244
VI References   244

I. Introduction

Aldolase (fructose-1,6-bisphosphate-D-glyceraldehyde-3-phosphate lyase, EC 4.1.2.13) catalyzes the cleavage of fructose-1,6-bisphosphate (FBP) into dihydroxyacetone-1-phosphate (DHAP) and glyceraldehyde-3-phosphate (GAP). There are two types of aldolases (Rutter 1964). Type I is typified by the mammalian muscle enzyme; type II is from yeast. The latter aldolases are principally characterized by inhibition with ethylenediamine tetraacetate (EDTA), by stimulation with divalent metal ions, by activation with K^+, and by requirement of SH reagents; almost the reverse is true of type I aldolases. Type II aldolases occur, in general, in phylogenetically lower organisms such as bacteria, fungi (Rutter 1964), and planktonic algae (Antia 1967). In contrast, type I aldolases are found in green algae, protozoa, higher plants, and animals (Rutter 1964). However, the protists *Chlamydomonas* (Russell and Gibbs 1967; Guerrini et al. 1971) and *Euglena* (Rutter 1964) possess both types.

Ikawa et al. (1972) found that all the marine brown and red algae tested also contain both types of this enzyme, and these authors investigated the properties of some of these aldolases using partially purified enzyme preparations.

II. Enzyme extraction

Aldolase extracts may be prepared from various marine algae in the following way (Ikawa et al. 1972). Grind approximately 100 g of fresh algal fronds that have been cleaned practically free of macroscopic epiphytes in a chilled mortar or a wooden grinder (Type 18 Ishikawa Co.), for 20–50 min together with a small amount of quartz sand and 60–100 ml of 0.5 M TRIS-HCl buffer, pH 8.2, containing 5 mM 2-mercaptoethanol (ME), 1% polyvinylpyrrolidone K 90, and 1 drop of Tween 80. In some cases it is necessary to use more polyvinylpyrrolidone, and in other cases the addition of 10–20% final concentration of glycerol and/or 1 mM EDTA and/or 10 mM $MgSO_4$ (e.g., for extraction of ribulose-1,5-bisphosphate carboxylase from brown algae) is required for optimal enzyme activities. Squeeze the homogenate through several layers of nylon cloth, and centrifuge the extract at 16 000 × g for 20 min. Solid $(NH_4)_2SO_4$ is added to give a saturation of 25%. After the preparation has stood for 20–30 min at 0 C, the precipitate containing polyvinylpyrrolidone is removed by centrifugation and discarded. Additional solid $(NH_4)_2SO_4$ is added to the supernatant solution to bring the concentration to 75–90% saturation. Leave for 20–30 min at 0 C. Centrifuge at 16 000 × g for 20 min and discard the supernatant. Dissolve the precipitate in about 10 ml H_2O and dialyze overnight in a cold room against 10 volumes of 50 mM TRIS-HCl buffer, pH 7.8, containing 30 mM ME. The dialyzed solution contains the crude aldolase.

III. Enzyme assay

Aldolase activity in algal extracts can be measured by the method of Beck (1955), which is a modification of that described by Sibley and Lehninger (1949).

A test tube containing 1.0 ml of TRIS-HCl buffer (100 μmol, pH 8.6 or 7.5), 0.25 ml of FBP (12.5 μmol), 0.25 ml of hydrazine sulfate (55 μmol), 0.2 ml of enzyme extract, and 0.8 ml H_2O is incubated at 30 C for appropriate periods (30 min in general). The reaction is stopped by the addition of 2.0 ml of 10% trichloroacetic acid. In the blank test, FBP is added after the trichloroacetic acid treatment. Then 1-ml aliquots of the filtrate of the above mixture are transferred to test tubes, and 1.0 ml of 0.75 N NaOH is added to each. After the tubes have stood at room temperature for 10 min, 1.0 ml of 2,4-dinitrophenylhydrazine (5 μmol in 2 N HCl) is added, and the tubes are incubated at 38 C for 10 min. Subsequently, 7 ml of 0.75 N NaOH is added, and after 10 min the absorbance at 540 nm is measured.

For more highly purified aldolase preparations, the method by Wu and Racker (1959) may be used to determine aldolase activity.

IV. Purification and properties

A. Ulva pertusa

An $(NH_4)_2SO_4$ fraction of *U. pertusa* aldolase was applied to a Sephadex G-200 column (3.8 × 79 cm) and eluted with 25 mM TRIS-HCl buffer, pH 7.5, containing 1 mM ME at a flow rate of 40 ml h^{-1}; 10-ml fractions of the eluate were collected. The aldolase activity in each tube was assayed with or without 20 mM EDTA in the reaction mixture. The aldolase activities were found in fractions 25–32 and formed a sharp peak irrespective of the presence or absence of EDTA. No other peak with aldolase activity was found.

This aldolase showed a broad optimum activity from pH 6.5 to 8.8. The enzyme was slightly activated by ME (1 mM) and by cysteine (1 mM), but strikingly inhibited by p-chloromercuribenzoate (p-CMB, 0.1 mM). EDTA (20 mM) and K^+ (0.1 M) did not inhibit or activate the enzyme. The aldolase is, therefore, type I. The high sensitivity of the *Ulva* aldolase to p-CMB resembles that from *Chlamydomonas* (Russell and Gibbs 1967), but is different from muscle type I aldolase, which is insensitive to this substance (Rutter 1964).

B. Hizikia fusiformis

An $(NH_4)_2SO_4$ fraction of *H. fusiformis* was fractionated by gel filtration on Sephadex G-200 column in the same way as *U. pertusa*. Aldolase activity in each fraction was measured similarly, with or without 20 mM EDTA. In contrast to *Ulva* aldolase, however, two completely separate aldolase peaks (one covering fractions 23–38, and the other covering fractions 44–57) were obtained in the experiment without EDTA, and only the former peak was detected in the presence of EDTA. This result indicates that aldolase of the latter peak is strongly inhibited by EDTA. Thus, these aldolases belong to types I and II, respectively.

The two forms of aldolase were then purified further by chromatography on a DEAE-cellulose column (2.0 × 25 cm) in which a gradient elution was made with 0.05–0.2 M phosphate buffers, pH 7.0. Each form of aldolase fraction obtained seemed to be a single component, judging from the coincidence of their activities and the protein peaks. Both types of aldolases were strongly inhibited by p-CMB (0.1 mM), as was the *Ulva* aldolase (type I), and similarly were moderately inhibited by some divalent metal ions, such as Co^{2+} and Cu^{2+} (1 mM), but the type II aldolase, which had been completely inactivated by EDTA (0.5 mM), was reactivated by addition of particular divalent metal ions such as Co^{2+}, Fe^{2+}, and especially Zn^{2+}. Potassium ions also strongly activated this aldolase. The enzyme showed a narrow pH optimum, pH 7.5–7.8. These properties of the brown algal type II aldolase closely resemble those of the yeast aldolase (type II), in which is contained a significant quantity of tightly bound zinc (Harris et al. 1969). In contrast, the type II aldolases of *Anacystis* (Willard and Gibbs 1968), *Chlamydomonas* (Russell and Gibbs 1967), and *Euglena* (Rutter 1964) require Fe^{2+} for their activities.

C. Porphyra suborbiculata

By entirely similar fractionation procedures to those for green and brown algal aldolases, two types of aldolases were obtained from the red alga *P. suborbiculata*. Most properties of these aldolases resemble those from *H. fusiformis*.

D. Oscillatoria sp.

From this freshwater blue-green alga extraction of aldolase was attempted in the same way as with marine algae. Only type I aldolase was obtained. In contrast to the type I aldolase of other algal classes, aldolase of this blue-green alga exhibited an absolute requirement for

Table 24-1. *Distribution of aldolases in marine algae*

Species	Enzyme activity (units g^{-1} fresh wt.)^a	Percent inhibition with 20 mM EDTA
Green algae		
Ulva pertusa	16.5	0
Cladophora japonica	19.0	0
Codium latum	3.8	0
Brown algae		
Spatoglossum pacificum	11.7	19
Dictyota dichotoma	0.9	15
Ishige foliacea	3.0	62
Colpomenia sinuosa	0.7	51
Colpomenia bullosa	3.7	13
Eisenia bicyclis	1.1	28
Undaria pinnatiforme	4.3	27
Hizikia fusiformis	1.8	23
Sargassum horneri	5.1	30
Sargassum thunbergii	2.0	25
Red algae		
Porphyra suborbiculata	42.6	33
Galaxaura falcata	3.2	48
Gelidium amansii	12.6	25
Serraticardia maxima	11.4	18
Corallina pilulifera	16.4	31
Halymenia acuminata	4.7	14
Gloiopeltis complanata	7.5	35
Gracilaria gigas	6.4	27
Gymnogongrus flabelliformis	10.3	66
Gigartina intermedia	1.9	67
Gigartina tenella	9.6	29
Gigartina mamillosa	17.3	49
Chondrus ocellatus	9.4	44
Blue-green alga (freshwater)		
Oscillatoria sp.	—	100

^{a} Measured in the absence of EDTA. Unit activity is defined as the amount of enzyme that splits 1 μmol FBP per 60 min under the assay conditions described by Sibley and Lehninger (1949).

Fe^{2+} and cysteine for its activation, as has been reported for *Anacystis nidulans* (Willard and Gibbs 1968, 1975).

V. Distribution of aldolases in marine algae

Ikawa et al. (1972) measured the aldolase activities in crude or (NH$_4$)$_2$SO$_4$ fractions from a variety of marine algae including green, brown, and red algae in the presence or absence of EDTA (20 mM). The values from the former reaction mixtures apparently represent activities of type I aldolases; those from the latter apparently give the sum of both type I and II activities. The results are presented in Table 24-1. Molecular weights of several marine algal aldolases have been determined by Ikawa et al. (1972) by gel filtration on Sephadex G-200 columns.

VI. References

Antia, N. J. 1967. Comparative studies on aldolase activity in marine planktonic algae, and their evolutionary significance. *J. Phycol.* 3, 81–5.

Beck, W. S. 1955. Determination of triose phosphates and proposed modifications in the aldolase method of Sibley and Lehninger. *J. Biol. Chem.* 212, 847–57.

Guerrini, A. M., Cremona, T., and Preddie, E. C. 1971. The aldolases of *Chlamydomonas reinhardii. Arch. Biochem. Biophys.* 146, 249–55.

Harris, C. E., Kobes, R. D., Teller, D. C., and Rutter, W. J. 1969. The molecular characteristics of yeast aldolase. *Biochemistry* 8, 2442–54.

Ikawa, T., Asami, S., and Nisizawa, K. 1972. Comparative studies on fructose diphosphate aldolases mainly in marine algae. *Proc. Int. Seaweed Symp.* 7, 526–31.

Russell, G. K., and Gibbs, M. 1967. Partial purification and characterization of two fructose diphosphate aldolases from *Chlamydomonas mundana. Biochim. Biophys. Acta* 132, 145–54.

Rutter, W. J. 1964. Evolution of aldolase. *Fed. Proc.* 23, 1248–57.

Sibley, J. A., and Lehninger, A. L. 1949. Determination of aldolase in animal tissues. *J. Biol. Chem.* 177, 859–72.

Willard, J. M., and Gibbs, M. 1968. Purification and characterization of the fructose diphosphate aldolases from *Anacystis nidulans* and *Saprospira thermalis. Biochim. Biophys. Acta* 151, 438–48.

Willard, J. M., and Gibbs, M. 1975. Fructose diphosphate aldolase from blue-green algae. *Methods Enzymol.* 42, 228–34.

Wu, R., and Racker, E. 1959. Regulatory mechanisms in carbohydrate metabolism. III. Limiting fractors in glycolysis of ascites tumor cells. *J. Biol. Chem.* 234, 1029–35.

25: The glucosyltransferases of red algae

JEROME F. FREDRICK

The Research Laboratories,
Dodge Chemical Company, Bronx, New York 10469

CONTENTS

I	Introduction	*page* 246
II	Materials	247
III	Methods	248
	A. Preparation of purified extracts	248
	B. Polyacrylamide gel electrophoresis	248
	C. Visualization techniques	250
IV	Discussion	251
V	Advanced techniques	252
VI	References	252

I. Introduction

The synthesis of the storage glucan, floridean starch, is the cumulative result of the actions of three groups of enzymes generally known as "glucosyltransferases" (Fredrick 1967). The product is a highly branched glucan containing only α-1,4- and α-1,6-glucosyl linkages. The glucosyltransferases include: phosphorylases (α-1,4-glucan: orthophosphate glucosyltransferase, EC 2.4.1.1); synthetases or "synthases" (ADP-glucose: α-1,4-glucan α-4-glucosyltransferase, EC 2.4.1.b); and branching enzymes, Q and b.e. (α-1,4-glucan: α-1,4-glucan 6-glycosyltransferase, EC 2.4.1.18).

These enzymes exist in multiple molecular forms (isoenzymes) in all classes of algae (Fredrick 1967, 1968a, 1968b, 1971). In the red algae, and particularly in *Rhodymenia pertusa,* two synthetase isozymes, one phosphorylase, and three branching isozymes are present (Fredrick 1967, 1971). Both synthetase isozymes have been shown to be capable of utilizing ADP glucose (ADPG) and UDP glucose (UDPG) as substrates for the lengthening of already formed linear glucans containing α-1,4-glucosyl bonds. Of the branching isozymes, two have the usual classic Q enzyme action (able to branch linear glucans such as amylose to form moderately branched polyglucans such as amylopectin); the other isozyme is capable of introducing further α-1,6 branch points into already moderately branched amylopectins, thereby forming highly branched glycogens and phytoglycogens (Fredrick 1971).

The three groups of isozymes have very similar properties and exhibit serological identity reactions (Fredrick 1976). Phosphorylase and synthetase have identical amino acid sequences in their respective phosphorylated hexapeptides (Nolan et al. 1964; Larner and Sanger 1965). Branching enzyme and phosphorylase also have some structural similarity of their molecules (Fredrick 1961, 1976, 1977). Because of this, classic methods of enzyme isolation usually result in the coprecipitation of two or more of the enzymes. It has been found that polyacrylamide gel electrophoresis is the most successful method (Frederick 1962) for resolving them.

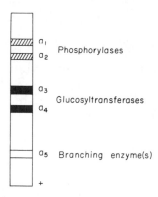

Fig. 25-1. Typical separation of algal glucosyltransferases on polyacrylamide gel. *Rhodymenia pertusa* contains only one phosphorylase (a_2); otherwise the pattern is identical with the one shown. Note that the branching isozymes travel as a single band in this method (disk columnar electrophoresis). (From Fredrick 1967.)

II. Materials

The screening properties and sieving effects of polyacrylamide gel (Ornstein 1964) are made use of in either columns or slab forms (Raymond 1962). Phosphorylase isozymes and synthetase isozymes are readily separated and isolated by polyacrylamide gel electrophoresis (Fredrick 1962). The method, however, does not isolate the individual branching isozymes. These isozymes move as one band even when discontinuous buffers are used (Fig. 25-1). The Q branching isozymes of *Rhodymenia pertusa* differ from the b.e. isozyme of this red alga only in electrical charge and are "charge isomers" (Fredrick 1971). In order to isolate the individual branching isozymes, it is necessary to use an orthogonal technique (Raymond and Aurell 1962), which entails the use of two-dimensional gel electrophoresis with two different polyacrylamide concentrations (Fredrick 1967). This technique separates three distinct branching isozymes in purified extracts of *R. pertusa*, which can then be eluted individually for further studies.

In all these studies, the E-C Vertical Gel Cell (E-C Apparatus Corp.) has been used. This gives a gel slab area of 12 × 17 cm, and the thickness of the gel slab can be varied from 3 to 6 mm by the use of 3-mm Lucite spacer rails on the outer edges of the cooling plate (Raymond 1962). The use of the 6-mm polyacrylamide gel is recommended for preparative work and for the elution of each individual isozyme in sufficient quantity for more advanced studies.

III. Methods

A. Preparation of purified extracts

Algae, 100 g, are washed for 2 h in running 8 C H_2O, drained on paper toweling, and washed in cold (5 C) double-distilled H_2O for 1 h. The algae are macerated in a cold Waring Blendor until a thick slurry results. This is transferred to a chilled mortar, and 5 g of fine quartz sand is added. Then 2 ml of 1.8% (w/v) solution of $NaHCO_3$ is added for each gram of wet algae. After being ground for 30 min, the resulting slurry is passed through two layers of cheesecloth and allowed to stand at 5 C for 1 h. The supernatant liquid is centrifuged for 15 min at $500 \times g$. The supernatant is neutralized with 0.1 M HCl to pH 7.0–7.2 and chilled to 4 C. Solid $(NH_4)_2SO_4$ is added to 0.8 saturation and the mixture allowed to stand at 2–3 C for 1 h. It is centrifuged at $1000 \times g$ for 5 min, and the resulting precipitate is dissolved in 0.1 saturated $(NH_4)_2SO_4$ and neutralized to pH 7. The $(NH_4)_2SO_4$ concentration is now raised to 0.35 saturation and the mixture allowed to remain at 0–5 C for 30 min. The precipitate is centrifuged off and resuspended in 0.01 M TRIS-HCl buffer (pH 7.3). It is dialyzed in Visking tubing against six to eight changes of buffer at 5 C for a 12-h period. The opalescent solution, called the "phosphorylase" fraction (Fredrick and Gentile 1960), contains all three groups of isozymes and is used for the polyacrylamide gel separations.

B. Polyacrylamide gel electrophoresis

TRIS-HCl (0.5 M, pH 7.2–7.4) is the most effective buffer for separating the glucosyltransferases using the slab method in the cell described by Raymond (1962). The cell is available commercially as the E-C 470 vertical cell, and excellent separations are obtained in 5–7% (w/v) acrylamide in the TRIS buffer. Both phosphorylase and synthetase isozymes are separated in 2–3 h using 200 V and 100 mA. Continuous circulation of water at 5 C through the cooling plates ensures active isozymes. The 5% gel will separate the phosphorylase isozymes and the synthetase isozymes, but will give only partial separation of the three branching isozymes of *R. pertusa*.

The orthogonal method (two-dimensional gel electrophoresis using two different concentrations of polyacrylamide as shown in Fig. 25-2) gives separations of the branching isozymes when the second gel concentration is greater than 8%. Using 8% gel gives separation of the Q branching isozymes from the b.e. branching isozyme. Figure 25-3 shows the Q isozymes moving as a single spot toward the anode, while the b.e. isozyme moves by itself.

For orthogonal separations, a strip 2–3 cm wide is excised after the

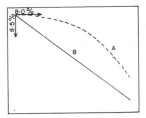

Fig. 25-2. Orthogonal electrophoresis in 8% polyacrylamide in one direction and in 8.5% polyacrylamide in the other. Glucosyltransferases of the same molecular size are found on line *B;* those of polymeric size assume positions on line *A.* Note that if the pattern of Fig. 25-3 is compared, the synthetase isozymes of *Rhodymenia pertusa* (a_3 and a_4) would seem to be of identical molecular size. (From Fredrick 1968a.)

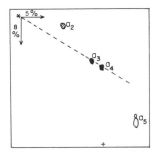

Fig. 25-3. Orthogonal electrophoresis of the glucosyltransferases of *Rhodymenia pertusa.* Note that the two Q branching isozymes (a_5, lower spot) travel as a single spot in the 8% gel, but the b.e. (a_5, upper spot) isozyme has been separated from these other branching isozymes. (From Fredrick 1967.)

separation in the lower gel concentration and embedded at the slot position (horizontally) in the higher gel concentration before gelation. The technique is generally described by Raymond and Aurell (1962), and specifically for algae by Fredrick (1971). The three branching isozymes of *R. pertusa* are separated completely after 2 h at 400 V and 120 mA. Using this orthogonal method, the three branching isozymes of this alga were shown to be charge isomers (Fig. 25-4). These can be individually isolated and their physical characteristics studied (Fredrick 1971).

For preparative work, separations using the 6-mm-thick gels were found to yield good quantities of each of the isozymes. Elution from these 6-mm gels was achieved by excising the particular areas of the gels, lyophilizing the segment, and extracting with 1.0 M TRIS-HCl buffer (pH 7.6) at 5 C. Up to 50 mg of protein can be obtained in this manner (Fredrick 1974).

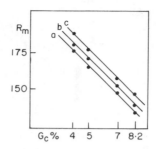

Fig. 25-4. Mobilities of the branching isozymes of *Rhodymenia pertusa* plotted against the gel concentration. This indicates that although the Q (*b* and *c*) and b.e. (*a*) isozymes differ in their action on glucan substrates, they are essentially of the same molecular size and differ only in electrical charge (charge isomers). (From Fredrick 1971.)

C. Visualization techniques

1. Protein-staining methods. For locating the isozymes, the gels are placed into a staining tray and covered with dye solution (2 g amido black in 1 l of methanol:water:glacial acetic acid, 5:5:1). After 1 h the dye solution is carefully poured off and the gel is destained in a mixture of acetic acid:methanol:water (1:5:5), either electrophoretically in 30 min or through several changes of the destaining reagent.

Another protein stain that gives permanent staining of the protein is the reactive dye, Procion Blue MRS (0.5–1.0%) in a 45:10:45 mixture of water:acetic acid:methanol. This stain forms covalent bonds with the protein component of the isozymes.

2. Substrate-staining techniques. For phosphorylases and synthetases, the gels may be incubated in their respective buffered substrates (dipotassium glucose-1-phosphate and the sodium salt of ADPG or UDPG). The reaction mixtures should contain about 1% maltoheptaose (Whelan et al. 1953) as primer to initiate the reaction. Incubations at 25–30 C for 2–3 h give good yields of linear iodine-staining glucan. The gels can then be washed briefly and placed into the iodine reagent described by Krisman (1962), in which the glucans stain a deep blue to violet.

A modification of the Gomori technique (Fredrick 1963; Davis et al. 1967) can be used to make the in situ action of the phosphorylases visible. In this method, a soluble calcium salt is added to the incubation mixture. After incubation, the gels are placed into a dilute $AgNO_3$ solution, irradiated with ultraviolet light for 15–20 min, and thoroughly washed in running H_2O. The phosphate released from the glucose-1-phosphate is precipitated by the Ca^{2+} at the site of the enzyme. The insoluble calcium phosphate is converted to the silver

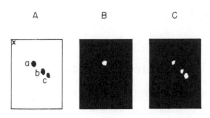

Fig. 25-5. The branching isozymes of *Rhodymenia pertusa* and their action on different substrates: *A*, stained with amido black; *B*, amylopectin incorporated into gel; *C*, amylose incorporated into gel. Note that isozyme *a* is active on both substrates (b.e. isozyme), whereas isozymes *b* and *c* are active only on amylose. (From Fredrick 1968b.)

salt, which is reduced by the ultraviolet irradiation to metallic black silver. The gels are then fixed in sodium thiosulfate, and the phosphorylase bands show up as sharp brown-black areas in the clear gels.

To allow study of the branching isozymes, the gels are placed into intimate contact (in the form of a gel sandwich) with thin polyacrylamide gels cast with either amylose or amylopectin as substrates incorporated into the gels. After incubation, the substrate-containing gels are separated and stained in iodine solution (Fredrick 1968a). The areas of branching enzyme activity show up as either red or violet bands against a light blue background (Fig. 25-5).

IV. Discussion

There is a paucity of data on the glucosyltransferases of algae, particularly the red algae. Some histochemical studies on phosphorylase were done a number of years ago (Yin 1948). Most of the work with glucosyltransferases has concentrated on the synthetases and not on the phosphorylases and branching enzymes. Preiss and Greenberg (1967) have studied the synthetases of the green alga, *Chlorella pyrenoidosa,* and describe techniques for obtaining active soluble and particulate preparations of the synthetases of this alga.

In the red algae, besides the polyacrylamide studies described with *Rhodymenia pertusa,* some work has been done with the synthetases of *Serraticardia maxima* (Nagashima et al. 1968, 1969, 1971). These methods do not involve electrophoresis, but rather use classic methods of enzyme isolation. The studies are dependent on the incorporation of labeled substrate (U[14]CDPG) into the floridean starch.

It is encouraging to see the use of polyacrylamide gel electrophoresis for taxonomic studies in algae. Techniques have been described using polyacrylamide gel electrophoresis of proteins other than glucosyltransferases in blue-green and green algae (Derbyshire and Whitton 1968; Thomas and Brown 1970), and these may prove useful for studies of the evolutionary processes in thallophytes (Holton 1973).

V. Advanced techniques

The use of concanavalin-Sepharose columns for the isolation and purification of the a_2 phosphorylase isozyme of algae has proved successful (Fredrick 1975). As the a_2 isozyme is present in *Rhodymenia pertusa*, this method may be of value in isolating sufficiently pure isozyme for kinetic studies, etc.

The various affinity chromatography methods using the principle of hydrophobic groups in the column have been employed by Shaltiel and Er-el (1973) and Shaltiel (1974) for separating mixtures of phosphorylase and synthetase.

VI. References

Davis, C. H., Schliselfeld, L. H., Wolff, D. P., Leavitt, C. A., and Krebs, E. G. 1967. Interrelationships among glycogen phosphorylase isozymes. *J. Biol. Chem.* 242, 4824–33.

Derbyshire, E., and Whitton, B. A. 1968. A disc electrophoretic study of proteins of blue-green algae. *Phytochemistry* 7, 1355–60.

Fredrick, J. F. 1961. Immunochemical studies of the phosphorylases of Cyanophyceae. *Phyton* 16, 21–6.

Fredrick, J. F. 1962. Multiple molecular forms of 4-glucosyltransferase (phosphorylase) in *Oscillatoria princeps*. *Phytochemistry* 1, 153–7.

Fredrick, J. F. 1963. An algal α-glucan phosphorylase which requires adenosine-5-phosphate as coenzyme. *Phytochemistry* 2, 413–5.

Fredrick, J. F. 1967. Glucosyltransferase isozymes in algae. *Phytochemistry* 6, 1041–6.

Fredrick, J. F. 1968a. Multiple forms of polyglucoside-branching enzyme in the algae. *Physiol. Plant.* 21, 176–82.

Fredrick, J. F. 1968b. Glucosyltransferase isozymes in algae. II. Properties of branching enzymes. *Phytochemistry* 7, 931–6.

Fredrick, J. F. 1971. Polyglucan branching isoenzymes of algae. *Physiol. Plant.* 24, 55–8.

Fredrick, J. F. 1974. Formation of primer-independent phosphorylases by synthetic coupling of glucans to primer-requiring isozyme. *Plant Sci. Lett.* 3, 183–9.

Fredrick, J. F. 1975. Affinity chromatography studies of the *de novo* glucan synthesizing phosphorylase isozyme of blue-green algae. *Plant Sci. Lett.* 5, 131–5.

Fredrick, J. F. 1976. *Cyanidium caldarium* as a bridge alga between Cyanophyceae and Rhodophyceae: Evidence from immunodiffusion studies. *Plant Cell Physiol.* 17, 317–22.

Fredrick, J. F. 1977. Possible common origin of glucosyl-transferases in *Oscillatoria princeps*. *Phytochemistry* 16, 55–7.

Fredrick, J. F., and Gentile, A. C. 1960. The effect of 3-amino-1,2,4-triazole on phosphorylase of *Oscillatoria*. *Arch. Biochem. Biophys.* 86, 30–3.

Holton, R. W. 1973. Electrophoresis and the taxonomy of algae. *Bull. Torrey Bot. Club* 100, 297–303.

Krisman, C. R. 1962. A method for the colorimetric estimation of glycogen with iodine. *Anal. Biochem.* 4, 17–23.

Larner, J., and Sanger, F. 1965. The amino acid sequence of the phosphorylation site of muscle uridine diphosphoglucose α-1,4-glucan, α-4-transferase. *J. Mol. Biol.* 11, 491–500.

Nagashima, H., Nakamura, S., and Nisizawa, K. 1968. Biosynthesis of floridean starch by chloroplast preparations from a marine red alga, *Serraticardia maxima*. *Bot. Mag. Tokyo* 81, 411–3.

Nagashima, H., Nakamura, S., Nisizawa, K., and Hori, T. 1971. Enzymic synthesis of floridean starch in a red alga, *Serraticardia maxima*. *Plant Cell Physiol.* 12, 243–53.

Nagashima, H., Ozaka, H., Nakamura, S., and Nisizawa, K. 1969. Physiological studies on floridean starch, floridoside, and trehalose in a red alga, *Serraticardia maxima*. *Bot. Mag. Tokyo* 82, 462–73.

Nolan, C., Novoa, W. B., Krebs, E. G., and Fischer, E. H. 1964. Further studies on the site phosphorylated in the phosphorylase *b* to *a* reaction. *Biochemistry* 3, 542–51.

Ornstein, L. 1964. Disc electrophoresis. I. Background and theory. In Fredrick, J. F. (ed.). *Gel Electrophoresis*, pp. 321–49. New York Academy of Sciences Press, New York.

Preiss, J., and Greenberg, E. 1967. Biosynthesis of starch in *Chlorella pyrenoidosa*. I. Purification and properties of the adenosine diphosphoglucose α-1, 4-glucan, α-4-glucosyltransferase from *Chlorella*. *Arch. Biochem. Biophys.* 118, 702–8.

Raymond, S. 1962. A convenient apparatus for vertical gel electrophoresis. *Clin. Chem.* 8, 455–70.

Raymond, S., and Aurell, B. 1962. Two-dimensional gel electrophoresis. *Science* 138, 152–3.

Shaltiel, S. 1974. Hydrophobic chromatography. *Methods Enzymol.* 34, 126–40.

Shaltiel, S., and Er-el, Z. 1973. Hydrophobic chromatography: Use for purification of glycogen synthetase. *Proc. Natl. Acad. Sci. U.S.A.* 70, 778–81.

Thomas, D. L., and Brown, M. R. 1970. New taxonomic criteria in the classification of *Chlorococcum* species. III. Isozyme analysis. *J. Phycol.* 6, 293–9.

Whelan, W. J., Bailey, J. M., and Roberts, P. J. P. 1953. The mechanism of carbohydrase action. I. The preparation and properties of maltodextrin substrates. *J. Chem. Soc.* 260, 1293–8.

Yin, H. C. 1948. Phosphorylase in plastids. *Nature* 162, 928.

26: Anion-activated adenosine triphosphatases

PAUL G. FALKOWSKI

Oceanographic Sciences Division,
Brookhaven National Laboratory, Upton, New York 11973

CONTENTS

I	Introduction	*page* 256
II	Extraction of membrane-bound ATPases	256
	A. Disruption methods	256
	B. Enrichment methods	257
III	Methods of assay	258
IV	Problems frequently encountered	259
V	Acknowledgments	259
VI	References	259

I. Introduction

In the two decades following Skou's discovery of the $(Na^+ + K^+)$-activated transport ATPase (adenosine triphosphatase, EC 3.6.1.3), extensive research efforts have been made to isolate and characterize membrane-bound enzymes capable of translocating anions, as well as cations, across cell membranes (Skou 1957; Hemmingsen 1971; Hodges et al. 1972; Stam et al. 1973; Karlsson and Kylin 1974; Kuiper et al. 1974; Maslowski and Komoszyński 1974; Falkowski 1975a, 1975b; Cambraia et al. 1976). Because these enzymes are associated with phospholipid moieties (Grisham and Barnett 1972; Goldman and Albers 1973; Simpkins and Hokin 1973), many conventional biochemical techniques, useful in the purification of soluble enzymes, have not been particularly successful when applied to membrane-bound ATPases. In general, the majority of problems in handling these delicate lipoproteins are encountered in the early stages of isolation from cells with initially low enzyme activity. This chapter will describe some general approaches used in extracting and assaying anion-activated ATPases from marine phytoplankton.

II. Extraction of membrane-bound ATPases

The extraction of ATPases from the plasmalemma consists of the following steps: (1) physical disruption of the cell in a suitable buffer, (2) separation of the membrane fraction from the rest of the cellular material, and (3) solubilization of the membrane proteins.

A. Disruption methods

After 6–10 l of a culture has been harvested by continuous centrifugation (10 000–12 000 × g at ca. 100 ml min⁻¹), the cells must be lysed in a suitable buffer. Two acceptable buffers are 1–5 mM tris(hydroxymethyl)aminomethane–acetate (TRIS-acetate), at pH 7.5–8.2 or 1–5 mM TRIS–2(N-morpholino)-ethanesulfonic acid (TRIS-MES), at pH 7.2–7.5. Both should contain ca. 5 mM ethylenediaminetetraacetic acid, 1 mM glutathione, dithiothreitol, or mercaptoethanol. Some investigators include sucrose (100–300 mM) in the buffer; however, it is not clear that this increases the efficiency of the extraction.

One of the gentlest and simplest methods used for disrupting cells is osmotic shock. Unfortunately, this technique is limited to relatively few species of marine phytoplankton, including certain pennate diatoms (e.g., *Nitzschia alba*) and naked flagellates. Most centric diatoms, however, even those weakly silicified species such as *Chaetoceros* sp., are not disturbed by this method. For such osmotically resistant cells, sonication is usually an effective disrupting technique. The cell pellet should be resuspended in ice-cold buffer in a 50-ml stainless-steel centrifuge tube, immersed in a beaker of crushed ice, and sonicated at intervals of less than 1 min using a needle probe. The suspension is examined microscopically until no intact cells are observed. Species such as *Skeletonema costatum* may be so silicified that even after 8–10 min at maximum intensity, sonication has not disrupted the majority of the cells. In such rare cases it may be easiest to grind the cells by hand with a glass fiber filter. Grinding should, however, be avoided if possible; the process may disturb the integrity of these delicate proteins.

B. Enrichment methods

After the cells are disrupted, it is necessary to prepare a membrane fraction containing the bulk ATPase activity. This step consists of a low-speed centrifugation (10 000–15 000 × g) to remove larger inclusions and debris. The supernatant is then applied to a zonal rotor or a discontinuous sucrose gradient (15, 20, 30, 45, and 50% sucrose + 1 mM $MgSO_4$ and 1 mM TRIS-acetate or TRIS-MES buffer) and sedimented at 80 000–110 000 × g (Hokin et al. 1973; Hodges and Leonard 1974). In cases where enzyme activity is low in the starting material, it may be desirable to stimulate the formation of vesicles by the addition of ca. 18 mM NaI to the low-speed supernatant (Uesugi et al. 1971). This particular salt often induces the formation of membrane vesicles and has been used with some success in the preparation of the (NO_3^-, Cl^-)-activated ATPase from *S. costatum* (Falkowski 1975b).

The pellet obtained at 80 000–110 000 × g (or the layers if sucrose gradients are used) contains membrane fragments and ATPase activity. To solubilize these membrane-bound enzymes, small quantities of detergents (e.g., Triton X-100, sodium dodecyl sulfate, deoxycholate) are added. These are most effective at low concentrations (0.01–0.1%); however, they are difficult to remove from the preparation and should not be used if solubilized enzymes are not required. It should be stressed that the detergent-solubilized proteins are no longer in a native state; as a result, specific enzyme activity may be artificially enhanced or suppressed (Askari 1974).

III. Methods of assay

Three basic methods have been used to assay membrane-bound ATPases: the inorganic phosphate analysis, the coupled enzyme assay, and the radioisotopic assay. They are described in increasing order of their sensitivity in detecting ATPase activity.

The inorganic phosphate method of Fiske and Subbarow (1925) as adapted by Lindeman (1958) has the advantage' of being rapid and simple, although its sensitivity depends on high specific activity of the enzyme preparation. First, 5 μg of protein from the enzyme extracts is incubated in 1.5 ml of a solution containing TRIS-acetate or TRIS-MES buffer, 100 μmol; $MgCl_2$, 50 μmol; TRIS-ATP (Sigma), 50 μmol; dithiothreitol, 1 μmol; and varying concentrations of the activating anions (NO_3^-, HCO_3^-, Cl^-, etc.). The reaction mixture is incubated at 20–30 C for up to 30 min, after which 1.0 ml of 60% trichloroacetic acid is added. The precipitated protein is centrifuged at 10 000 × g. Then 2 ml of the supernatant is collected and assayed for inorganic phosphate (Lindeman 1958; Strickland and Parsons 1972). The inorganic phosphate reaction is both time- and temperature-dependent and, therefore, may be difficult to control experimentally. Further, reactive phosphate remaining in the membrane preparation may lead to errors in the calculation of kinetic constants.

ATPases may be coupled in vitro to pyruvate kinase (EC 2.7.1.40) and lactate dehydrogenase (EC 1.1.1.27); the activity of the enzyme can be determined by following the change in absorption at 340 nm or fluorescence, owing to the stoichiometric oxidation of NADH. The same concentration of substrates is used as in the Fiske and Subbarow method, with the addition of 5 mM phosphoenolpyruvate, 0.2 mM NADH, and excess lactate dehydrogenase with pyruvate kinase. The ATPase activity is followed from 0 to 0.2 absorbance units in a recording spectrophotometer or in a fluorometer with a 7-37 excitation filter and a Wratten 8 emission filter. Because of the wide range of metabolic modifiers known to affect pyruvate kinase, this procedure is not useful for studies of nucleotide specificity, Na^+ vs. K^+ activation, or Arrhenius interactions.

The most sensitive assay for the determination of ATPase activity is with the use of γ-^{32}P-ATP. In this procedure the final concentrations in the assay mixture are 2 mM TRIS-ATP; 2 mM $MgCl_2$; 10 μM dithiothreitol; 1.5 mM TRIS-acetate buffer at pH 7.9; varying concentrations of activating anions; 2–3 × 10^6 cpm γ-^{32}P-ATP. The mixture is preincubated for 10 min, and the reaction is initiated by the addition of ca. 5 μg protein from the membrane extract. The final volume should be between 0.5 and 1.25 ml. After 15–30 min the reaction is terminated by the addition of a 4% solution of $(NH_4)_6Mo_7O_{24} \cdot 4H_2O$ in 10 N H_2SO_4 (Goldman and Albers 1973). Then 500 μl of *iso*butanol is added to each tube and the solutions are shaken for 1 min. A 200-μl aliquot of the *iso*butanol phase is removed and counted in 10 ml of a dioxane-naphthalene cocktail (e.g., Aquafluor, New England Nuclear) in a liquid scintillation counter. This assay is by far the most accurate and most sensitive procedure for ATPases; however, because of the expense, danger, and the short half-life of ^{32}P (14.3 days) it is not always feasible.

IV. Problems frequently encountered

Frequently, anion-activated ATPases account for only 10% of the total ATPase activity, yet these are the very enzymes one wishes to study. It is sometimes possible to separate the proteins using sodium dodecyl sulfate gel electrophoresis (Hokin et al. 1973), but some enzymes may be inactivated. The solution to this problem is surely held in the promise of affinity chromatography. Unfortunately, at present, this technique is not suitable for most ATPases.

Finally, it becomes desirable to demonstrate that the enzyme functions in the active transport of the activating ions. This problem may be partially resolved by comparing physiological features of the transport process in vivo (e.g., pH profile, temperature coefficient, inhibitor studies) directly with the ATPase in vitro (Fisher et al. 1970; Falkowski 1975a, 1975b; Falkowski and Stone 1975). Recently, however, purified $(Na^+ + K^+)$-ATPase has been reinserted into an artificial phospholipid bilayer to demonstrate active transport directly (Hilden and Hokin 1975).

V. Acknowledgments

I thank Dr. V. Palaty for teaching me all I know about cell membranes, and the late Dr. E. Bruce Tregunna for his encouragement and ready smile.

VI. References

Askari, A. (ed.). 1974. Properties and functions of $(Na^+ + K^+)$-activated adenosinetriphosphatase. *Ann. N.Y. Acad. Sci.* 242, 1–741.
Cambraia, J., Balke, N. E., and Hodges, T. K. 1976. Cation and anion-sensitive ATPases of the plasma membrane of oat roots. *Plant Physiol. Suppl.* 57, 84.

Falkowski, P. G. 1975a. Nitrate uptake in marine phytoplankton: Comparison of half-saturation constants from seven species. *Limnol. Oceanogr.* 20, 412–17.

Falkowski, P. G. 1975b. Nitrate uptake in marine phytoplankton: (Nitrate, chloride)-activated adenosine triphosphatase from *Skeletonema costatum* (Bacillariophyceae). *J. Phycol.* 11, 323–6.

Falkowski, P. G., and Stone, D. P. 1975. Nitrate uptake in marine phytoplankton: Energy sources and the interaction with carbon fixation. *Marine Biol.* 32, 77–84.

Fisher, J. D., Hansen, D., and Hodges, T. K. 1970. Correlation between ion fluxes and ion-stimulated adenosine triphosphatase activity of plant roots. *Plant Physiol.* 46, 812–4.

Fiske, C. H., and Subbarow, Y. 1925. The colorimetric determination of phosphorus. *J. Biol. Chem.* 66, 375–400.

Goldman, S. S., and Albers, R. W. 1973. Sodium-potassium-activated adenosine triphosphastase. IX. The role of phospholipids. *J. Biol. Chem.* 248, 867–74.

Grisham, C. M., and Barnett, R. E. 1972. The interrelationship of membrane and protein structure in the functioning of the $(Na^+ + K^+)$-activated ATPase. *Biochim. Biophys. Acta* 266, 613–24.

Hemmingsen, B. B. 1971. A mono-silicic acid stimulated adenosinetriphosphatase from protoplasts of the apochlorotic diatom *Nitzschia alba*. Ph.D. thesis, University of California, San Diego. 132 pp.

Hilden, S., and Hokin, L. E. 1975. Active potassium transport coupled to active sodium transport in vesicles reconstituted from purified sodium and potassium ion-activated adenosine triphosphatase from the rectal gland of *Squalus acanthias*. *J. Biol. Chem.* 250, 6296–303.

Hodges, T. K., and Leonard, R. T. 1974. Purification of a plasma membrane-bound adenosine triphosphatase from plant roots. *Methods Enzymol.* 32, 392–406.

Hodges, T. K., Leonard, R. T., Bracker, C. E., and Keenan, T. W. 1972. Purification of an ion-stimulated adenosine triphosphatase from plant roots: Association with plasma membranes. *Proc. Natl. Acad. Sci. U.S.A.* 69, 3307–11.

Hokin, L. E., Dahl, J. L., Deupree, J. D., Dixon, J. F., Hackney, J. F., and Perdue, J. F. 1973. Studies on the characterization of the sodium-potassium transport adenosine triphosphatase. X. Purification of the enzyme from the rectal gland of *Squalus acanthias*. *J. Biol. Chem.* 248, 2593–605.

Karlsson, J., and Kylin, A. 1974. Properties of Mg^{2+}-stimulated and $(Na^+ + K^+)$-activated adenosine-5'-triphosphatase from sugar beet cotyledons. *Physiol. Plant.* 32, 136–42.

Kuiper, P. J. C., Kähr, M., Stuiver, C. E. E., and Kylin, A. 1974. Lipid composition of whole roots and of Ca^{2+}, Mg^{2+}-activated adenosine triphosphatases from wheat and oat as related to mineral nutrition. *Physiol. Plant.* 32, 33–6.

Lindeman, W. 1958. Observations on the behaviour of phosphate compounds in *Chlorella* at the transition from dark to light. *Proc. 2nd Int. Conf. Peaceful Uses Atomic Energy* 24, 8–15.

Maslowski, P., and Komoszyński, M. 1974. Purification and properties of adenosinetriphosphatase from *Zea mays* seedling microsomes. *Phytochemistry* 13, 89–92.

Simpkins, H., and Hokin, L. E. 1973. Studies on the characterization of the sodium-potassium transport adenosinetriphosphatase. XIII. On the organization and role of phospholipids in the purified enzyme. *Arch. Biochem. Biophys.* 159, 897–902.

Skou, J. C. 1957. The influence of some cations on an adenosine triphosphatase from peripheral nerves. *Biochim. Biophys. Acta* 23, 394–401.

Stam, A. C., Jr., Weglicki, W. B., Gertz, E. W., and Sonnenblick, E. H. 1973. A calcium-stimulated, oubain-inhibited ATPase in a myocardial fraction enriched with sarcolemma. *Biochim. Biophys. Acta* 298, 927–31.

Strickland, J. D. H., and Parsons, T. R. 1972. *A Practical Handbook of Seawater Analysis.* Bull. Fish. Res. Board Can. 167, 311 pp.

Uesugi, S., Dulak, N. C., Dixon, J. F., Hexum, T. D., Dahl, J. L., Perdue, J. F., and Hokin, L. E. 1971. Studies on the characterization of the sodium-potassium transport adenosine triphosphatase. VI. Large-scale partial purification and properties of a lubrol-solubilized bovine brain enzyme. *J. Biol. Chem.* 246, 531–43.

27: Extracellular acid phosphatase of Ochromonas danica

N. J. PATNI AND S. AARONSON

Biology Department, Queens College,
City University of New York, Flushing, New York 11367

CONTENTS

I	Introduction	page 264
II	Equipment and supplies	264
	A. Materials	264
	B. Reagents	264
	C. Enzyme source	264
III	Assay method	265
	A. Limitations and variations	265
IV	Acknowledgments	266
V	References	266

I. Introduction

Acid phosphatase or orthophosphoric phosphohydrolase (EC 3.1.3.2) hydrolyzes a wide variety of orthophosphate esters. Depending on the substrate used, acid phosphatase can be assayed by the formation of phenol from phenylphosphate, p-nitrophenol from p-nitrophenylphosphate, inorganic phosphate from many phosphate esters, as well as 3-O-methylfluorescein from its phosphate ester. The described assay is based on the formation of p-nitrophenol from the hydrolysis of p-nitrophenylphosphate, because of ease and convenience; p-nitrophenol is measured spectrophotometrically at 410 nm.

II. Equipment and supplies

A. Materials

All glassware should be cleaned with chromic–sulfuric acid solution, washed many times, rinsed thoroughly (at least six to eight times) with glass-distilled water, and dried. Matched spectrophotometer cells (cuvettes) should be cleaned with detergent and dried with 95% alcohol or acetone. Spectrophotometer is set with wavelength at 410 nm.

B. Reagents

1. Sodium acetate buffer, 0.1 M, pH 4.8. Mix 30 ml of 0.2 M sodium acetate (16.4 g of $C_2H_3O_2Na$ or 27.2 g of $C_2H_3O_2Na\cdot3H_2O$ in 1000 ml) and 20 ml of 0.2 M acetic acid (11.55 ml of glacial acetic acid in 1000 ml) and bring to 100 ml with distilled water.

2. p-Nitrophenylphosphate (15 mM) in buffer. Dissolve 39.4 mg of p-nitrophenylphosphate (Sigma or ICN Pharmaceuticals) in 10 ml of 0.1 M sodium acetate buffer.

3. NaOH, 0.1 N. Dissolve 4 g of NaOH in 1 l of distilled water.

C. Enzyme source

Ochromonas danica Pringsheim L 933/2 of Culture Centre of Algae and Protozoa (Cambridge) is maintained and grown as described earlier (Patni and Aaronson 1974). Supernatant, obtained after harvesting cells from the exponential phase by centrifuging at 4340 × *g* for 20 min in a Sorval RC-2B at 4 C, is used for the assay of extracellular acid phosphatase secreted into the medium. Purification of the secreted extracellular acid phosphatase has been described by Patni and Aaronson (1974).

III. Assay method

The following reagents are present in total volume of 1.0 ml: 0.5 ml of 15 mM *p*-nitrophenylphosphate in acetate buffer, pH 4.8; 0.3 ml of acetate buffer, pH 4.8; and 0.2 ml of culture supernatant as enzyme source (20–50 μg total protein). Incubate the reaction mixture in duplicate for 30 min at 37 C. For an assay control, incubate 0.5 ml of *p*-nitrophenylphosphate in acetate buffer and 0.3 ml of buffer without enzyme under similar conditions. The reaction is terminated by adding 5 ml of 0.1 N NaOH. Add 0.2 ml of enzyme to the control and read the absorbance of *p*-nitrophenol at 410 nm against control. The molar extinction coefficient used for *p*-nitrophenol under these conditions is 1.75×10^4 cm^2 mol^{-1}. One unit of enzyme activity is defined as the quantity of enzyme required to liberate 1 μmol of *p*-nitrophenol per minute. The specific activity is expressed as enzyme units per milligram of protein. Protein is estimated either by the method of Lowry et al. (1951) with bovine serum albumin as standard, or spectrophotometrically as described by Layne (1957).

A. Limitations and variations

The rate of reaction is linear with respect to time and enzyme concentration under these conditions. For samples having high enzyme activity, the enzyme should be diluted with buffer or time of incubation should be decreased. The rate of enzyme activity is linear for ca. 24 h. For samples having very low acid phosphatase activity, fluorogenic assays should be used (Rotman et al. 1963; Perry 1972).

For enzyme activity toward different phosphate esters, free inorganic phosphate can be determined by the method of Fiske and Subbarow (1925) or by the method of Chen et al. (1956) as modified by Ames and Dubin (1960). However, liberation of free *p*-nitrophenol and free inorganic phosphate from phenolic phosphate esters like *p*-nitrophenylphosphate should be assayed simultaneously to see if there is any phosphotransferase activity in the enzyme preparations.

Acid phosphatase activity could be assayed at different pH values by substituting the 0.1 M sodium acetate buffer with the following buffers of the same molarity: glycine-HCl (pH 1.5–3.0); citric acid–sodium citrate (pH 3.0–6.2); succinic acid–TRIS (pH 3.5–5.5); succinic acid–NaOH (pH 3.8–6.0); histidine·HCl (pH 6.0–6.5); or TRIS-maleate (pH 5.4–7.0).

IV. Acknowledgments

This work was supported by grant BMS 74-08918 from the National Science Foundation to S.A. and grant NIH 5-S05-RR-07064 from the National Institute of Health to Queens College.

V. References

Ames, B. N., and Dubin, D. T. 1960. The role of polyamines in the neutralization of bacteriophage deoxyribonucleic acid. *J. Biol. Chem.* 235, 769–75.

Chen, P. S., Toribara, T. Y., and Warner, H. 1956. Microdetermination of phosphorus. *Anal. Chem.* 28, 1756–8.

Fiske, C., and Subbarow, Y. 1925. The colorimetric determination of phosphorus. *J. Biol. Chem.* 66, 375–400.

Layne, E. 1957. Spectrophotometric and turbidimetric methods for measuring proteins. *Methods Enzymol.* 3, 451–4.

Lowry, O. H., Rosebrough, N. J., Farr, A. L., and Randall, R. J. 1951. Protein measurement with the Folin phenol reagent. *J. Biol. Chem.* 193, 265–75.

Patni, N. J., and Aaronson, S. 1974. Partial characterization of the intra- and extracellular acid phosphatase of an alga, *Ochromonas danica. J. Gen. Microbiol.* 83, 9–20.

Perry, M. J. 1972. Alkaline phosphatase activity in subtropical central north Pacific waters using a sensitive fluorometric method. *Marine Biol.* 15, 113–9.

Rotman, B., Zderic, J. A., and Edelstein, M. 1963. Fluorogenic substrates for β-galactosidases and phosphatases derived from fluorescein (3,6-dihydroxyfluoran) and its monomethyl ether. *Proc. Natl. Acad. Sci. U.S.A.* 50, 1–6.

Section IV

Physiological and biochemical processes

28: Determination of photosynthetic rates and ^{14}C photoassimilatory products of brown seaweeds

BRUNO P. KREMER

*Botanisches Institut der Universität Köln,
Gyrhofstrasse 15, D-5000 Köln 41, Germany*

CONTENTS

I	**Introduction**	*page* **270**
II	**Materials**	**270**
	A. Technical equipment	270
	B. Test organisms	271
III	**Methods**	**271**
	A. ^{14}C incubation	271
	B. Analytical procedures	273
IV	**Discussion**	**280**
V	**References**	**280**

I. Introduction

Photosynthetic carbon assimilation is a basic process in all photoautotrophic organisms. This chapter deals with some selected methods for the determination of (1) the rates of photosynthesis expressed as net carbon uptake using light-dependent incorporation of ^{14}C and (2) the typical photoassimilates after relatively long-term incubations (≥ 10 min photosynthesis) in a $^{14}CO_2$ medium, particularly evaluating accumulated end products (photosynthates) rather than intermediate compounds.

The objectives of the study in hand largely determine which of the steps outlined below are required. These methods are designed for use with minimal reference to original literature; however, it may be generally profitable to consult the more comprehensive specialized surveys (Paech and Tracey 1955, 1956; Ruhland 1960; Linskens and Tracey 1962; San Pietro 1971, 1972).

II. Materials

A. Technical equipment

Because most of the methods described in this chapter are based on the application and analysis of compounds labeled with radioisotopes, it is absolutely necessary that the user handle radioactive materials with appropriate caution, particularly with respect to uncontrolled contamination.

1. Glassware. Usual laboratory equipment, except plastic materials, including radiochemically clean pipettes, tubes, beakers, etc., is required. More specialized articles are mentioned in the description of the procedures.

2. Apparatus. Working with radiotracers usually requires a radioisotope laboratory and some relatively expensive equipment for detecting and quantitatively measuring radioactivity. Liquid scintillation techniques are now widely used. A wide-window methane flow counter coupled with a linear rate meter is very useful for counting single radioactive spots on chromatograms or for scanning a one-dimensional radiochromatogram. A simple β detector for continuously monitoring ^{14}C activity during incubation procedures or any other analytical step involving labeled materials is recommended as well.

B. *Test organisms*

The methods of ^{14}C incubation and subsequent analysis of the ^{14}C-photosynthate have often been used in investigations on certain brown algal species such as *Giffordia mitchellae* (Harv.) Ham., *Laminaria hyperborea* (Gunn.) Fosl., *Laminaria saccharina* (L.) Lam., *Pelvetia canaliculata* (L.) Dcne. et Thur., *Fucus vesiculosus* L., *Fucus serratus* L., and others (Bidwell 1958; Bidwell et al. 1958; Yamaguchi et al. 1966; Kremer and Willenbrink 1972; Schmitz et al. 1972; Kremer 1973, 1975a; Weidner et al. 1975). These techniques or their modifications are not restricted to the Phaeophyceae, but are applicable to representatives of other algal classes (Craigie et al. 1968; Nagashima et al. 1969).

III. Methods

A. ^{14}C *incubation*

The rate of net photosynthetic carbon assimilation in benthic macroalgae can be very precisely measured by light-dependent incorporation of radiocarbon (Šesták et al. 1971). Because photosynthesis is a process that is influenced by a variety of environmental and biological factors (Heath 1969; Marcelle 1975), important parameters such as irradiance, temperature, pH range, salinity, and total CO_2 concentration should be recorded routinely, especially when photosynthetic rates are to be expressed in absolute terms. This complex of factors affecting efficiency of photosynthesis cannot be detailed further in this chapter. A summarizing review on these topics is still lacking; special aspects of seawater chemistry have been treated by Strickland and Parsons (1968), whereas problems arising from underwater light measurements have been discussed by Jerlov and Nygård (1969) or Burr and Duncan (1972).

1. In situ incubation under field conditions. To determine the rate of photosynthesis under natural conditions, selected parts of the thallus are placed either in closed glass bottles (Wallentinus 1975) or in clear polyethylene bags of an appropriate size tightly fixed around the thallus area (Parker 1965; Towle and Pearse 1973; Schmitz and Srivastava 1975; Schmitz and Lobban 1976). Radioactive compounds can be added directly to these containers or can be injected through a rub-

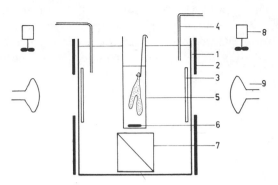

Fig. 28-1. Diagram of an apparatus for photosynthetic incubation experiments. *1*, glass aquarium; *2*, cover; *3*, infrared filter; *4*, tubes to thermostatically controlled bath; *5*, incubation chamber containing $H^{14}CO_3^-$-seawater medium; *6*, magnetic stirring bar; *7*, magnetic stirrer; *8*, ventilation; *9*, tungsten bulb.

ber serum stopper incorporated in the apparatus. For rate determinations, about 5 μCi $Na_2^{14}CO_3$ or $NaH^{14}CO_3$ per liter of seawater medium will give good results in incubation periods exceeding 10 min up to several hours. If a radiochromatographic analysis of the photosynthates is desired, the specific ^{14}C activity should be increased. Calculation of the assimilation rates achieved under the appropriate incubation conditions is based (a) on the volume of the incubation container used and (b) on total carbon uptake in relation to $^{12}CO_2$ and $^{14}CO_2$ content of the medium. Total inorganic carbon concentration (CO_2, HCO_3^-, H_2CO_3) in seawater is determined according to Strickland and Parsons (1968).

2. Incubation under laboratory conditions. Thalli collected on the shores and transported to the laboratory should be adapted carefully to the intended test conditions prior to use, the more important parameters of the natural environment being simulated as far as possible. Before samples are exposed to the ^{14}C-medium in the laboratory, they are usually preincubated in a nonradioactive medium for at least 15 min to provide steady-state conditions of photosynthesis (Steemann Nielsen 1942).

a. $H^{14}CO_3^-$-seawater. A diagrammatic representation of an incubation apparatus coupled with an external thermostat for maintaining constant temperature even during longer incubation periods is given in Fig. 28-1. Light sources (slide projectors or tungsten bulbs) can be varied by a rheostat if minor changes in the spectral emission can be tolerated. Specific ^{14}C activity of the seawater medium used depends on the objectives of the experiment: low concentrations, about 5–10

μCi NaH^{14}CO$_3$, for rate determinations; higher concentrations, up to 1 mCi/50 ml, for particular biochemical investigations.

In special investigations on long-distance translocation in members of the Laminariales some slightly modified incubation techniques have been adopted (Hellebust and Haug 1972; Nicholson and Briggs 1972; Schmitz et al. 1972).

b. ^{14}CO$_2$-atmosphere. An interesting modification of the incubation conditions arises from the ecological question whether an alga may also photosynthesize ^{14}CO$_2$ when exposed to a radioactive gas phase (Bidwell and Craigie 1963; Kremer and Schmitz 1973). In such experiments ^{14}CO$_2$ can be released from Na$_2$^{14}CO$_3$ or NaH^{14}CO$_3$ by addition of a few milliliters of 10% HClO$_4$ in a closed incubation chamber in which the thallus samples have been suspended. In similar experiments also infrared gas analyzers have been used (Johnson et al. 1974; Brinkhuis et al. 1976).

B. *Analytical procedures*

1. Sample fixation. There are several techniques for terminating metabolic reactions at the end of an experiment. The choice of the procedure depends on the analytical steps to follow: If the assimilation rates are to be referred to chlorophyll or protein content or to a dry-weight basis, thallus samples have to be deep-frozen on dry ice (-76 C) or in liquid N$_2$ (-196 C) and subsequently freeze-dried. Other reference systems such as thallus fresh weight or surface area allow sample fixation in cold or boiling ethanol (80%). Samples should be thoroughly rinsed with seawater or freshwater to remove adhering inorganic ^{14}C-carbon.

2. Determination of ^{14}C uptake
a. Solubilization of total tissue. For rapid assay of total ^{14}C incorporated in the thallus even of solid Phaeophyceae, the routine method of Lobban (1974) is generally recommended. Up to 250 mg fresh blade or equivalents of dry matter is mixed together with 0.2 ml 60% HClO$_4$ in a scintillation vial. Then 0.4 ml H$_2$O$_2$ is added, the vial is capped and heated to 70–80 C. The mixture is cooled to room temperature; 6 ml ethylene glycolmonoethyl ether (= 2-ethoxyethanol) and 10 ml scintillation cocktail (6.0 g PPO in 1 l toluene) are added. The sample is immediately counted.

b. Combustion of tissue to ^{14}CO$_2$. Freeze-dried tissue can be directly combusted in an oxygen atmosphere in a closed volume. Carbon dioxide released from ^{14}C-labeled organic material is absorbed by ethylamine or phenylethylamine (Kalberer and Rutschmann 1961). Oxidation takes place in special reaction columns or directly in scintillation vials. Compared with solubilization of tissue, this method is relatively slow and requires special equipment. Technical details are described in the operation manuals provided by the suppliers (e.g., Oxymat/Intertechnique; Sample Oxidizer 305/Packard; Micro-Mat/Berthold-Friesecke). A simple method of this type has been described by Burnison and Perez (1974).

c. Liquid counting of extracts. Liquid scintillation counting, either of tissue extracts or of tissue converted to a form that is measurable in a scintillation liquid, is now a widely applied method of choice. Many instrument systems (e.g., Beckman, Packard, Tracerlab) are available. Most provide for the correction of number of counts per minute recorded to absolute activity (disintegrations per minute, dpm) by the external standardization method, because counting efficiency is often reduced by chemical and physical quenching effects. Scintillation cocktails often used are composed of 4 g p-terphenyl and 100 mg di-methyl-POPOP (= 2,2'-p-phenylen-bis-[4-methyl-5-phenyloxazol]) in 1 l toluene, or (for aqueous extracts) of 12 g butyl-PBD (= 2-[4-t-methylphenyl]-5-[4-biphenylyl]-1,2,3-oxadiazol) in 1 l toluene and 1 l methanol.

3. Fractionation of products of photosynthesis. In long-term experiments an increasing amount of assimilated ^{14}C is converted to polymeric compounds of structure and storage (Bidwell 1958; Yamaguchi 1966; Hellebust and Haug 1972; Kremer 1975a). Analytical differentiation between the major groups of assimilates can be easily achieved by means of their solubility.

a. Soluble photosynthates. Grinding the thallus material with quartz sand in a mortar or other procedures of homogenization (blenders or dismembrators) followed by repeated extraction with hot or cold ethanol (80%, 50%) under vigorous shaking provide exhaustive extraction of low-molecular-weight assimilates including phosphate esters, amino acids, sugars, polyols, and organic acids (Bidwell et al. 1958; Yamaguchi et al. 1966; Hellebust and Haug 1972; Kremer and Willenbrink 1972; Schmitz et al. 1972). This material extractable in ethanol is generally regarded as the crude extract.

b. Insoluble compounds. The polymeric compounds in Phaeophyceae comprising mainly proteins, laminaran, alginic acid, and some further polysaccharides (Percival 1970), apart from nucleic acids, can be further separated by fractionation. Polymeric carbohydrates can be isolated from ethanol-extracted thallus material by treatment with dilute HCl, pH 2 (acid extract = "laminaran fraction"); then 0.5% Na_2CO_3 (alkali extract = "alginic acid fraction"); and finally in 1–4% NaOH or 24% KOH (= "cellulose fraction"). This or slightly modified flow schemes of extractions have been used by Bidwell et al. (1958, 1972), Yamaguchi et al. (1966), Craigie et al. (1971), Hellebust and Haug (1972).

4. Radiochromatography of crude extracts. Low-molecular-weight soluble assimilates formed from $^{14}CO_2$ during longer periods of photosynthesis can be characterized by several chromatographic techniques. Because a complex variety of single compounds is generally found even after short incubation times, a prefractionation of the crude (ethanol) extracts is desirable.

a. Ion-exchange chromatography. Column chromatography using 1×10-cm columns of certain ion exchangers, such as Amberlite, Dowex, Lewatit, or Permutit, provide effective prefractionation into neutral, basic, and acidic compounds. Besides others (Craigie 1963; Nagashima et al. 1969; Atkins and Canvin 1971; Hofer 1974), the procedure described by Splittstoesser (1969) has often been used. An aliquot of crude extract is run into a column of (1) Dowex 50 W $\times$ 8 (H^+-form). The neutral and acidic fraction is passed through and washed with 50 ml H_2O onto a column of (2) Dowex 1 $\times$ 10 ($HCOO^-$ form) directly connected with column 1. The effluent of column 2 represents the neutral fraction. The adsorbed basic fraction is eluted from column 1 with 50 ml 10 N HCl, the acidic material from column 2 with 50 ml 8 N formic acid. Columns are regenerated by washing with H_2O until the effluent is neutral.

b. Paper chromatography. The ^{14}C-labeled constituents of each fraction can be further fractionated by one- or two-dimensional paper chromatography (PC) and/or thin-layer chromatography (TLC). For a preparative scale, PC is generally recommended, as paper chromatograms allow loading with considerably greater volumes of extracts. Sheets of 60×60-cm (two-dimensional PC) or 60×15-cm (one-dimensional PC) of Schleicher and Schüll 2043b equivalent to Whatman No. 1 are used. Table 28-1 gives a survey of some common solvent systems. Spots or bands of the separated ^{14}C-labeled compounds are localized either by contact autoradiography (exposure of the chromatogram against X-ray film or X-ray paper) or by scanning with a flow counter coupled with a linear rate meter.

c. Thin-layer chromatography. If the crude extracts or fractions are sufficiently ^{14}C-labeled (more than ca. 20 000 dpm/0.1 ml liquid), a very effective separation by TLC as proposed by Feige et al. (1969) can be carried out. Solvent systems to be used are included in Table 28-2. The diagrammatic representation in Fig. 28-2 shows the position of frequently occurring metabolites of plant extracts after two-dimensional TLC on normal or microcrystalline cellulose on 20×20-cm

Table 28-1. *Some solvent systems for separation of photosynthates by paper chromatography (PC) and thin-layer chromatography (TLC)*

Composition	Separation of	PC	TLC	Reference
i-Butyric acid:1 N NH_4OH:EDTA = 100:60:46.7 (v/v/w)	Crude extracts	x		Tyszkiewicz (1962)
n-Butanol:propionic acid:H_2O = 375:180:245	Crude extracts	x		Tyszkiewicz (1962)
i-Butyric acid:n-butanol:i-propanol:n-propanol:H_2O:NH_4OH:EDTA = 100:3:3:14:38:4:50	Crude extracts		x	Feige et al. (1969)
n-Butanol:n-propanol:n-propionic acid:H_2O = 100:44:71:93	Crude extracts		x	Feige et al. (1969)
n-Butanol:acetic acid:H_2O = 5:1:4 (upper phase)	Crude extracts		x	Feige et al. (1969)
n-Butanol:ethanol (abs.):H_2O = 45:5:50	Amino acids	x	x	Quillet (1957)
n-Butanol:acetic acid:H_2O = 4:1:1	Amino acids	x	x	Reed (1950)
Ethanol:NH_4OH = 95:5	Organic acids	x	x	Wallen and Geen (1971)
Butyl formate:formic acid:H_2O = 10:4:1	Organic acids	x	x	Blundstone (1963)
Ethyl acetate:acetic acid:formic acid:H_2O = 18:3:1:4	Sugars	x	x	Percival and Smestad (1972)
Ethyl methylketone:acetic acid:0.75 M boric acid = 40:10:9	Sugars and polyols	x	x	Kremer (1975b)
Ethyl methylketone:acetic acid:0.70 M boric acid = 100:10:5	Polyols	x		Kremer (1975b)

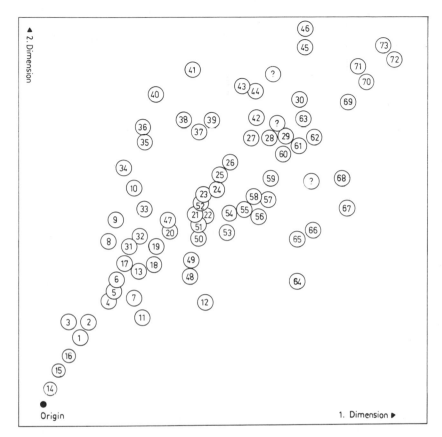

Fig. 28-2. Two-dimensional thin-layer chromatogram on 20 × 20-cm plates, according to Feige et al. (1969). Nomogram for the position of frequently occurring metabolites.

1–13. Phosphate esters: *1,* fructose-1,6-dP; *2,* uridine-diphosphoglucose; *3,* ribulose-1,5-dP; *4,* glucose-6-P; *5,* fructose-6-P; *6,* fructose-1-P; *7,* ribose-5-P; *8,* 3-phosphoglycerate; *9,* phosphoglycolate; *10,* phosphoenolpyruvate; *11,* ATP; *12,* NADH + FADH; *13,* dihydroxyacetone-P.

14–30. Carbohydrates: *14–16,* oligosaccharides; *17,* lactose + raffinose; *18,* maltose; *19,* trehalose; *20,* sucrose; *21,* glucose; *22,* galactose; *23,* fructose; *24,* hexitols (mannitol, dulcitol, sorbitol); *25,* xylose + arabinose; *26,* pentitols (arabitol, xylitol, ribitol); *27,* erythritol; *28,* glyceraldehyde; *29,* glycolaldehyde; *30,* glycerol.

31–46. Organic acids: *31,* galacturonate; *32,* glucuronate; *33,* orotic acid; *34,* tartarate; *35,* aconitate; *36,* citrate; *37,* ketoglutarate; *38,* malate; *39,* glycerate; *40,* glyoxylate; *41,* fumarate; *42,* oxaloacetate; *43,* glycolate; *44,* pyruvate; *45,* succinate; *46,* lactate.

47–73. Amino acids: *47,* aspartate; *48,* cystine; *49,* diaminopimelic acid; *50,* taurine; *51,* phosphoethylamine; *52,* glutamate; *53,* asparagine; *54,* serine; *55,* glycine; *56,* glutamine; *57,* hydroxyproline; *58,* citrulline; *59,* threonine; *60,* alanine; *61,* β-alanine; *62,* proline; *63,* tyrosine; *64,* cysteine; *65,* lysine; *66,* ornithine; *67,* histidine; *68,* arginine; *69,* valine; *70,* aminobutyric acid; *71,* methionine; *72,* leucine + *i*-leucine; *73,* phenylalanine. (With kind permission of Dr. G. B. Feige, Köln.)

Table 28-2. *Concentrations and amounts of reference substances in test solutions for chromatography*

Compound	Concentration (mg ml^{-1}) required for		Amount (μl) to apply in	
	PC	TLC	PC	TLC
Amino acids	10	1	1	1
Organic acids	50	5	5	1
Sugars/polyols	20	2	2	2
Phosphate esters	50	10	2	2

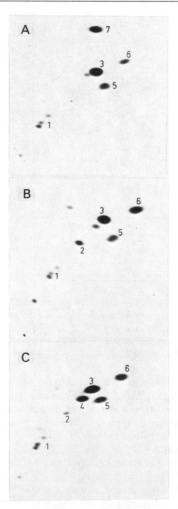

Fig. 28-3. ^{14}C assimilation patterns of some marine brown algae after two-dimensional thin-layer chromatography, according to Feige et al. (1969). *A, Giffordia mitchellae; B, Fucus serratus; C, Pelvetia canaliculata.* Some major photosynthates: *1*, sugar phosphates; *2*, aspartate; *3*, mannitol; *4*, volemitol; *5*, serine/glycine; *6*, alanine; *7*, glycolate.

Table 28-3. *Frequently used detection reagents for tracing reference substances on chromatograms*

Compound	Spray reagent	Reference
Amino acids	Ninhydrin/n-butanol	Consden et al. (1944)·
Organic acids	Aniline-glucose	Linskens (1959)
	Bromophenol blue	Blundstone (1963)
Keto acids	Dinitrophenyl hydrazine	Mehlitz et al. (1963)
Aldoses	Aniline-phthalate	Partridge (1948)
Ketoses	Urea-HCl	Dedonder (1952)
Polyols	Periodate-benzidine	Cifonelli and Smith (1954)
Phosphate esters	Perchlorate–ammonium molybdate	Hanes and Isherwood (1949)

plates. Figure 28-3 gives some samples of autoradiograms obtained by this method.

5. Identification of single photosynthates

a. Co-chromatography. The position of a ^{14}C-labeled assimilate on a thin-layer chromatogram (Fig. 28-2) developed in the solvent systems of Feige et al. (1969) gives preliminary information on the nature of the compound. For further characterization and identification, spots are scraped from the thin-layer or cut from the paper chromatogram and eluted with H_2O or 50% EtOH. Substances recovered in this manner are rechromatographed in any of the solvents listed in Table 28-1 together with authentic reference substances. Compounds supposed to be involved in an assimilate pattern are added either to the origin or in a separate channel of further test chromatograms. Concentrations and amounts of test solutions to be applied are given in Table 28-2. Coincident R_f values of unknown assimilates and co-chromatographed reference substances in various solvent systems are a generally accepted test of identity.

b. Specific color reactions. The various metabolites, such as amino acids, sugars and derivatives, phosphate esters, and organic acids, can be detected on the chromatograms by several specific spray reagents and the position of the spots compared with the ^{14}C-labeled photosynthates located by autoradiography or scanning. Table 28-3 gives some practical and widely used reagents for detecting specific compounds. Detailed information on their preparation should be taken from original literature or from standard manuals of chromatography (Linskens 1959; Stahl 1969).

IV. Discussion

The procedure of $^{14}CO_2$ assimilation in the light or in the dark yields reliable values of rates of net carbon fixation under the appropriate incubation conditions. The rates, often expressed as amounts of carbon fixed per hour and surface area or pigment concentrations, are easily reproduced provided homogeneous and comparable plant material of the same developmental stages have been used. The methods for the analysis of single major products of photosynthesis are in general applicable for the characterization of any algal constituent. Techniques other than the simple chromatographic methods are usually less effective and require more time and equipment. For isolation, purification, and identification of any algal constituent involved in intermediary metabolism or in storage, the chromatographic procedures briefly presented here give best results.

V. References

Atkins, C. A., and Canvin, D. T. 1971. Analysis of ^{14}C-labeled acidic photosynthetic products by ion-exchange chromatography. *Photosynthetica* 5, 341–51.

Bidwell, R. G. S. 1958. Photosynthesis and metabolism of marine algae. II. A survey of rates and products of photosynthesis in $C^{14}O_2$. *Can. J. Bot.* 36, 337–49.

Bidwell, R. G. S., and Craigie, J. S. 1963. A note on the greatly reduced ability of *Fucus vesiculosus* to absorb or to evolve CO_2 when not submerged. *Can. J. Bot.* 41, 179–82.

Bidwell, R. G. S., Craigie, J. S., and Krotkov, G. 1958. Photosynthesis and metabolism in marine algae. III. Distribution of photosynthetic carbon from $C^{14}O_2$ in *Fucus vesiculosus*. *Can. J. Bot.* 36, 581–90.

Bidwell, R. G. S., Percival, E., and Smestad, B. 1972. Photosynthesis and metabolism of marine algae. VIII. Incorporation of ^{14}C into the polysaccharides metabolized by *Fucus vesiculosus* during pulse labeling experiments. *Can. J. Bot.* 50, 191–7.

Blundstone, H. A. W. 1963. Paper chromatography of organic acids. *Nature (London)* 197, 377.

Brinkhuis, B. H., Tempel, N. R., and Jones, R. F. 1976. Photosynthesis and respiration of exposed salt-marsh fucoids. *Marine Biol.* 34, 349–59.

Burnison, B. K., and Perez, K. T. 1974. A simple method for the dry combustion of ^{14}C-labeled materials *Ecology* 55, 899–902.

Burr, A. H., and Duncan, M. J. 1972. Portable spectroradiometer for underwater environments. *Limnol. Oceanogr.* 17, 466–75.

Cifonelli, J. A., and Smith, F. 1954. Detection of glycosides and other carbohydrate compounds on paper chromatograms. *Anal. Chem.* 26, 1132–4.

Consden, R., Gordon, A. H., and Martin, A. J. P. 1944. Qualitative analysis of proteins: A partition chromatographic method using paper. *Biochem. J.* 38, 224–32.

Craigie, J. S. 1963. Dark fixation of C^{14}-bicarbonate by marine algae. *Can. J. Bot.* 41, 317–25.

Craigie, J. S., Leigh, C., Chen, L. C.-M., and McLachlan, J. 1971. Pigments, polysaccharides and the photosynthetic products of *Phaeosaccion collinsii*. *Can. J. Bot.* 49, 1067–74.

Craigie, J. S., McLachlan, J., and Tocher, R. D. 1968. Some neutral constituents of the Rhodophyceae with special reference to the occurrence of the floridosides. *Can. J. Bot.* 46, 605–11.

Dedonder, R. 1952. Glucides et composés voisins. *Bull. Soc. Chim. Fr.* 19, 874–83.

Feige, B., Gimmler, H., Jeschke, W. D., and Simonis, W. 1969. Eine Methode zur dünnschichtchromatographischen Auftrennung von ^{14}C- and ^{32}P-markierten Stoffwechselprodukten. *J. Chromatogr.* 41, 80–90.

Hanes, C. S., and Isherwood, F. A. 1949. Separation of the phosphoric esters on the filter paper chromatogram. *Nature* 164, 1107–12.

Heath, O. V. S. 1969. *The Physiological Aspects of Photosynthesis*. Heinemann, London. 310 pp.

Hellebust, J. A., and Haug, A. 1972. Photosynthesis, translocation, and alginic acid synthesis in *Laminaria digitata* and *Laminaria hyperborea*. *Can. J. Bot.* 50, 169–76.

Hofer, H. W. 1974. Separation of glycolytic metabolites by column chromatography. *Anal. Biochem.* 61, 54–61.

Jerlov, N. G., and Nygård, K. 1969. A quanta and energy meter for photosynthetic studies. *Københavens Univ. Inst. Fys. Oceanogr. Rep.* 10, 1–19.

Johnson, W. S., Gigon, A., Gulmon, S. L., and Mooney, H. A. 1974. Comparative photosynthetic capacities of intertidal algae under exposed and submerged conditions. *Ecology* 55, 450–3.

Kalberer, F., and Rutschmann, J. 1961. Eine Schnellmethode zur Bestimmung von Tritium, Radiokohlenstoff und Radioschwefel in beliebigem organischem Probenmaterial mittels des Flüssigkeits-Scintillations-Zählers. *Helv. Chim. Acta* 44, 1956–66.

Kremer, B. P. 1973. Untersuchungen zur Physiologie von Volemit in der marinen Braunalge *Pelvetia canaliculata*. *Marine Biol.* 22, 31–5.

Kremer, B. P. 1975a. Mannitmetabolismus in der marinen Braunalge *Fucus serratus*. *Z. Pflanzenphysiol.* 74, 255–63.

Kremer, B. P. 1975b. Separation of isomeric pentitols and hexitols by paper and thin-layer chromatography. *J. Chromatogr.* 110, 171–3.

Kremer, B. P., and Schmitz, K. 1973. CO_2-Fixierung und Stofftransport in benthischen marinen Algen. IV. Zur ^{14}C-Assimilation einiger litoraler Braunalgen im submersen und emersen Zustand. *Z. Pflanzenphysiol.* 68, 357–63.

Kremer, B. P., and Willenbrink, J. 1972. CO_2-Fixierung und Stofftransport in benthischen marinen Algen. I. Zur Kinetik der $^{14}CO_2$-Assimilation bei *Laminaria saccharina*. *Planta* 103, 55–64.

Linskens, H. F. (ed.). 1959. *Papierchromatographie in der Botanik*, 2nd ed. Springer-Verlag, Berlin. 408 pp.

Linskens, H. F., and Tracey, M. V. (eds.). 1962. *Modern Methods of Plant Analysis,* vol. V. Springer-Verlag, Berlin. 535 pp.

Lobban, C. S. 1974. A simple, rapid method of solubilizing algal tissue for scintillation counting. *Limnol. Oceanogr.* 19, 356–9.

Marcelle, R. (ed.). 1975. *Environmental and Biological Control of Photosynthesis.* Junk, The Hague. 408 pp.

Mehlitz, A., Gierschner, K., and Minas, T. 1963. Dünnschichtchromatographische Trennung von 2,4-Dinitrophenylhydrazonen. *Chem. Ztg.* 87, 573–6.

Nagashima, H., Ozaki, H., Nakamura, S., and Nisizawa, K. 1969. Physiological studies on floridean starch, floridoside and trehalose in a red alga, *Serraticardia maxima. Bot. Mag.* 82, 462–73.

Nicholson, N. L., and Briggs, W. R. 1972. Translocation of photosynthate in the brown alga *Nereocystis. Am. J. Bot.* 59, 97–106.

Paech, K., and Tracey, M. V. (eds.). 1955. *Modern Methods of Plant Analysis,* vol. II. Springer-Verlag, Berlin. 626 pp.

Paech, K., and Tracey, M. V. (eds.). 1956. *Modern Methods of Plant Analysis,* vol. I. Springer-Verlag, Berlin. 542 pp.

Parker, B. C. 1965. Translocation in the giant kelp *Macrocystis.* I. Rates, direction, quantity of C^{14}-labeled products and fluorescein. *J. Phycol.* 1, 41–6.

Partridge, S. M. 1948. Filter-partition chromatography of sugars. *Biochem. J.* 42, 238–48.

Percival, E. 1970. Algal polysaccharides. In Pigman, W., and Horton, D. (eds.). *The Carbohydrates,* vol. IIB, pp. 537–68. Academic Press, London.

Percival, E., and Smestad, B. 1972. Photosynthetic studies on *Ulva lactuca. Phytochemistry* 11, 1967–72.

Quillet, M. 1957. Volémitol et mannitol chez les Phéophycées. *Bull. Lab. Marit. Dinard* 43, 119–24.

Reed, L. J. 1950. The occurrence of γ-aminobutyric acid in yeast extract: Its isolation and identification. *J. Biol. Chem.* 183, 451–8.

Ruhland, W. (ed.). 1960. *Encyclopedia of Plant Physiology,* vol. V, pts. 1 and 2. Springer-Verlag, Berlin. 1013 pp.; 868 pp.

San Pietro, A. (ed.). 1971. *Methods in Enzymology,* vol. 23, *Photosynthesis A.* Academic Press, New York. 743 pp.

San Pietro, A. (ed.). 1972. *Methods in Enzymology,* vol. 24, *Photosynthesis and Nitrogen Fixation B.* Academic Press, New York. 526 pp.

Schmitz, K., and Lobban, C. S. 1976. A survey of translocation in Laminariales (Phaeophyceae). *Marine Biol.* 36, 207–16.

Schmitz, K., Lüning, K., and Willenbrink, J. 1972. CO_2-Fixierung und Stofftransport in benthischen marinen Algen. II. Zum Ferntransport ^{14}C-markierter Assimilate bei *Laminaria hyperborea* und *Laminaria saccharina. Z. Pflanzenphysiol.* 67, 418–29.

Schmitz, K., and Srivastava, L. M. 1975. On the fine structure of sieve tubes and the physiology of assimilate transport in *Alaria marginata. Can. J. Bot.* 53, 861–76.

Šesták, Z., Čatský, J., and Jarvis, P. G. 1971. *Plant Photosynthetic Production: Manual of Methods.* Junk, The Hague. 818 pp.

Splittstoesser, W. E. 1969. Arginine metabolism by pumpkin seedlings: Separation of plant extracts by ion exchange resins. *Plant Cell Physiol.* 10, 87–94.

Stahl, H. 1969. *Thin-layer Chromatography: A Laboratory Handbook.* Springer-Verlag, Berlin. 1041 pp.

Steemann Nielsen, E. 1942. Der Mechanismus der Photosynthese. *Dan Bot. Ark.* 11, 1–95.

Strickland, J. D. H., and Parsons, T. R. 1968. *A Practical Handbook of Seawater Analysis.* Bull Fish. Res. Board Can. 167, 27–34.

Towle, D. W., and Pearse, J. S. 1973. Production of the giant kelp, *Macrocystis,* estimated by in situ incorporation of ^{14}C in polyethylene bags. *Limnol. Oceanogr.* 18, 155–9.

Tyszkiewicz, E. 1962. An improved solvent system for the paper chromatography of phosphate esters. *Anal. Biochem.* 3, 164–6.

Wallen, D. G., and Geen, G. H. 1971. Light quality in relation to growth, photosynthetic rates and carbon metabolism in two species of marine plankton algae. *Marine Biol.* 10, 34–43.

Wallentinus, J. 1975. Primary production of macroalgae measured by the ^{14}C-method. *Merentutkimuslaitoksen. Julk. Havsforskningsinst. Skr.* 239, 72–7.

Weidner, M., Nordhorn, G., Kremer, B. P., and Küppers, U. 1975. Untersuchungen zur Photosynthese bei *Giffordia mitchellae. Z. Pflanzenphysiol.* 76, 423–43.

Yamaguchi, T., Ikawa, T., and Nisizawa, K. 1966. Incorporation of radioactive carbon from $H^{14}CO_3$ into sugar constituents by a brown alga, *Eisenia bicyclis,* during photosynthesis and its fate in the dark. *Plant Cell Physiol.* 7, 217–29.

29: Polarographic measurements of photosynthesis and respiration

ALAN D. JASSBY

Lawrence Berkeley Laboratory,
University of California, Berkeley, California 94720

CONTENTS

I	**Introduction**	*page* **286**
II	**Materials and test organisms**	**287**
	A. Culture vessel	287
	B. Oxygen measurement	289
	C. Stirring	289
	D. Aeration	289
	E. Temperature control	290
	F. Illumination	290
	G. Test organisms	290
III	**Method**	**290**
	A. Preparation of culture	290
	B. Oxygen exchange	291
IV	**Sample data**	**292**
V	**Discussion**	**293**
	A. Uses of the apparatus	293
	B. Difficulties	294
	C. Modifications	295
VI	**Acknowledgments**	**295**
VII	**References**	**295**

I. Introduction

The use of the oxygen electrode for the investigation of algal photosynthetic activity offers several unique advantages over alternate methods. Calculation of the derivative of continuous recordings of pO_2 allows O_2-exchange rates of algal cultures to be obtained with higher resolution and less labor than methods that require discrete measurements (e.g., ^{14}C uptake and manometric or chemical O_2 determination). Continuous recording also is possible with pCO_2 and pH electrodes. However, the nonlinear response of pCO_2 electrodes, and the necessity of titration to calibrate pH changes against alkalinity changes, renders pO_2 recording the simplest and most accurate electrometric determination of photosynthetic activity. Although infrared analyzers can be used to monitor continuously the CO_2 content of a gas stream passing through or over a well-stirred algal culture, lack of equilibration between the inorganic carbon of the gas stream and culture medium during non-steady-state conditions again presents difficulties in calibration that are avoided with pO_2 recording.

The pO_2 electrode has enjoyed widespread use in cell physiology during the past several decades, especially since the advent of the Clark-type probe in which the platinum electrode is protected by a membrane (Clark et al. 1953). Details of pO_2 electrode construction and calibration have been described by a number of authors (Beechey and Ribbons 1972; Lessler 1972) and need not be repeated here. Fork (1972) discusses various modifications of the rate-measuring electrode in which the O_2-exchange rate is produced as a direct output by placing the algae in immediate contact with the electrode. Several chamber designs exist for containing illuminated algal suspensions when using the pO_2 electrode. Particularly noteworthy is the sophisticated apparatus of Bartoš et al. (1975).

The methods described here are directed toward an examination of the diel patterns of O_2 exchange. The relation between algal photosynthesis and ambient irradiance, such as that expressed by "light-saturation curves" (Jassby and Platt 1976), depends to a great extent on the previous light history of the cells, among other factors. Because the effect of light history on light-saturation curves is not known precisely, it is impossible to derive the diel pattern of photosynthesis from a light-saturation curve, even when the daily pattern of irradiance is known. Severe gaps exist in our knowledge of the manner in which algal oxygen exchange changes throughout a 24-h period, and at present there is no substitute for direct measurement of these changes. With higher plants, the use of differential infrared gas analyzers (Lange et al. 1970) already has closed many of these gaps. Although several investigators have examined diel photosynthesis in algae (Verduin 1957; Vollenweider and Nauwerck 1961; Harris 1973), the techniques used usually involve a laborious number of discrete measurements of pH changes, ^{14}C uptake, or O_2 changes.

A technique for automatic recording of O_2 exchange on a 24-h basis is described below. Inherent in any technique that depends on variations of O_2 or carbon to measure photosynthesis is the difficulty that the O_2 and carbon concentrations themselves affect photosynthesis (as well as respiration). Much of this effect appears to be attributable to the photorespiratory pathways of algae (Tolbert 1974). In most natural waters, daily changes in the extracellular O_2 levels are not significant in the photic zone. It thus is necessary to minimize the gas changes that can occur in an algal suspension over a 24-h period, while still permitting exchange rates to be calculated from pO_2 changes in the medium. This is particularly important with the dense cultures necessary for short-term pO_2 measurements.

II. Materials and test organisms

Only the basic design of the apparatus is dealt with in detail. Several modifications are possible, and will be discussed briefly later, but the materials listed in this section refer only to the simplest form of the O_2-exchange apparatus.

A. Culture vessel

A glass culture vessel with a water-cooled jacket and connections for the pO_2 electrode, air input and exhaust, and inoculation or sampling is required (Fig. 29-1). The expense of having a glassblower construct the entire vessel from a simple container (e.g., a reagent bottle) generally is little more than the cost of purchasing and modifying most of the specialized commercial culture containers. A 1-l vessel was used in these experiments, but the exact size is not defined by the method.

The water jacket extends halfway up from the bottom of the container, leaving sufficient room for a glass sleeve on the side of the container to contain the pO_2 electrode. The axis of the sleeve is oriented so that the probe points slightly upward into the medium, minimizing

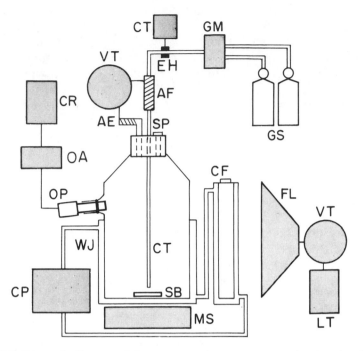

Fig. 29-1. Schematic diagram of the oxygen-exchange apparatus showing the air exhaust (*AE*), air filter (*AF*), chemical filter (*CF*), circulating pump (*CP*), chart recorder (*CR*), controlling timer and capillary tube (*CT*), electric hosecock (*EH*), floodlight (*FL*), gas-metering system (*GM*), gas supply (*GS*), light timer (*LT*), magnetic stirrer (*MS*), oxygen analyzer (*OA*), oxygen probe (*OP*), stirring bar (*SB*), sampling port (*SP*), variable transformer (*VT*), and water jacket (*WJ*).

the accumulation of air bubbles on the probe surface. The sleeve and container connect in a gradual curve, also to prevent air bubbles from being trapped in the sleeve. The diameter of the sleeve depends, of course, on the pO_2 probe chosen by the experimenter, the probe being sealed against the sleeve with Teflon O-rings fixed in an annular bulge in the sleeve.

A neoprene stopper at the top of the container is perforated with three holes. A capillary tube through which air is injected extends through the central opening to the bottom of the vessel, leaving adequate space for a magnetic stirring bar. The air is exhausted through an adjacent bent-glass tube of larger bore. The exhaust tube is plugged with cotton-wool and wrapped with heating tape connected to a variable transformer. The third opening contains a length of glass tubing equal to the height of the neoprene stopper. This opening is capped with a rubber serum bottle stopper through which samples may be extracted with a syringe. Because almost all rubber products contain substances that can modify algal metabolism, it is preferable to avoid contact between any of the stoppers and the medium.

B. Oxygen measurement

The pO_2 electrode (Radiometer, model E5046) is covered by a polypropylene membrane of 20 μm thickness and is connected to an analyzer that provides the proper polarizing voltage of 0.63 V and amplifies the current generated by the probe (Radiometer, model PHM71 Mk2, with a pO_2 module, Radiometer, model PHA934). The analyzer output is recorded continuously on a single channel of a dual-channel chart recorder (Gould, model 220). The remaining channel is used for recording accessory information such as irradiance.

C. Stirring

The medium is agitated by a magnetic stirrer on which the culture vessel rests. The stirring motor must provide speeds high enough that the pO_2 reading becomes independent of stirring speed. Most large commercial magnetic stirrers are suitable (e.g., Thermolyne, model SL-7225, with a heat barrier; Thermolyne, model PT-10000).

D. Aeration

Gas is supplied by commercially available cylinders of air and CO_2, mixed in predetermined amounts in a self-contained metering system (National, model 3634) before delivery to the vessel. By using air supplies from which CO_2 has been removed, it is possible to examine the effects of different carbonate alkalinities on O_2 exchange over the wide range found in natural freshwaters. The gas mixture passes through a glass drying tube packed with cotton-wool and surrounded by heating tape connected to the same variable transformer as the exhaust tube. The filter attaches to the air inlet with autoclavable polyvinylchloride tubing.

An electric hosecock (Fisher, model 5-846-5) on thin-walled surgical tubing between the filter and the gas-metering system contains a solenoid that opens or closes the clamping bar in response to an external signal. The gas flow thus can be turned on or off automatically in a programmed sequence with an appropriate controller.

The controlling timer is a device constructed specifically for this apparatus. The periods during which the gas flow is on or off are adjusted independently over the range from 0.25 to 256 min in multiples of 2, following which the preset on and off times cycle indefinitely. Manual control also is possible. The controller is based on an astable multivibrator driving a cascade of binary counters with appropriate logic. A detailed circuit diagram is available upon request.

E. Temperature control

Water is passed through a cooling jacket by means of a constant-temperature circulating pump (Haake, model FK). The pump also cools the chemical filter used to modify the spectrum of the light source (see below). It is important to maintain culture temperatures precisely, the pO_2 electrode having a temperature coefficient of approximately 3% $°C^{-1}$ with the $20\mu m$ polypropylene membrane.

F. Illumination

The culture vessel is illuminated by projector floodlights (Sylvania, 150 W). The light emitted by incandescent lamps must be filtered of its infrared energy to provide a source with a spectrum resembling natural sunlight. For this purpose, a water-jacketed rectangular container 2 cm in thickness is interposed between the lamp and the culture vessel. The chemical filter in the container is identical to that used by Jitts et al. (1964). Light of different irradiances is chosen by means of a variable transformer, and the diel light-dark cycle is controlled by an automatic timer (General Electric, model CR121). A spectroradiometer (ISCO, model SR) is used to determine the spectrum of photosynthetically active radiation (380–725 nm) incident on the culture vessel, and a pyranometer (Eppley, model 8-48) monitors total irradiance.

G. Test organisms

The apparatus has been tested with the following organisms isolated from the coastal waters of Nova Scotia: *Dunaliella tertiolecta* Butcher, *Monochrysis lutheri* Droop, and *Phaeodactylum tricornutum* Bohlin. The organisms are maintained axenically on f/2 media (McLachlan 1973) prepared with Bedford Basin seawater. Illumination is provided by cool-white fluorescence on a 12:12 h light-dark cycle and the maintenance temperature is held constant at 15 C. Sterile transfer of each species is carried out on a weekly basis.

III. Method

A. Preparation of culture

The pO_2 electrode is removed from its sleeve, the sleeve itself is plugged with cotton-wool, and aluminum foil is wrapped around the external sleeve opening and the port for the rubber serum stopper. The gas-metering system is disconnected from the air filter, the circulating pump is separated from the water jacket, and all heating tape is removed. The vessel is filled to the base of the glass sleeve with f/2 medium, and the container, filter, and a glass syringe are sterilized in an autoclave. After autoclaving, the air supply, water jacket, and heating tape are reconnected. The circulating pump temperature, light-dark cycle, and the irradiance then are selected, the last with the aid of the spectroradiometer probe at the surface of the culture vessel. The gas supply is adjusted to aerate vigorously, and a 20-cm³ aliquot of stock culture is transferred aseptically into the culture vessel with the sterile syringe.

The organisms are permitted to multiply until a concentration is reached that allows precise measurement of oxygen exchange in time intervals of less than 1 h. If we make the reasonable assumption that pO_2 readings have a precision of approximately ± 0.1 mg O_2 l^{-1}, then (by assuming propagation of errors with zero covariance) a rate of 1.0 mg l^{-1} h^{-1} can be detected with a coefficient of variation of 14%. Optimal rates of photosynthesis generally are less than 25 μg C (μg chl a)$^{-1}$ h^{-1} (Platt and Jassby 1976), or 67 μg O_2 (μg chl a)$^{-1}$ h^{-1} assuming a photosynthetic quotient of unity. Then a minimum of 15 μg chl a is required to achieve coefficients of variation of 14% for hourly measurements at *optimal* irradiance, and usually much more will be necessary. Samples are withdrawn from the sampling port at regular intervals to determine when sufficient biomass is obtained.

B. Oxygen exchange

Because of the linearity of the pO_2 electrode, calibration is accomplished simply by immersion in two culture filtrates, one of which has had the oxygen removed by vigorous N_2 bubbling, the other of which is saturated with air. The probe is prepared by rinsing quickly in ethanol followed by filter-sterilized water. The aluminum foil and cotton-wool are removed from the glass sleeve, and the pO_2 electrode is fixed into its O-ring seal. Additional f/2 medium, which has been sterilized in a large glass syringe, is introduced through the rubber port until the culture level is well above the pO_2 electrode. Sterile conditions are not ensured when the probe is inserted, but contamination is negligible for several days.

The gas-flow timer usually is set so that the flow is on for 16 min and off for 32 min, allowing single O_2-exchange measurements 30 times each day. The maximum resolution depends, of course, on the chlorophyll levels and illumination. When the gas flow is on, aeration quickly brings pO_2 levels to approximately saturation. When the flow is off, the pO_2 changes in the vessel are recorded on the chart recorder and a net oxygen-exchange rate during the 32-min period is determined.

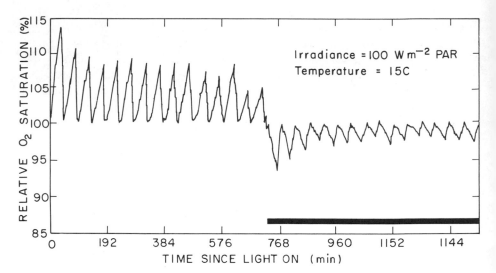

Fig. 29-2. Tracing of a 24-h O_2-exchange record for *Phaeodactylum tricornutum*. Aeration is on and off, sequentially, for 32 and 16 min, respectively. The O_2-exchange rate is proportional to the difference between the maximum and minimum O_2 levels in any 32-min interval. The dark bar signifies the duration of the dark period.

By matching the time during which aeration is off to the maximum oxygen-exchange rate, the pO_2 levels never deviate more than 20% from saturation. The Warburg effect of photosynthetic suppression by O_2 thus is minimized. Without the frequent return of pO_2 to saturation conditions, the chlorophyll concentrations required for precise measurements every hour or so would entail severe O_2 supersaturation and depletion over the light and dark periods, respectively. On the other hand, minimizing the deviation from saturation decreases the precision with which exchange rates can be determined, especially during periods of low oxygen exchange. The exact settings for the flow timer thus must be decided on the basis of a trade-off between temporal resolution, on the one hand, and precision and pO_2 deviations, on the other.

IV. Sample data

Figure 29-2 illustrates a 24-h recording of O_2 concentrations from a culture of *Phaeodactylum tricornutum*. The irradiance level was 100 W m^{-2} of photosynthetically active radiation (PAR) and the culture was grown and monitored at 15 C. Figure 29-3 shows data from a culture of *Dunaliella tertiolecta* grown and monitored at 14 C. The raw recording has been converted to exchange rates, and each point is plotted in

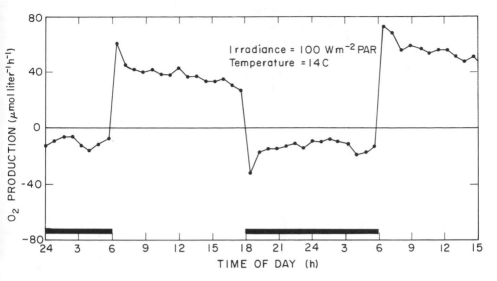

Fig. 29-3. Diel O_2-exchange pattern for *Dunaliella tertiolecta* derived from a recording such as in Fig. 29-2. Dark bar represents the dark period.

the center of the 32-min interval for which it has been computed. Irradiance incident on the culture vessel again is 100 W m^{-2} PAR.

V. Discussion

A. Uses of the apparatus

Figure 29-3 is a good example of the phenomena that can be observed with this technique. Transient maxima in photosynthesis when the light turns on, and in respiration when the light extinguishes, are seen almost always, a feature known from other diel studies on aquatic communities (Beyers 1966). Oxygen exchange decreases in absolute magnitude during both day and night. These phenomena may be attributable in part to a buildup of photosynthate during the day and a gradual decrease at night, resulting in, respectively, increases and decreases in respiratory O_2 demands. Diel rhythms in chlorophyll and photosynthetic enzyme levels also probably play an important role (Sournia 1974). The local maximum in respiration sometimes seen several hours before the light period commences usually correlates with a sudden increase in cell number in samples collected through the sampling port. Either the dividing cells or young cells appear to require increased amounts of O_2. The ratio of net photosynthesis to night respiration is 2.9 for this culture.

The techniques described here offer novel opportunities for investigating some significant ecological problems. Algae preconditioned in continuous culture under defined conditions can be subjected systematically to different levels and patterns of daily irradiance, providing information on the effect of changing meteorological conditions. Monitoring of diel patterns during conditions of tropical, temperate, and polar light and temperature can yield information on the diel oxygen-exchange strategies employed by phytoplankton in extreme environmental conditions. Introduction of nutrients at various times of the light-dark cycle can demonstrate how the timing of daily cycles in nutrient regeneration affects algal growth.

B. Difficulties

One of the major problems associated with any pO_2 method is the need for high culture densities to obtain short-term measurements. The technique is unsuitable for samples collected from most natural environments or from the early stages of a batch culture. Concentration by centrifugation is one preparation technique for these cases, but some cells invariably are damaged. Problems of severe nutrient deficiency and algae-produced inhibitors at high densities are alleviated somewhat by preconditioning in continuous culture.

The probe membrane tends to become colonized by algae within several days after the electrode is inserted. At this point, the oxygen-exchange measurements are not representative of the entire culture, and the experiment must be terminated. If care is taken, significant contamination of the culture by bacteria does not appear before this time.

The addition of fresh medium to immerse the electrode at the start of an experiment has unknown effects on the algae already in the vessel. The problem easily can be avoided by growing a sufficient volume of algal culture in a larger vessel. These organisms then are transferred to the empty experimental vessel in which the pO_2 electrode already is inserted for the start of an experiment. More glassware is required and extra care is necessary to prevent contamination when transferring the culture, but sudden changes in nutrient conditions are avoided.

The final major difficulty concerns the conversion of oxygen-exchange rates to productivity and respiration in terms of carbon. For a variety of reasons (photorespiration, excretion, different respiratory substrates), the photosynthetic quotient and respiratory quotient rarely are equal to unity. Occasionally, a photosynthetic quotient as high as 3 has been observed (Fogg 1969). There is no unequivocal way to convert O_2 exchange to losses or gains of cellular organic carbon.

C. Modifications

A number of obvious minor modifications can be made to the vessel. A thin rectangular vessel would allow more even incident irradiance on the culture. The optical density at chosen wavelengths could then be monitored continuously with the spectroradiometer probe positioned against the side of the vessel opposite the light source. A constant correction for the absorbance of the vessel walls and water jacket would be necessary, but optical densities at appropriate wavelengths often correlate well with cell concentrations (Sorokin 1973).

The vessel is easily altered for continuous culture. The sampling port tube is extended and connected to a peristaltic pump and nutrient reservoir. An additional tube fixed to the glass wall above the pO_2 electrode serves as an overflow to maintain constant water levels.

A more extensive modification of the gas flow control would improve the design substantially. A feedback from the pO_2 analyzer to the electric hosecock could turn the gas flow on when pO_2 levels deviated from saturation by 10%, and off when pO_2 levels reached saturation. A controlling timer then would be unnecessary. This modification has not yet been tested.

VI. Acknowledgments

I am indebted to E. Larson for construction of the gas-flow timer and to Dr. T. Platt for helpful discussions concerning the technique. Dr. K. Denman made numerous comments during the course of this research. This work was supported by Environment Canada through the Marine Ecology Laboratory, Dartmouth, Nova Scotia, and by the U.S. Energy Research and Development Administration.

VII. References

Bartoš, J., Berková, E., and Šetlík, I. 1975. A versatile chamber for gas exchange measurements in suspensions of algae and chloroplasts. *Photosynthetica* 9, 395–406.

Beechey, R. B., and Ribbons, D. W. 1972. Oxygen electrode measurements. *Methods Microbiol.* 6B, 25–53.

Beyers, R. J. 1966. The pattern of photosynthesis and respiration in laboratory microecosystems. In Goldman, C. R. (ed.). *Primary Productivity in Aquatic Environments*, pp. 61–74. University of California, Berkeley.

Clark, L. C., Jr., Wold, R., Granger, D., and Taylor, Z. 1953. Continuous recording of blood oxygen tensions by polarography. *J. Appl. Physiol.* 6, 189–93.

296 *Physiological and biochemical processes*

Fogg, G. E. 1969. Oxygen versus ^{14}C-methodology. In Vollenweider, R. A. (ed.). *A Manual on Methods for Measuring Primary Production in Aquatic Environments*, pp. 76–8. Blackwell, Oxford.

Fork, D. C. 1972. Oxygen electrode. *Methods Enzymol.* 24, 113–22.

Harris, G. P. 1973. Diel and annual cycles of net plankton photosynthesis in Lake Ontario. *J. Fish. Res. Board Can.* 30, 1779–87.

Jassby, A. D., and Platt, T. 1976. Mathematical formulation of the relationship between photosynthesis and light for phytoplankton. *Limnol. Oceanogr.* 21, 540–7.

Jitts, H. R., McAllister, C. D., Stephens, K., and Strickland, J. D. H. 1964. The cell division rates of some marine phytoplankters as a function of light and temperature. *J. Fish. Res. Board Can.* 21. 139–57.

Lange, O. L., Schulze, E. D., and Koch, W. 1970. Evaluation of photosynthesis measurements taken in the field. In Šetlík, I. (ed.). *Prediction and Measurement of Photosynthetic Productivity*, pp. 339–52. Centre for Agricultural Publishing and Documentation, Wageningen.

Lessler, M. A. 1972. Macro- and micro-oxygen electrode techniques for cell measurement. *Methods Cell Physiol.* 5, 199–218.

McLachlan, J. 1973. Growth media–marine. In Stein, J. R. (ed.). *Handbook of Phycological Methods: Culture Methods and Growth Measurements*, pp. 25–51. Cambridge University Press, London.

Platt, T., and Jassby, A. D. 1976. The relationship between photosynthesis and light for natural assemblages of coastal marine phytoplankton. *J. Phycol.* 12, 421–30.

Sorokin, C. 1973. Dry weight, packed cell volume and optical density. In Stein, J. R. (ed). *Handbook of Phycological Methods: Culture Methods and Growth Measurements*, pp. 321–43. Cambridge University Press, London.

Sournia, A. 1974. Circadian periodicities in natural populations of marine phytoplankton. *Adv. Marine Biol.* 12, 325–89.

Tolbert, N. E. 1974. Photorespiration. In Stewart, W. D. P. (ed.). *Algal Physiology and Biochemistry*, pp. 474–504. Blackwell, Oxford.

Verduin, J. 1957. Daytime variations in phytoplankton photosynthesis. *Limnol. Oceanogr*, 2, 333–6.

Vollenweider, R. A., and Nauwerck, A. 1961. Some observations on the C 14 method for measuring primary production. *Verh. Int. Ver. Theor. Angew. Limnol.* 14, 134–9.

30: Polarographic measurements of respiration following light-dark transitions

ALAN D. JASSBY

Lawrence Berkeley Laboratory,
University of California, Berkeley, California 94720

CONTENTS

I Introduction   page 298
II Materials and test organisms   299
III Method   299
IV Sample data   300
V Discussion   301
VI Acknowledgments   303
VII References   303

I. Introduction

Endogenous or dark respiratory O_2 exchange, as opposed to photo-respiration, refers to the O_2 consumption associated primarily with oxidative phosphorylation. Substrate oxidation in algae takes place via glycolysis, the pentose phosphate cycle, the glyoxylate cycle, and the tricarboxylic acid cycle as in higher plants, but novel pathways of mitochondrial electron transport are suspected in certain algae (Lloyd 1974).

The presence of photosynthetic O_2 production and photorespiratory O_2 consumption precludes direct measurement of endogenous respiration in the light using a pO_2 electrode. Some investigators have attempted to estimate the levels of endogenous respiration under illumination by measuring the dark O_2 consumption in samples recently removed from a source of light. However, the processes of endogenous respiration are not independent of other metabolic pathways in the cell, particularly those pathways associated with photosynthesis. It has been suggested, for example, that rates of glycolysis are strongly affected by chloroplast activity (Jackson and Volk 1970). In addition, certain wavelengths of light may affect endogenous respiration directly (Kowallik 1967). At present, there is no way to estimate accurately endogenous respiration in the light using only pO_2 electrodes.

Oxygen consumption in the dark depends on the previous light history in several ways. The duration, spectrum, and magnitude of the light, as well as other factors, determine the type and amount of photosynthate produced. Subsequent respiration in the dark will be affected by the metabolism of the photosynthate and by certain diel rhythms (e.g., of mitochondrial morphology) in cell cultures synchronized by the light-dark cycle (Lloyd 1974). The previous light history thus may affect the ensuing dark respiration for many hours after a light-dark transition.

Transient phenomena in O_2 exchange also are noted for approximately 10 min after a light-dark transition (Ried 1968). The structure of the transients can be complex, and only part of it is attributable to endogenous respiration. After long periods of irradiation (greater than several minutes), the main transient phenomenon is a slowly declining stimulation of O_2 uptake, for 5–10 min, which appears to be linked to an energy-consuming process following photosynthesis, but cannot be attributed to endogenous respiration.

If one is interested in measurements of endogenous respiration over periods of 1 h or more, the technique described in Chp. 29 is appropriate. Effects of O_2 on respiration are mitigated by the approximately constant O_2 conditions, and semicontinuous measurements of O_2 consumption in the dark are recorded automatically. It must be noted, however, that the first measurement after the light-dark transition will be affected by the transient stimulation of O_2 uptake mentioned above. The extent of this effect will depend on the time constant of the transient, its initial magnitude after darkening, and the duration of the dark O_2 consumption measurement.

Occasionally, short-term endogenous respiration measurements of 1 h or less are required for cells removed sequentially from an illuminated culture into darkness. As pointed out, these measurements cannot be identified quantitatively with endogenous respiration in the light, but they do have value in qualitative descriptions of endogenous respiration changes during the life cycle of synchronized cells. For these short-term measurements, the transient stimulation of O_2 uptake may lead to significant overestimates of endogenous respiration. A technique for estimating the size of the transient, and avoiding overestimates of endogenous respiration, is described below.

II. Materials and test organisms

The algae are contained in a 1.5-cm³ water-jacketed cell with a capillary bore stopper (Gilson, model OX-15253). A Clark-type electrode with a fast response time (Gilson, model OX-15259) monitors pO_2 levels in the small container. The electrode current is amplified and recorded with an oxygraph (Gilson, model K-ICRP-C) containing a heated stylus recording system.

Special magnetic stirring bars (Gilson, model OX-15252) are required to fit the 1.5-cm³ container. Otherwise, the materials needed for stirring, illumination, temperature control, and the culture of test organisms are as described in Chp. 29. A source of air connected to a syringe needle is required to return samples to pO_2 saturation when it is desirable to repeat measurements on the same sample.

III. Method

Algal cultures are grown axenically under defined illumination and temperature until sufficient biomass is attained (Chp. 29). The 1.5-cm³ vessel is autoclaved and connected to the calibrated pO_2 electrode and constant-temperature circulating pump. Temperature and illu-

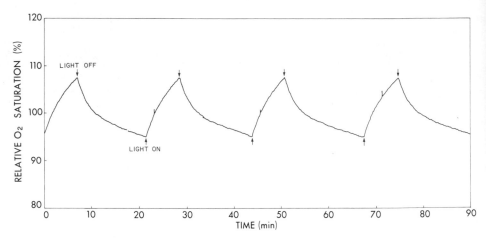

Fig. 30-1. Tracing of an O_2-exchange record for *Dunaliella tertiolecta* at 18 C with short-term alternation of light (150 W m^{-2} PAR) and dark periods.

mination in the vessel are adjusted to conform to the culture conditions. A 1.5-cm³ aliquot is transferred from the culture to the small vessel and aerated through the syringe needle to maintain saturation until a measurement is desired.

To determine dark O_2 consumption rates, the stopper is inserted into the vessel, the lights are turned off, and pO_2 levels are monitored until the initial transient stimulation of O_2 uptake is over (5–10 min) and an approximate steady-state rate is achieved. The response time of the electrode (90% of final value in 10 sec) is fast enough to follow the transient without significant error, but the response time can be decreased by stretching the membrane carefully so that it provides a thinner diffusion barrier between the medium and electrode. This procedure is desirable especially with low temperatures, where the response time can increase to an undesirable level.

IV. Sample data

Oxygen exchange in aliquots of a *Dunaliella tertiolecta* culture is illustrated in Fig. 30-1. The temperature was constant at 18 C and the irradiance level incident on the 1.5-cm³ container was 150 W m^{-2} of photosynthetically active radiation (PAR). The transient stimulation of O_2 uptake after a light-dark transition is marked in these cultures. The sequence of light and dark periods also illustrates the reproducibility of this phenomenon.

The initial slope of O_2 consumption (measured over 1 min after darkening) is approximately five times the final steady-state slope reached before the light is turned on again. Steady-state values of O_2 consumption are achieved within 10 min of the light-dark transition.

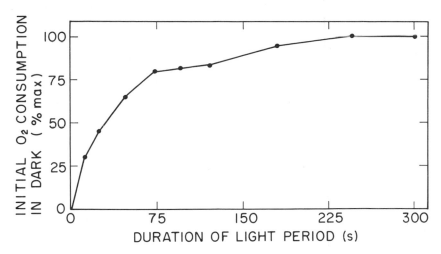

Fig. 30-2. Dependence of the initial O_2-consumption rate (measured over 1 min after darkening), following a light-dark transition, on the duration of the preceding light period. The *Dunaliella tertiolecta* culture is the same as for Fig. 30-1.

The magnitude of the initial slope after darkening depends on the duration of the preceding light period. For the population in Fig. 30-1, approximately 4 min of light is required to reach maximum values for the initial slope (Fig. 30-2). The initial slope has a higher sensitivity to O_2 levels and a lower Q_{10} than the final steady-state slope (Figs. 30-3 and 30-4).

V. Discussion

The high temporal resolution afforded by this technique enables endogenous respiration to be measured accurately. The dependency of the initial slope on the preceding light duration, O_2 levels, and temperature indicates that it represents a phenomenon different from endogenous respiration (as measured by the final steady-state slope). Normally, endogenous respiration would be estimated by measuring the O_2 levels at the start of the dark period ($t = 0$) and after some time interval ($t = T$). The data of Fig. 30-1 enable us to determine the overestimate in endogenous respiration that could occur.

Let R_D equal the endogenous respiration rate represented by the final slope; let R_L equal the additional rate of O_2 consumption that occurs immediately after a light-dark transition; and let R equal the total O_2 consumption rate during the dark period. A good description of the relative O_2 consumption rate after darkening in Fig. 30-1 is given by:

$$R = R_L \, e^{-t/\tau} + R_D \qquad (1)$$

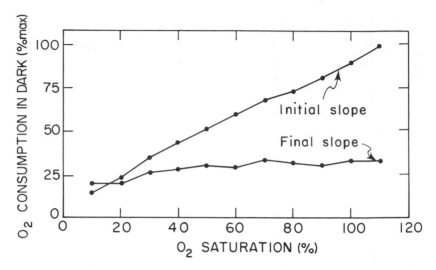

Fig. 30-3. Dependence of the initial O_2-consumption rate (measured over 1 min after darkening) on O_2 levels. The *Dunaliella tertiolecta* culture is the same as for Fig. 30-1. Points are plotted at the O_2 levels corresponding to the concentrations reached at the light-dark transition. The duration of the preceding light period is approximately 5 min for each point.

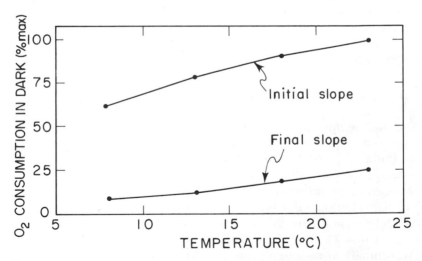

Fig. 30-4. Dependence of the initial O_2-consumption rate (measured over 1 min after darkening) on temperature. The *Dunaliella tertiolecta* culture is the same as for Fig. 30-1. Culture aliquots are transferred from the main culture vessel (which is at 18 C) to the 1.5-cm³ cell, where they are maintained under identical light conditions, but different temperatures, for 30 min before a measurement.

where $\tau = 3$ min, and $R_L/R_D = 4$. The amount of O_2 consumed between $t = 0$ and $t = T$ is:

$$\int_0^T R dt = R_L \tau (1 - e^{-T/\tau}) + R_D T \tag{2}$$

If the O_2 decrease were attributed entirely to endogenous respiration, the apparent endogenous respiration rate R_D' would be:

$$R_D' = R_D \left[1 + \frac{R_L \tau}{R_D T} (1 - e^{-T/\tau}) \right] \tag{3}$$

When $T = 15$, 30, 60, and 120 min. $R_D'/R_D = 1.8$, 1.4, 1.2, and 1.1, respectively. Thus, overestimates of R_D by 20% still occur for $T = 1$ h.

The obvious way to avoid this problem with discrete O_2 measurements is to wait 10 min after darkening before measuring initial O_2 levels. This requires the assumption that the time constant of the transient is never much more than the value observed in these experiments ($\tau = 3$ min). Although this assumption has proved consistently valid, continuous high-resolution monitoring still is recommended, for the transients themselves provide interesting information on O_2-consuming processes within the cell. The correct interpretation of the transient stimulation of O_2 uptake is not known precisely, but the available evidence suggests that it is associated with the reducing end of photosystem I and possibly with photorespiration (Jackson and Volk 1970), decaying as some light-produced substrate is used up.

VI. Acknowledgments

Discussions with Dr. T. Platt were invaluable during the course of this research. This work was supported by Environment Canada through the Marine Ecology Laboratory, Dartmouth, Nova Scotia, and by the U.S. Energy Research and Development Administration.

VII. References

Jackson, W. A., and Volk, R. J. 1970. Photorespiration. *Annu. Rev. Plant Physiol.* 21, 385–432.

Kowallik, W. 1967. Action spectrum for an enhancement of endogenous respiration by light in *Chlorella. Plant Physiol.* 42, 672–6.

Lloyd, D. 1974. Dark respiration. In Stewart, W. D. P. (ed.). *Algal Physiology and Biochemistry*, pp. 505–29. Blackwell, Oxford.

Ried, A. 1968. Interactions between photosynthesis and respiration in *Chlorella.* I. Type of transients of oxygen exchange after short light exposures. *Biochim. Biophys. Acta* 153, 653–63.

31: Hill reaction and photophosphorylation with chloroplast preparations from Chlamydomonas reinhardi

STEPHEN LIEN

Department of Plant Sciences,
Indiana University, Bloomington, Indiana 47401

CONTENTS

I	Introduction	*page* 306
II	Materials and test organism	306
	A. Chemicals	306
	B. Test organism	307
III	Methods	307
	A. Preparation of chloroplast fragments	307
	B. Assay of Hill reaction	309
	C. Assay of photophosphorylation	312
IV	Sample data	314
V	References	315

I. Introduction

Several major advances in our knowledge of photosynthesis were made possible through the use of unicellular algae as experimental material. For example, the extensive use of *Chlorella* and *Scenedesmus* in the earlier works on the pathway of photosynthetic CO_2 fixation was instrumental in the elucidation of the Calvin cycle. The demonstration of the Emerson enhancement effect and the subsequent recognition of the two photochemical systems functioning cooperatively in O_2-evolving organisms were greatly facilitated through the use of various algal species (notably belonging to the genera *Anabaena, Anacystis, Chlorella, Chroococcus, Phormidium,* and *Porphyridium*) having diverse pigment composition. In recent years, a large number of mutant strains isolated from several algal species (in particular *Chlamydomonas reinhardi, Scenedesmus obliquus,* and *Euglena gracilis*) has become widely available for research on biochemical, biophysical, physiological, and developmental aspects of photosynthesis. As a result, a greatly increased usage of these algal materials in the near future can be expected. It is the purpose of this chapter to describe (1) the basic technique for preparing cell-free chloroplast membrane preparations suitable for the measurement of Hill reaction and photophosphorylation, and (2) assay procedures for Hill reaction and photophosphorylation.

II. Materials and test organism

Detailed information about materials and equipment is given under Methods (III).

A. Chemicals

All chemicals used here are analytical grade reagents obtained from commercial sources, except that *p*-benzoquinone (PBQ) should be purified by sublimation of the commercial product before use.

[306]

B. Test organism

Chlamydomonas reinhardi strain 137c. The stock cultures of *C. reinhardi* are maintained on appropriate agar media in screw-cap culture tubes as described by Levine (1971). To obtain the *experimental culture* for daily use, the cells are most conveniently grown in 500-ml Erlenmeyer flasks containing 300 ml of the TRIS-acetate-phosphate (TAP) medium of Gorman and Levine (1965) at a desired temperature (usually 20–25 C). Adequate agitation and aeration of the cultures are provided by placing the culture flasks on rotary shakers (ca. 180 rpm). Illumination of the cultures is provided from above by fluorescent lamps that give approximately 400–600 ft-c at the surface of the shaker platform. To obtain consistent results, the cultures should be harvested at a cell density not exceeding 5×10^6 ml^{-1}, which corresponds to approximately 12–15 μg chl ml^{-1} of culture of wild-type strain. It should be noted that the chlorophyll content of certain mutant strains may vary greatly from that of the wild-type cells, and therefore the cell density of the culture should be determined by direct counting.

III. Methods

A. Preparation of chloroplast fragments

1. Reagents. Buffer I. Tricine-NaOH, 50 mM, pH 7.5; KCl, 10 mM. Buffer II. HEPES-NaOH, 50 mM, pH 7.5; sorbitol, 0.4 M; KCl, 50 mM; MgCl$_2$, 2 mM; bovine serum albumin (defatted), 2 mg ml^{-1}.

2. Procedure I. A fast method for preparing cell-free chloroplast membrane fragments by sonication.

a. Harvesting and washing of cells. Cell suspension (ca. 300 ml) from one culture flask is centrifuged at $3000 \times g$ for 5 min. The cell pellets are washed twice by resuspending in 100 ml of buffer I and centrifuging at $3000 \times g$ for 5 min. The washed cells are resuspended in ca. 10 ml of buffer I at room temperature.

b. Determination of chlorophyll and adjustment of cell density. An aliquot (0.2–0.4 ml) of the cell suspension from III.A.2.a is transferred into a 12-ml glass conical centrifuge tube (prechilled in −20 C freezer). Allow the suspension to become frozen in the freezer (or by dipping into dry ice–acetone) and then thaw at room temperature. Adjust the total volume of the sample to 2.0 ml by adding distilled H$_2$O. Thoroughly mix the contents of the tube with 8.0 ml of acetone. Allow the mixture to stand in darkness for 5 min at room temperature. Place the centrifuge tube in a desk-top clinical centrifuge operating at medium speed for at least 5 min to obtain a clear, green supernatant solution. The chlorophyll concentration of the supernatant solution is determined spectrophotometrically according to the method of Arnon (1949) or Chp. 6. Adjust the cell density of the remaining cell suspension of III.A.2.a to give a final total chlorophyll concentration of 0.1 mg chl ml^{-1} by dilution with buffer I, and store the cell suspension at room temperature until use. If the sample is to be stored for an extended period (>30 min), it is necessary to keep the cells in suspension by gentle stirring.

c. Disruption of cells by sonication. A 7.0-ml portion of diluted cell suspension from III.A.2.b is transferred into a. 10-ml capacity glass beaker placed in an ice-water bath. Insert the probe of a sonicator (Sonic Dismembrator, Quigley-Rochester) into the sample until the tip of the probe is 1 cm below the sample surface. Adjust the power output level to 80% of maximal power and sonicate the sample for a fixed duration of 10–30 sec. (The optimal duration of sonication should be determined empirically for a given type of sonicator–probe–sample size combination). After sonication, the cell-free preparation (the degree of cell breakage can be determined by examining a drop of the sonicated sample under a microscope) is transferred into a test tube and stored in an ice bath and kept in darkness (up to 2–3 h).

3. Procedure II. Preparation of washed and photochemically active chloroplast membrane fragments by a Yeda press (model 92-50, Yeda Research and Development Co.)

a. Harvesting and washing of cells. Approximately 600 ml of culture (from two flasks) is harvested and washed as described under III.A.2.a. The twice-washed cell pellet is resuspended in 10 ml of buffer II. Determine the chlorophyll concentration of the suspension (III.A.2.b). Adjust the cell density of the suspension, by diluting with an appropriate volume of buffer II, to a final chlorophyll concentration of 500 μg chl ml^{-1}. Store the cell suspension in an ice bath.

b. Disruption of cells by Yeda press. Load the cell suspension from III.A.3.a into the sample chamber of a prechilled (4 C) Yeda press. Pressurize the sample chamber with N$_2$ gas from a high-pressure nitrogen tank (via a conventional high-pressure gas regulator valve) to 800 lb in.$^{-2}$ While maintaining the same pressure reading, slowly release the needle valve of the sample outlet port until a steady flow of the sample (2–3 drops sec^{-1}) is obtained. Collect the entire sample into an ice-cold glass container. Allow the sample to stay in the ice bath for 3 min and then reprocess it through the Yeda press at the same pressure setting.

c. Isolation and washing of chloroplast membrane fragments by differential centrifugation. Centrifuge the Yeda-press-treated sample from III.A.3.b at 1500 $\times$ *g* for 90 sec to remove unbroken cells. Carefully decant the dark green supernatant into a centrifuge tube and centrifuge at 10 000 × g for 20 min. Remove the pale green supernatant and resuspend the dark green pellet in 10 ml of buffer II. Avoid taking the white starchlike precipitate at the bottom of the pellet into suspension. Using a tissue homogenizer, gently disperse the resuspended green pellet into suspension and dilute with buffer II to a total volume of 20 ml and then centrifuge again at 10 000 × g for 20 min. Discard the supernatant and resuspend the green pellet in a small amount (ca 2 ml) of buffer II. Disperse the sample with a small tissue homogenizer and withdraw a 0.1-ml sample for determination of total chlorophylls, as described above (III.A.2.b), but omit the freeze-thaw treatment, which is required only for whole cells. Dilute the sample with buffer II to give a final chlorophyll concentration of 0.5–1.0 mg chl ml^{-1} and store in an ice bath; protect the preparation from exposure to room light until used for assays.

B. Assay of Hill reaction

As in the case of chloroplast preparations from higher plants, a properly isolated chloroplast membrane preparation from algal sources can carry out a variety of light-induced oxidation-reduction reactions in the presence of suitable electron donor and acceptor compounds. When water serves as the ultimate source of electrons for the reduction of electron acceptors other than CO_2, the reaction is known as the Hill reaction and is normally accompanied by a net evolution of O_2.

Hill reaction: $$H_2O + \underset{\text{(oxidized)}}{\text{acceptor}} \xrightarrow[\text{chloroplast}]{h\nu} \underset{\text{(reduced)}}{\text{acceptor}} + \tfrac{1}{2} O_2 \qquad (1)$$

However, if the reduced form of the acceptor (such as the viologen dyes and low-potential quinones) are autooxidizable, a net uptake of O_2 from the medium can be observed, especially if the breakdown of H_2O_2, the autooxidation product of the reduced acceptor, is prevented. The light-dependent uptake of O_2 by illuminated chloroplast preparations is known as the Mehler reaction, and is composed of a light-dependent Hill reaction and a dark autooxidation of the reduced endogenous and/or exogenous acceptor.

$$H_2O + A \xrightarrow[\text{chloroplast}]{h\nu} AH_2 + \tfrac{1}{2} O_2 \qquad (2)$$

$$AH_2 + O_2 \longrightarrow A + H_2O_2 \qquad (3)$$

Mehler reaction:
(net of equations 2 and 3).

$$H_2O + \tfrac{1}{2} O_2 \xrightarrow[\text{chloroplast}]{h\nu} H_2O_2 \qquad (4)$$

According to equations 1 and 4, the Hill activity of chloroplast preparations can be assayed by various means, including (1) chemical or spectrophotometric analysis of the redox state of the acceptor molecules, and (2) manometric or polarographic determination of the O_2 tension of the sample. In general, the manometric method is somewhat tedious and poor in sensitivity as well as in time resolution. Although the spectrophotometric assay of the Hill reaction is potentially unsurpassed for the time resolution (valuable for kinetic analysis), it is applicable only to reactions involving a limited number of Hill oxidants having large change in the extinction coefficient accompanying a change in their redox state. In recent years, the use of polarographic determination of O_2 as a means of assaying the Hill reaction has become increasingly popular because of the ready availability of the Clark-type O_2 electrode, which is easy to operate and provides good sensitivity and high specificity for O_2 even in the presence of a complex mixture of electrolytes, organic compounds, and biological materials. Furthermore, the moderately fast response time (typically 10–20 sec) of the Clark-type O_2 electrode allows direct recording of the time courses of the Hill reaction, which is useful in kinetic analysis. The procedures described below are designed for assay of the Hill reaction and related electron transport reactions of chloroplast membrane preparations with a Clark-type O_2 electrode as the basic analytical tool.

1. Reagents. (a) HEPES-NaOH buffer, 0.5 M, pH 8.0 (b) Ascorbate-DCPIP (prepare fresh daily). 300 mg sodium ascorbate plus 4.4 mg 2,6-dichlorophenol-indophenol (Na salt) dissolved with distilled H_2O to a final volume of 10 ml. (c) PBQ solution (prepare fresh daily). 16.2 mg of p-benzoquinone (purified by sublimation) dissolved in distilled H_2O to a final volume of 5.0 ml. Keep in darkness. (d) Ferricyanide solution (prepare weekly). 80 mg potassium ferricyanide dissolved in distilled H_2O to a final volume of 10 ml. (e) Methylviologen (MV) solution. 6.4 mg methylviologen (dichloride salt) dissolved in distilled H_2O, 10 ml total volume. (f) KCN solution. 0.06 M in distilled H_2O (prepare daily and keep tightly sealed when not in use). (g) Sorbitol solution. 1.5 M in distilled H_2O. (h) KCl solution. 1.5 M in distilled H_2O.

2. Equipment. (a) One Clark-type O_2 electrode (such as model 5331 oxygen probe from Yellow Springs Instrument Co.), together with an appropriate electrode polarization and signal-amplification system. (b) One water-jacketed reaction vessel to accommodate the electrode and to provide a sealable reaction chamber of 1- to 2-ml capacity. The reaction vessel is preferably made of glass and has provision for illumination of sample. (c) One circulating constant-temperature water bath. (d) One magnetic stirrer and glass- or Teflon-coated small stirring bar. (e) One strip chart recorder compatible with the electrode-amplifier system. (f) Light source for illumination of sample.

3. Assay procedures. The Hill activity of the chloroplast preparation is assayed as light-induced O_2 evolution or consumption in samples containing the appropriate reagents for the Hill reaction. Upon introduction of the complete reaction mixture (as listed below) into the sample chamber of the electrode vessel, the entire sample mixture is sealed against the external air space and is kept under constant stirring by the magnetic stirring system. Record the O_2 tension of the sample on the strip chart recorder in the absence of illumination (encase the electrode vessel assembly in a light-tight box or cover it with black felt cloth) for 1–2 min. While continuing recording, admit the illuminating light to the sample chamber for a desirable duration (routinely 2–5 min) so that a suitable deflection in the O_2-tension trace is obtained. Terminate the illumination and continue the recording for an additional 1–2 min. The rate of O_2 evolution or consumption by the sample is calculated from the difference in the slope of the recorder tracing during illumination and during the dark period of the same sample. To obtain the absolute rate of O_2 evolution or consumption in terms of micromoles of oxygen per hour per milligram of chlorophyll, it is necessary to know (a) the chlorophyll concentration of the sample and (b) the combined sensitivity of the electrode-amplifier-recorder system to O_2 tension in the sample (see Delieu and Walker 1971).

The reagents to be used for the assay of various types of Hill reactions with chloroplast membrane preparation from *C. reinhardi* are listed below. (The amount of each reagent to be added is based on a final sample volume of 1.5 ml and should be proportionally adjusted when a different final sample volume is employed for the assay.)

a. Reaction 1: $H_2O \rightarrow$ PBQ (assayed as O_2 evolution). The reduction of PBQ and the concomitant O_2 evolution are absolutely dependent on a functional photosystem II. This reaction does not exhibit an absolute requirement for photosystem I or for an efficient redox coupling between the two photosystems.

The reaction mixture contains: (i) HEPES buffer, 0.1 ml (solution a). (ii) KCl, 0.05 ml (solution h). (iii) Distilled H_2O, 0.5 ml. (iv) 0.3 ml of cell-free chloroplast membrane preparation of procedure I (III.A.2), or 0.25 ml of sorbitol (solution g) plus 0.03–0.05 ml of washed chloroplast membranes prepared by Yeda press (procedure II, III.A.3) containing 30 μg of chlorophyll. (v) PBQ solution, 0.1 ml (solution c). (vi) Add distilled H_2O to a final volume of 1.50 ml.

b. Reaction 2: $H_2O \rightarrow$ ferricyanide (assayed as O_2 evolution). O_2 evolution with ferricyanide is absolutely dependent on a functional photosystem II. This reaction shows a stronger dependence on the activity of photosystem I and a redox coupling between the two photosystems when compared with the PBQ reaction. However, an absolute requirement for photosystem I activity has not been established.

The reaction mixture contains: (i)–(iv) are identical to reaction 1 (III.B.3.a). (v) Potassium ferricyanide, 0.05 ml (solution d). (vi) Add distilled H_2O to a final volume of 1.50 ml.

c. Reaction 3: $H_2O \rightarrow$ MV (assayed as O_2 consumption). Owing to the low redox potential of methylviologen, its photoreduction is absolutely dependent on photosystem I. In the absence of an artificial electron donor, a sustained consumption of O_2 (owing to reoxidation of reduced MV by O_2) is dependent on a functional photosystem II and an efficient redox coupling between the two photosystems. This reaction involves all segments of the photosynthetic electron transport system from the water-splitting site of photosystem II to the reducing end of photosystem I.

The reaction mixture contains: (i)–(iv) are identical to reaction 1 (III.B.3.a). (v) Methylviologen, 0.08 ml (solution e). (vi) KCN, 0.05 ml (solution f). (vii) Add distilled H_2O to a final volume of 1.50 ml.

d. Reaction 4: Ascorbate-DCPIP $\rightarrow$ MV (assayed as O_2 consumption). This reaction is dependent only on the electron transport reactions closely associated with photosystem I. The electrons supplied by photosystem II in reaction 3 are replaced by the artificial electron donor system of ascorbate-DCPIP.

The reaction mixture contains: (i)–(iv) are identical to reaction 1 (III.B.3.a). (v) Ascorbate-DCPIP, 0.05 ml (solution b). (vi) Add distilled H_2O to a final volume of 1.50 ml.

C. Assay of photophosphorylation

The chloroplast membrane fragments prepared by the two procedures described above (III.A.2 and 3) exhibit good electron transport activities. However, unlike the well-coupled chloroplast membrane preparations from higher plants, the rate of photosynthetic electron transport reactions of the preparations from *C. reinhardi* is not tightly regulated by the phosphorylation reactions. Neither the addition of uncouplers such as *p*-trifluoromethoxycarbonylcyanidephenylhydrazone (FCCP), gramicidin, or methylamine, nor the inclusion of the phosphorylation substrates in the reaction mixture caused a significant stimulation of the rate of electron transport. Nonetheless, these chloroplast membrane preparations are active in photophosphorylation. Reasonably good rates of a light-induced ATP formation can be demonstrated to accompany the noncyclic flow of electrons from water to ferricyanide, PBQ, methylviologen, ferredoxin, or to accompany the phenazine methosulfate (PMS)-mediated cyclic flow of electrons around photosystem I.

1. Reagents. In addition to the solutions listed under III.B.1, the following reagents are needed for the assay of photophosphorylation: (i) $MgCl_2$, 30 mM in distilled H_2O. (j) PMS solution, 1.5 mM (keep in

Table 31-1. *Reagents for photophosphorylation experiments*

Photophosphorylation associated with	Basic substrates[a]	Acceptor or cofactor[b]	H_2O + chloroplast membrane preparation[c]
Noncyclic electron flow			
1. $H_2O \rightarrow PBQ$	1.0 ml	0.10 ml of c	0.40 ml
2. $H_2O \rightarrow$ ferricyanide	1.0 ml	0.05 ml of d	0.45 ml
3. $H_2O \rightarrow MV$	1.0 ml	0.08 ml of e	0.42 ml
Cyclic electron flow with			
4. PMS	1.0 ml	0.03 ml of j	0.47 ml

[a] Basic substrate solution as prepared above (III.C.2.a).
[b] Solutions as prepared above (III.B.1 and III.C.1).
[c] Containing 30–45 μg of total chlorophyll.

darkness, prepare weekly). Dissolve 4.6 mg of phenazine methosulfate in distilled H_2O to a final volume of 10 ml. (k) ADP solution, 15 mM, pH 7.0. Store frozen (use grade 1 adenosine 5-diphosphate, sodium salt from Sigma). (l) Potassium phosphate solution, 15 mM, pH 7.9 (make 100-ml batch and store frozen in tightly capped tubes each containing 2–3 ml). (m) Carrier-free $^{32}P_i$ in 1 N HCl (0.5–5 mCi ml^{-1}). (n) NaOH, 1 N, in distilled H_2O. (o) TCA-solution: 20% w/v of trichloroacetic acid in distilled H_2O.

2. Assay procedures

a. Preparation of the basic phosphorylation substrates. (The amount of each reagent is specified for assay of 10 samples). Pipette into a 25-ml Erlenmeyer flask the following reagents: (i) 1.0 ml HEPES buffer (solution a); (ii) 0.5 ml KCl (solution h); (iii) 1.0 ml $MgCl_2$ (solution i); (iv) 1.0 ml ADP (solution k); (v) 1.0 ml potassium phosphate (solution l); (vi) a few microliters of carrier-free $^{32}P_i$ (solution m), such that the final specific activity of the phosphate is ca. 0.5–1×10^6 cpm μmol^{-1} of P_i, and then follow by an equal volume of NaOH (solution n); and (vii) distilled H_2O to a total volume of 10 ml. Mix the content before use.

b. Preparation of the complete reaction mixture. Accurately transfer 1.0 ml of the basic phosphorylation substrate solution into each of small Pyrex test tubes (ca. 8 × 75 mm) and then introduce the rest of the reagents (Table 31-1) into each tube. The chloroplast membrane preparation should be added to the mixture just prior to the illumination.

c. Illumination of samples. Sample tubes containing the entire reaction mixture are placed in a constant-temperature water bath (25 C) with a transparent side window and illuminated for 2 min with white light from two 150-W projector floodlight lamps placed at a distance

Table 31-2. *Rates of Hill reaction with Chlamydomonas reinhardi chloroplast preparations* (μmol O_2 [evolution or consumption] h^{-1} mg^{-1} chlorophyll)

Reaction	Cell-free preparation by sonication	Washed membranes by Yeda press
$H_2O \rightarrow PBQ$	120–180	100–150
$H_2O \rightarrow$ ferricyanide	90–150	40–70
$H_2O \rightarrow MV$	100–120	60–90
Ascorbate-DCPIP $\rightarrow MV$	600–1000	600–1000

Table 31-3. *Rates of photophosphorylation with Chlamydomonas reinhardi chloroplast preparations* (μmol ATP h^{-1} mg^{-1} chlorophyll)

Reaction	Cell-free preparation by sonication	Washed membranes by Yeda press
$H_2O \rightarrow PBQ$	25–40	20–30
$H_2O \rightarrow$ ferricyanide	40–100	30–60
$H_2O \rightarrow MV$	50–100	35–60
PMS (cyclic)	150–250	100–180

of 15 cm from the samples. (An evenly illuminated area large enough to accommodate six to eight tubes can be obtained.) At the end of the illumination 0.2 ml of TCA solution (solution o) is quickly added to each tube to stop the reaction. After mixing, the tubes are placed in an ice bath until analysis of $^{32}P_i$ incorporation is performed. Dark controls, having the same composition of reaction mixtures, are covered with aluminum foil and incubated for the same length of time in the constant-temperature bath. They are then analyzed in the same manner as the illuminated samples.

d. Analysis of $^{32}P_i$ incorporation. TCA-treated samples are centrifuged for 10 min at 7000 $\times$ g to precipitate the chloroplast membranes and proteins. An aliquot of the clarified supernatant solution (0.8 ml) from each sample is transferred into a larger test tube (18 $\times$ 150 mm) for extraction of inorganic phosphate using the procedure of Avron (1960).

IV. Sample data

Under the experimental conditions described above the typical rates of Hill reaction and photophosphorylation using the chloroplast membrane preparations from *C. reinhardi* are summarized in Tables 31-2 and 31-3.

V. References

Arnon, D. I. 1949. Copper enzymes in isolated chloroplasts: Polyphenoloxidase in *Beta vulgaris*. *Plant Physiol.* 24, 1–15.

Avron, M. 1960. Photophosphorylation by Swiss-chard chloroplasts. *Biochim. Biophys. Acta* 40, 257–72.

Delieu, T., and Walker, D. A. 1971. An improved cathode for the measurement of photosynthetic oxygen evolution by isolated chloroplasts. *New Phytol.* 71, 201–25.

Gorman, D. S., and Levine, R. P. 1965. Cytochrome f and plastocyanin: Their sequence in the photosynthetic electron transport chain of *Chlamydomonas reinhardi*. *Proc. Natl. Acad. Sci. U.S.A.* 54, 1665–9.

Levine, R. P. 1971. Preparation and properties of mutant strains of *Chlamydomonas reinhardi*. *Methods Enzymol.* 23A, 119–29.

32: Respiratory assimilation of
^{14}C-labeled substrates by a microalga

KEITH E. COOKSEY

*Rosenstiel School of Marine and Atmospheric Science,
University of Miami, Miami, Florida 33149*

CONTENTS

I	**Introduction**	*page* **318**
II	**Materials and organisms**	**318**
III	**Methods**	**319**
	A. Design of experiments.	319
IV	**Typical method and sample data**	**323**
	A. Uptake of ^{14}C-(U)-glucose by *Amphora coffeaeformis* var. *perpusilla*	323
	B. Fractionation of acetate-labeled cells of *Cocconeis diminuta*	324
V	**Discussion**	**325**
	A. Potential problems	325
	B. Alternative procedures	325
VI	**Acknowledgment**	**326**
VII	**References**	**326**

I. Introduction

During the last 40 years the uptake of organic compounds by microalgae has been studied widely, but the role this process plays in the carbon economy of algal cells in nature is still controversial. Much of the work during this period concerned well-known laboratory green algae (Danforth 1962). The organisms of interest to the ecologist were studied rarely. Workers who did concern themselves with the ecological aspects of substrate assimilation seldom considered the metabolism of the compounds taken up or their respiratory dissimilation (Wright and Hobbie 1965; Bunt, 1969). This chapter covers methods for the measurement of substrate assimilation and CO_2 production. It also provides a method for preliminary investigation of the metabolism of assimilated compounds.

In general, phycologists are most interested in mediated transport of solutes into algal cells because this process is able to take place at the low substrate concentrations that are environmentally significant. The apparatus to carry out this transport is either constitutive or inducible under the conditions of the experiment. In the former case, cells exposed to an exogenous compound for the first time take it up without lag; in the latter case, however, a short time elapses before significant uptake occurs. The process of induction of the transport system and its specificity when induced can be studied by measuring uptake rates of ^{14}C-labeled solutes in the presence of metabolic inhibitors and solute analogues.

II. Materials and organisms

In this chapter I shall discuss specifically the uptake of ^{14}C-(U)-glucose and ^{14}C-(2)-acetate by the diatoms *Amphora coffeaeformis* var. *perpusilla* (Grunow) Cleve and *Cocconeis diminuta* Pantocsek, respectively. *A. coffeaeformis* is a facultatively heterotrophic pennate diatom (Cooksey and Chansang 1977). *C. diminuta* is an obligately autotrophic pennate diatom (Cooksey 1972). Both organisms were isolated from the sediments of Biscayne Bay, Florida.

[318]

III. Methods

A. Design of experiments

1. Medium. Growth medium is generally the best suspending fluid for microalgae in an assimilation experiment. A list of such media for both freshwater and marine algae has been published (McLachlan-1973; Nichols 1973). Modifications of the growth medium are necessary when the effect of inorganic ions on uptake is studied. These can be achieved by simple manipulations of the recipe. Often it is desirable to reduce the complexity of a growth medium for uptake studies, especially if subsequent experimentation involves extracting the cells with aqueous ethanol. In these cases, ions can be omitted sequentially from the medium until a minimal medium for uptake is obtained. It is unwise, however, to omit a buffer or to incubate cells below the salinity at which they will grow. It is also my experience that calcium ions are necessary for the uptake of solutes by *A. coffeaeformis* (Cooksey unpublished). A medium that will not support growth is necessary for experiments with resting cells.

2. The labeled substrate. The experiments described use ^{14}C-(U)-glucose and ^{14}C-(2)-acetate. Both substrates are available from most suppliers of radiochemicals at various specific activities and purities. It is wise to buy the purest compound available if no purification procedures are to be performed by the investigator. The specific activity of the substance used in the investigation should be considered carefully. Although it is tempting to buy the highest specific activity available, it is not advisable to do so unless the experimental procedure warrants it. Compounds of high specific activity undergo radiochemical decomposition more rapidly than those of lower specific activity and are thus more likely to contain radioactive breakdown products of the parent molecule. Manufacturers of labeled compounds add compounds such as ethanol as scavengers to retard this degradation process. For this reason it is also usual to store radiochemicals at low temperature. The actual temperature of storage depends on the nuclide and how it is packaged. The manufacturer's label gives the optimal storage temperature.

One of the requirements of an uptake experiment is that the specific activity of the substrate in the incubation medium be calculated. If we assume an uptake rate in terms of moles per cell per time, and a knowledge of the efficiency of the counting procedure to be used, the specific activity required to give radioactive counts significantly above the background of the experiment may be calculated. Usually, in short-term experiments, very small quantities of labeled compound are assimilated. Often the quantity is about 1% of the total in the incubation medium. Thus it becomes critically important to know the radiochemical purity of the substrate. Radiochemical purities of 95% are not uncommon in manufacturers catalogues, and the pitfall in using material of this purity, rather than 99%, is obvious. Total assimilation could be of the same order or less than the impurity in the substrate! It is essential, therefore, in all critical studies to determine radiochemical purity by radiochromatography (Smith 1969; Quayle 1972).

It is probably best to begin with a uniformly labeled substrate if it is available. Later, when the system is known and more detailed studies are planned, consideration should be given to the information that can be obtained by measuring the assimilation and respiration of a specifically labeled compound. For example, when ^{14}C-glucose was metabolized by a marine pseudomonad, it was found that only 38% of the carbon from positions 3 and 4 of the glucose molecule was retained by the cells and 37% was liberated as $^{14}CO_2$. Neither parameter was, therefore, a good measure of glucose uptake. On the other hand, 74% of the carbon from the 1 position was liberated as $^{14}CO_2$, making this a much better measure of glucose metabolism by the cells (Hamilton and Austin 1967).

The labeled compounds should be sterilized after dilution, not only for the control of experimental conditions, but also to prevent microbial breakdown when stock solutions are stored. Sterilization is best achieved by filtration through a filter (0.2 μm pore size). The plastic Swinnex adapters (Millipore) for use on disposable plastic syringes are particularly useful for this purpose (Hamilton 1973).

3. Preparation of cells. If experiments are to be reproducible and understandable in terms of the general metabolism of the organism, it is important to know the physiological status of the cells being investigated. Substrates metabolized by cells in early exponential phase usually have a different metabolic fate from those metabolized by cells in stationary phase. The former cells are the most metabolically active and are thus particularly suitable for use in assimilation experiments. A further point to consider is whether the uptake apparatus of the cells is constitutive or inducible. When cells have been exposed previously to the substrate, there should be no lag in the uptake of the same radioactive substrate if the cells take up the substance at all. If the cells have not been exposed previously, a lag of unpredictable length may occur before uptake begins.

4. Incubation conditions. Although it is not essential to work aseptically during assimilation experiments, all equipment and solutions should be sterile at the beginning of an experiment; the algal suspension should be axenic during any preincubation period. For respiratory uptake of substrates, cells are, of course, incubated in the dark (but see below). It is important that cells be adapted to the conditions of the experiment before the radioactive substrate is added. Usually a 30-min preincubation period is sufficient. Metabolic inhibitors should be added at the beginning of the preincubation period. During preliminary experiments it is wise to check that assimilation in terms of radioactivity is proportional to cell concentration. If it is not, perhaps the sample size is too great, leading to self-absorption in the counting procedure, or perhaps there is a minor radioactive contaminant of the substrate that is taken up at a different rate from the substrate itself. The latter problem is accentuated if the uptake of the major radioactive compound is small compared with the total radioactive material in the incubation mixture.

The conditions of temperature, aeration, and time of incubation must be determined by experiment. Usually an incubation period of 1–2 h is sufficient to achieve a linear uptake rate.

5. *Sampling technique.* The time course of assimilation of a substance into algal cells can be followed by taking sequential samples from the algal suspension. It is necessary, especially with diatoms, to stir the contents of the incubation flask with a magnetic stirring bar to facilitate representative sampling and provide aeration to the suspension. The sample volume should be at least 0.5 ml because pipetting error becomes significant when smaller samples are taken. The bore of the pipette should be reasonably large so that cell samples can be transferred with minimal adhesion to the pipette walls. It is convenient, when dealing with particularly sticky cells, to incubate small volumes of cell suspension in tubes and to harvest the entire contents of the tube at appropriate times. In this case the walls of the tube can be scraped with a rubber policeman to remove adhering algae. The cells must be separated from the medium in order to measure their radioactivity. This can be accomplished by centrifugation or by filtration, the former being used only when the packed cell volume of the cells is to be measured (Sorokin 1973). When washing by centrifugation, great care is needed not to lose cells when removing the supernatant liquid. Filtration can be accomplished on membrane or glass fiber filters. Samples can be pipetted directly into a filter holder containing an appropriate volume of wash liquid or diluent before vacuum is applied. This procedure provides a more uniform distribution of cells than does the direct application of the sample to the dry membrane filter. In our laboratory, cells on the filter are washed with five times the sample volume of medium. The wash solution contains unlabeled substrate at the concentration used in the experiment. It is not necessary for this wash liquid to be cold; indeed, cold washes have been known to lyse cells.

Sampling procedures when respiratory CO_2 is to be determined present more of a problem. Cells are incubated in center-well flasks of the type used in Warburg respirometry. The flasks are conveniently sealed with a serum cap. The center well usually contains 10% NaOH or an organic base such as Hyamine (New England Nuclear). Better absorption of the $^{14}CO_2$ can be achieved when a filter paper strip is placed in the center well of the flask. At the end of the incubation period it is essential, especially when dealing with marine media, to acidify the reaction of the medium to pH 2 by adding 0.1 volume of 50% trichloroacetic acid. After gentle agitation for 1 h, all the $^{14}CO_2$ in the incubation medium will be transferred to the center well. The paper or center-well contents can then be transferred with rinsings to a liquid scintillation vial for counting. Flasks are commercially available with disposable plastic center wells that can be added to the scintillation vial directly (Kontes). It is possible to count the radioactivity in the cells after the contents of the center well have been removed. However, it must be borne in mind that the radioactivity contained in the cells will be less than in those to which no trichloroacetic acid has been added, as the acid removes soluble compounds from the cells (III.A.7).

6. *Counting labeled cells and $^{14}CO_2$.* Radioactive cells can be counted in an end-window Geiger-Müller counter. The chief advantage of this type of counter is its economy of sample preparation and operation. Sample preparation consists only of gluing the algae-coated filter to the planchet. Liquid scintillation counting is more expensive; however, it is two to three times more efficient than the better end-window counters. For liquid counting, cells on cellulose acetate or glass fiber filters can be counted directly in a counting cocktail compatible with aqueous samples (e.g., Aquasol, New England Nuclear). Cocktails containing cellular solubilizing agents are useful too (e.g., Research Products International). Counting of $^{14}CO_2$ is best done in a liquid scintillation counter, but it can be counted as $Ba^{14}CO_3$ at infinite thickness in an end-window counter. Organic bases or $Na_2^{14}CO_3$ in NaOH can be counted directly. Detailed information can be found in the handbook provided with liquid scintillation spectrometers (e.g., Searle Analytic, *Preparation of Samples for Liquid Scintillation Counting*). Spurious counts will be observed if alkaline solutions of $NaH^{14}CO_3$ are counted before chemiluminescence has decayed.

7. *Methods for simple fractionation of labeled cells.* The procedure is basically that of Roberts et al. (1955). The cells should be washed twice with incubation medium before being subjected sequentially to the solvents listed below. The solvents and conditions of extraction are as follows (at each stage the extracted cells are recovered by centrifuga-

Table 32-1 *Fractionation of Cocconeis diminuta labeled with* ^{14}C-*(2)-acetate*

Cell fraction	Light (% of total)	Dark (% of total)
Soluble fraction	8.0	37.4
Lipids	62.6	25.0
Nucleic acid and carbohydrates	1.4	2.4
Protein and insoluble matter	28.0	35.1

Source: Data recalculated from Cooksey (1972).

tion): (a) 5% trichloroacetic acid for 30 min at 5 C; (b) 75% ethanol at 40–50 C for 30 min; (c) a mixture consisting of equal parts 75% ethanol and diethyl ether for 15 min at 40–50 C; (d) 5% trichloroacetic acid for 30 min at 100 C; (e) the precipitate from the extractions (a–d) is suspended in 1 N NH_4OH at 100 C until dissolved (diatom frustules are not dissolved).

Each extract is made to a known volume and assayed for radioactivity. It is necessary to start with at least 5 mg (wet weight) of algal cells for which 2 ml of each extractant is convenient. Fractions a–e correspond to the following: (a) metabolic intermediates, soluble fraction of the cell; (b) lipids and alcohol-soluble protein; (c) balance of the lipids; (d) nucleic acids and less soluble carbohydrates; (e) cell residue and protein. (Acid hydrolysis of this fraction, after removal of residual trichloroacetic acid, releases largely protein amino acids.)

An example of the use of this method is shown in Table 32-1. Before attempting this fractionation it is recommended that the original reference (Roberts et al. 1955) be consulted.

IV. Typical method and sample data

A. Uptake of ^{14}C-(U)-glucose by Amphora coffeaeformis var. perpusilla

A. coffeaeformis possesses a glucose uptake system that is induced by the sugar. Hellebust (1971) has shown that *Cyclotella cryptica* (Reimann, Lewin and Guillard) develops an uptake system in darkness in the absence of glucose. The following experiment shows that the synthesis of the uptake system in *A. coffeaeformis* appears to be controlled differently from that in *C. cryptica*.

Two cultures of *A. coffeaeformis* were grown at 5400 lux in modified ASP-2 medium (Provasoli et al. 1957; Cooksey and Chansang 1977). At the middle of the logarithmic phase of growth one culture was placed in darkness; the other was harvested by centrifugation and the

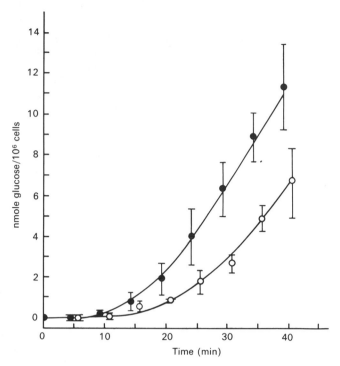

Fig. 32-1. Induction of glucose uptake in the dark in *Amphora coffeaeformis*. Black circles, autotrophically grown cells (controls); white circles, autotrophically grown cells preincubated in the dark for 40 h.

cells resuspended in 50 ml modified ASP-2 medium (2×10^6 cells ml^{-1}) to which 1 mM ^{14}C-glucose had been added (0.1 μCi μmol^{-1}). The mixture was incubated at 28 C in darkness and stirred magnetically. Then 2-ml samples were removed at 5-min intervals and the radioactivity in the cells determined. The other culture was treated similarly after incubation in darkness for 40 h. The results are shown in Fig. 32-1. Similar lag periods are shown by both cultures before significant glucose uptake occurs. Preincubation in darkness does not stimulate the synthesis of the uptake system in this organism.

B. Fractionation of acetate-labeled cells of Cocconeis diminuta

Cells grown in DC-C medium (Bunt 1969) at 25 C and 4000 lux were harvested and resuspended (2.5×10^6 cells ml^{-1}) in 48 ml of DC-C medium. Portions (24 ml) of this suspension were incubated in light (4000 lux) or darkness with 1 mM ^{14}C-(2)-acetate (0.25 μCi μmol^{-1}) in a final volume of 25 ml. After 4 h the cells from each incubation were harvested by centrifugation, washed twice with medium containing 1 mM acetate, and subfractionated (III.A.7). Note that although this organism is an obligate phototroph (Cooksey 1972), acetate is oxidized in the dark (respiration). The data in Table 32-1 show a greater incorporation into the soluble pool in the dark than in the light. Light favors direct assimilation of acetate into lipid.

V. Discussion

A. Potential problems

1. The respiratory uptake of ^{14}C-labeled organic substrates often results in the production of $^{14}CO_2$, which is found in the medium at the end of the incubation. Photosynthesis is a very rapid process. It needs only a few seconds of exposure to light for the cells to fix the respiratory $^{14}CO_2$ and obscure the incorporation pattern derived in the dark. Thus, it is important that all manipulation be carried out in dim light after the incubation has been completed.

2. A practical problem exists in dealing with some benthic diatoms and other attached forms because of their tendency to stick to glassware. Quantitative sampling and harvesting of these organisms is practically impossible. We have found that artificial seawater medium such as ASP-2 (Provasoli et al. 1957) with a reduced calcium concentration (0.25 mM) allows some organisms to stay in suspension rather than attach to the incubation flask. Most of the organisms we have tested grow normally at this calcium concentration, so we assume it is not detrimental to the cell.

3. When very low substrate concentrations are used in assimilation experiments designed to mimic natural conditions, the amount of radiolabeled substrate adhering to the cells becomes large compared with the amount taken up. This is especially true after short-term incubations. It then becomes necessary to use a dead-cell control for each sample time (i.e., heat-killed cells incubated under conditions identical to those in the experiment). The same problem exists when uptake rates at differing substrate concentrations are measured. In this case a dead-cell or zero-time control is necessary for each substrate concentration.

B. Alternative procedures

Although many algal cells assimilate organic substrates, the process is probably ecologically significant only if the cells can concentrate the substances against the concentration gradient that exists across the cell membrane. Accordingly, it is important to know if uptake consists of equilibration with exogenous solutes or concentration of these substances inside the cells. To make this differentiation, it is necessary to know the volume of the cells and the amount of substance transported, so that intracellular and extracellular concentrations can be compared. For the first of these, one can use the packed cell volume. This volume is an overestimate of the true volume of the cells by the amount of intercellular space. The space can be calculated by suspending the cells under investigation in a medium containing a radioactive substrate they cannot take up (e.g., ^{14}C-inulin). After centrifugation, the radioactivity in the cell pellet (unwashed) is measured. If the radioactivity per milliliter of fluid is known, the volume of the fluid in the intracellular space can be calculated.

The amount of radioactive substance taken up by the cells is calculated from the radioactivity of the cells and the specific activity of the substance in the incubation medium. Ideally, we should measure the radioactivity in the cells directly after transport has taken place and before metabolism has begun. This can be achieved rarely, so it has become usual practice to measure the accumulation of a nonmetabolizable analogue of the substance being investigated. For instance, glucose analogues used in transport studies are the various deoxyglucoses, α-methylglucose, 3-O-methylglucose, and in some circumstances, mannose. Commonly used analogues of amino acids are aminoisobutyric acid (neutral amino acid), cycloleucine (leucine), and α-methylglutamate (glutamate). An analogue, however, does not always mimic the normal substrate for transport, as 2-deoxyglucose is taken up by the glucose transport system in *Chlorella vulgaris* (Tanner et al. 1970), but not by that in *A. coffeaeformis* (Chansang and Cooksey 1977).

This chapter concerns respiratory uptake of substrates and thus, it is assumed, concerns only dark metabolism. Recent experiments in our laboratory (Chansang 1975) have indicated that the energy for the accumulation of mannose by the active glucose transport system of *A. coffeaeformis* is derived from oxidative reactions both in the dark and in the light. Anaerobiosis in the light prevented mannose accumulation.

VI. Acknowledgment

The author was a holder of National Science Foundation grant GB 31102 during the time this chapter was written.

VII. References

Bunt, J. S. 1969. Observations on photoheterotrophy in a marine diatom. *J. Phycol.* 5, 37–42.
Chansang, H. 1975. Growth physiology and glucose transport systems of the

pennate diatom, *Amphora coffeaeformis* var. *perpusilla* (Grunow) Cleve. Ph.D. dissertation, University of Miami, Coral Gables, Florida. 133 pp.

Chansang, H., and Cooksey, K. E. 1977. The glucose transport system of *Amphora coffeaeformis* (Grunow) Cleve. *J. Phycol.* 12:51–7.

Cooksey, K. E. 1972. The metabolism of organic acids by a marine pennate diatom. *Plant Physiol.* 50, 1–6.

Cooksey, K. E., and Chansang, H. 1977. Isolation and physiological studies on three isolates of *Amphora* (Bacillariophyceae). *J. Phycol.* 12, 455–60.

Danforth, W. F. 1962. Substrate assimilation and heterotrophy. In Lewin, R. A. (ed.). *Physiology and Biochemistry of Algae,* pp. 99–123. Academic Press, New York.

Hamilton, R. D. 1973. Sterilization. In Stein, J. R. (ed.). *Handbook of Phycological Methods: Culture Methods and Growth Measurements,* pp. 181–93. Cambridge University Press, London.

Hamilton, R. D., and Austin, K. 1967. Assay of relative heterotrophic potential in the sea: The use of specifically labelled glucose. *Can. J. Microbiol.* 13, 1165–73.

Hellebust, J. A. 1971. Kinetics of glucose transport and growth of *Cyclotella cryptica* Reimann, Lewin and Guillard. *J. Phycol.* 7, 1–4.

McLachlan, J. 1973. Growth media–marine. In Stein, J. R. (ed.). *Handbook of Phycological Methods: Culture Methods and Growth Measurements,* pp. 25–51. Cambridge University Press, London.

Nichols, H. W. 1973. Growth media–freshwater. In Stein, J. R. (ed.). *Handbook of Phycological Methods: Culture Methods and Growth Measurements,* pp. 7–24. Cambridge University Press, London.

Provasoli, L., McLaughlin, J. J. A., and Droop, M. R. 1957. The development of artificial media for marine algae. *Arch. Mikrobiol.* 25, 392–428.

Quayle, J. R. 1972. The use of isotopes in tracing metabolic pathways. *Methods Microbiol.* 6B, 157–83.

Roberts, R. B., Cowie, D. B., Abelson, P. H., Bolton, E. T., and Britten, R. J. 1955. *Studies in Biosynthesis in Escherichia coli.* Publ. No. 607. Carnegie Institution, Washington, D.C. 535 pp.

Smith, I. 1969. *Chromatographic and Electrophoretic Techniques,* vol. I, 3rd ed. Interscience, New York. 1080 pp.

Sorokin, C. 1973. Dry weight, packed cell volume and optical density. In Stein, J. R. (ed.). *Handbook of Phycological Methods: Culture Methods and Growth Measurements,* pp. 321–43. Cambridge University Press, London.

Tanner, W., Grünes, R., and Kandler, O. 1970. Spezifität und Turnover des induzierbaren Hexose-Aufnahmesystems von *Chlorella. Z. Pflanzenphysiol.* 62, 376–86.

Wright, R. T., and Hobbie, J. E. 1965. The uptake of organic solutes in lake water. *Limnol. Oceanogr.* 10, 22–8.

33: Respiratory processes in mitochondria

NEIL G. GRANT

*Department of Biology, The City College of the
City University of New York, New York, New York 10031*

CONTENTS

I	**Introduction**	*page*	**330**
II	**Materials**		**331**
	A. Mitochondria		331
	B. Chemicals and solutions		331
	C. Supplies		331
III	**Method**		**331**
	A. Estimation of O_2		331
	B. Determination of O_2 uptake rate and respiratory control		332
	C. Calculation of ADP/O ratios		332
IV	**Sample data**		**332**
V	**Discussion**		**334**
VI	**Acknowledgments**		**334**
VII	**References**		**334**

I. Introduction

Characterization of respiration of mitochondria is complicated. They are complex and diverse organelles (Lloyd 1974), and because so little is known about mitochondria from algae, they are probably even more diverse than we realize.

Mitochondrial respiration is investigated, in part, by using substrates that are oxidized and measuring the numbers of associated sites of oxidative phosphorylation. This is determined from the ratio of ATP formed per oxygen atom consumed (the ATP/O ratio) or, in the case of tightly coupled mitochondria, by the ADP/O ratio. Mitochondria from most organisms studied have three sites associated with NAD-linked substrates such as malate, and two sites associated with succinate.

An important consideration, however, in studying the respiratory processes of mitochondria from algae is the possibility of the presence of more than one pathway to O_2. For in addition to the cytochrome oxidase, which is apparently similar to that of mammals in being inhibited by cyanide, azide, and carbon monoxide, there is in *Chlorella* mitochondria another terminal oxidase, which appears to be similar to the so-called alternate oxidase of higher plants. Schonbaum et al. (1971) characterized a class of chemicals (substituted aromatic hydroxamic acids) that specifically inhibited cyanide-insensitive respiration in higher plant mitochondria. Grant and Hommersand (1974a, 1974b) showed that hydroxamic acids in a like manner inhibit cyanide-insensitive respiration in whole cells and mitochondria of *Chlorella protothecoides*.

Because the physiological function of the alternate oxidase is not understood in algae, I shall emphasize its characterization.

A rapid, sensitive, and versatile method for characterizing the kinetics of O_2 utilization employs the Clark-type O_2 electrode. There are many variations of this equipment and technique. Estabrook (1967) gives plans for the electrical circuitry associated with the electrode and for the reaction vessel. A suitable commercial instrument is available from Yellow Springs Instrument Co. The amplifier or circuit for the electrode is attached to a standard 10 mV full-scale potentiometric chart recorder.

[330]

II. Materials

A. Mitochondria

Mitochondria from such algae as *Chlorella* (Grant and Hommersand 1974b), *Euglena* (Sharpless and Butow 1970a, 1970b), or *Polytomella* (Lloyd and Chance 1968).

B. Chemicals and solutions

1. Reaction buffer (modified from Bonner 1967). Sucrose (0.25 M); potassium phosphate buffer, pH 7.2 (10 mM); KCl (10 mM); $MgCl_2$ (5 mM).

2. Substrates and cofactors, stock solutions. Succinic acid (0.8 M) neutralized with TRIS; L-malic acid (3.0 M) neutralized with TRIS; NADH (0.1 M) standardized optically, $\epsilon_{mM} = 6.22$ at 340 nm; ADP (150 mM) standardized optically, $\epsilon_{mM} = 15.4$ at 260 nm.

3. Inhibitors, stock solutions. (a) KCN (0.1 M) prepared fresh, neutralized before use, kept stoppered, *deadly poison.* (b) Antimycin (2 μM) dissolved in N,N-dimethylformamide, standardized optically, $\epsilon_{mM} = 4.8$ at 320 nm. (c) *m*CLAM (*m*-chlorobenzhydroxamic acid, prepared from *m*-chlorobenzoyl chloride and free hydroxylamine, Hauser and Renfrow 1943), 10 mM, or SHAM (salicylhydroxamic acid, Aldrich), 10 mM, dissolved in N,N-dimethylformamide.

C. Supplies

Syringe, microliter, 50-μl capacity, blunt-tipped.

III. Method

Keep the mitochondria in a centrifuge tube on ice until they are to be used. Add the required amount of reaction medium to the oxygen vessel (as algal mitochondria are likely to be obtained in small yield, use as small an amount of reaction mixture as feasible). Allow the mixture to come to equilibrium with the O_2 in the atmosphere and temperature, then close off from the atmosphere, ensuring that no bubbles of gas have been included.

A. Estimation of O_2

The amount of oxygen in the system can be estimated from tables in the *Handbook of Chemistry and Physics* (Weast 1969) or from Estabrook (1967). For example, at 25 C the concentration of oxygen in water at equilibrium with the atmosphere is 258 μM. If the volume of the vessel is 1.5 ml, then the amount of oxygen present is 387 nmol. A more accurate way of determining the amount of oxygen and also the characteristics of the monitor is to use known limiting amounts of NADH and mitochondria or submitochondrial particles that oxidize NADH, according to Estabrook (1967).

Set and note on the recorder the position of the pen at O_2 saturation and the position at zero O_2, which can be determined by adding a small amount of dry sodium dithionite (but see Laties 1974). The amount of O_2 present divided by the chart divisions between the two-positions of the pen gives the nanomoles of O_2 per chart division.

B. Determination of O_2 uptake rate and respiratory control

Add a small amount of mitochondria, about 50–100 μl, with a syringe or micropipette and observe the rate of O_2 uptake, which should be negligible for well-washed mitochondria. After a short while, add substrate to a concentration that is not rate limiting in a volume of about 0.01 of the vessel. That is, if the vessel is 1.5 ml, the addition should be about 15 μl. The resulting stimulated O_2 uptake rate is known as the substrate rate (Bonner 1974). After this stabilizes, add a known and limiting amount of ADP, at which time the mitochondria will be in state 3 (Chance and Williams 1955). If the mitochondria are tightly coupled, there will be a further stimulation of O_2 uptake rate while the ADP is converted to ATP, at which time the mitochondria will be in state 4 and rate of O_2 utilization should decrease. The respiratory control ratio (RC) is defined as the state 3 rate divided by the subsequent state 4 rate. The rate of respiration is customarily expressed as nanomoles of O_2 used per minute per milligram of mitochondrial protein.

C. Calculation of ADP/O ratios

The number of sites of phosphorylation between a substrate and O_2 may be estimated from the amount of O_2 utilized for a known, limiting amount of ADP (equation 1):

$$\frac{\text{natoms oxygen used in state 3}}{\text{nmol ADP added}} = \text{ADP/O} \qquad (1)$$

Note: natoms oxygen = 2 × nmol O_2.

IV. Sample data

Figure 33-1 illustrates a typical O_2 uptake trace and shows the calculation of respiratory control and ADP/O ratio.

The hydroxamic acid–sensitive pathway has the effect of lowering the ADP/O ratio, as shown in Fig. 33-2, where 1 mM mCLAM had been added before ADP (SHAM has a similar effect). Respiratory control has been increased over the noninhibited control, and in this case the ADP/O ratio indicates two sites of oxidative phosphorylation

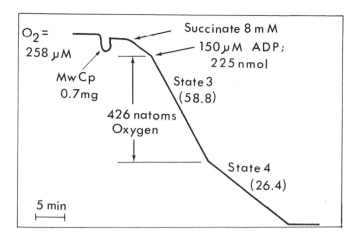

Fig. 33-1. Pattern of oxygen utilization by *Chlorella protothecoides* mitochondria showing the determination of respiratory control and ADP/O ratios. At the point indicated, 0.1 ml of mitochondria (*Mw Cp*), 0.7 mg protein, was added to 1.4 ml of reaction mixture (see text), 25 C. Succinate (15 μl) and ADP were added as indicated. Calculation of respiratory control and ADP/O ratios according to the method of Chance and Williams (1955) is explained in the text. The numbers beside the trace are the rates of oxygen uptake expressed as nanomoles of O_2 per minute per milligram of protein.

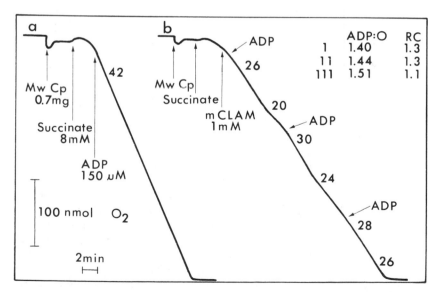

Fig. 33-2. Pattern of succinate oxidation. Conditions as in Fig. 33-1. *Curve a,* pattern of oxidation of succinate. *Curve b,* pattern of oxidation of succinate with *m*CLAM added before ADP. (From Grant and Hommersand 1974b.)

between succinate and oxygen. Respiratory control has not been observed in *Chlorella* mitochondria with the cytochrome oxidase pathway blocked by antimycin.

V. Discussion

This method is rapid, sensitive, and versatile. If the mitochondria, however, show no respiratory control, it is necessary to determine by other means whether they are capable of oxidative phosphorylation. The ATP produced may be assayed using the firefly luciferase assay (Wilson 1970; see Chp. 19). The inorganic phosphate consumed during oxidative phosphorylation, using a glucose-hexokinase trap to recycle the ATP, can be measured in a Warburg or similar manometer (Slater 1967a). The manometric method is time-consuming and uses a greater amount of mitochondria. An extensive discussion of the use of various classes of inhibitors and uncouplers on mitochondria is given in Slater (1967b).

VI. Acknowledgments

The bulk of this work was done while I was a doctoral candidate in the Department of Botany, University of North Carolina at Chapel Hill, supported by a National Aeronautics and Space Administration Predoctoral Fellowship and an American Cancer Society Institutional Subgrant to M. H. Hommersand and me. Additional work at The City College has been supported by a Faculty Senate Research Award.

VII. References

Bonner, W. D., Jr. 1967. A general method for the preparation of plant mitochondria. *Methods Enzymol.* 10, 126–33.

Bonner, W. D., Jr. 1974. Plant mitochondria. In San Pietro, A. (ed.). *Experimental Plant Physiology*, pp. 125–33. Mosby, St. Louis.

Chance, B., and Williams, G. R. 1955. Respiratory enzymes in oxidative phosphorylation. I. Kinetics of oxygen utilization. *J. Biol. Chem.* 217, 383–93.

Estabrook, R. W. 1967. Mitochondrial respiratory control and the polarographic measurement of ADP/O ratios. *Methods Enzymol.* 10, 41–7.

Grant, N. G., and Hommersand, M. H. 1974a. The respiratory chain of *Chlorella protothecoides*. I. Inhibitor responses and cytochrome components of whole cells. *Plant Physiol.* 54, 50–6.

Grant, N. G., and Hommersand, M. H. 1974b. The respiratory chain of *Chlorella protothecoides*. II. Isolation and characterization of mitochondria. *Plant Physiol.* 54, 57–9.

Hauser, C. R., and Renfrow, W. B., Jr. 1943. Benzohydroxamic acid. *Org. Synth. Collect.* 2, 67–8.

Laties, G. G. 1974. The respiration of plant storage organ tissue slices. In San Pietro, A. (ed.). *Experimental Plant Physiology,* pp. 118–24. Mosby, St. Louis.

Lloyd, D. 1974. *The Mitochondria of Microorganisms.* Academic Press, London. 753 pp.

Lloyd, D., and Chance, B. 1968. Electron transport in mitochondria isolated from the flagellate *Polytomella caeca. Biochem. J.* 107, 829–37.

Schonbaum, G. R., Bonner, W. D., Jr., Storey, B. T., and Bahr, J. T. 1971. Specific inhibition of the cyanide-insensitive respiratory pathway in plant mitochondria by hydroxamic acids. *Plant Physiol.* 47, 124–8.

Sharpless, T. K., and Butow, R. A. 1970a. Phosphorylation sites, cytochrome complement, and alternate pathways of coupled electron transport in *Euglena gracilis* mitochondria. *J. Biol. Chem.* 245, 50–7.

Sharpless, T. K., and Butow, R. A. 1970b. An inducible alternate terminal oxidase in *Euglena gracilis* mitochondria. *J. Biol. Chem.* 245, 58–70.

Slater, E. C. 1967a. Manometric methods and phosphate determination. *Methods Enzymol.* 10, 19–29.

Slater, E. C. 1967b. Application of inhibitors and uncouplers for a study of oxidative phosphorylation. *Methods Enzymol.* 10, 48–57.

Weast, C. R. (ed.). 1969. *Handbook of Chemistry and Physics.* Chemical Rubber Co., Cleveland.

Wilson, S. B. 1970. Energy conservation associated with cyanide-insensitive respiration in plant mitochondria. *Biochim. Biophys. Acta* 223, 383–7.

34: Determination of protein synthesis during synchronous growth

DENNIS J. O'KANE AND
RAYMOND F. JONES

Division of Biological Sciences,
State University of New York at Stony Brook, Stony Brook, New York 11794

CONTENTS

I	**Introduction**	*page* **338**
	A. Purpose	338
	B. General requirements and theory	338
II	**Materials and methods**	**339**
	A. Source and culture of cells	339
	B. Analytical procedures	340
	C. Precursor incorporation protocols	340
III	**Sample data**	**342**
	A. Comparison of filters	342
	B. Incorporation of precursors	343
IV	**Discussion**	**349**
	A. Changes in pool sizes during synchronous growth	349
	B. Incorporation of precursors	349
V	**Acknowledgments**	**350**
VI	**References**	**350**

I. Introduction

A. Purpose

Although radiolabeling with amino acids is a sensitive and generally applicable technique for determining the rate of protein biosynthesis, several factors may complicate the comparison of the apparent rates of synthesis in investigations of populations of cells at various stages of differentiation, such as occur during synchronous growth. The purpose of this chapter is twofold: (1) to present methods for labeling cells at different stages of the cell cycle using synchronous cultures of the unicellular alga *Chlamydomonas reinhardtii,* and (2) to comment on the filter method commonly used to determine the rate of protein synthesis (Mans and Novelli 1961).

B. General requirements and theory

Some comments on the requirements and methodology of precursor incorporation are appropriate.

1. The radiolabeled precursor. Before cells are labeled with any radioactive precursor, the purity of the precursor should be determined by two-dimensional chromatography and autoradiography. After storage for several months at − 20 C, significant contamination of ^{14}C-arginine by peptidelike compounds is evident, although neither citrulline nor ornithine is detectable. Although these contaminants do not alter the specific activity of free arginine, they would decrease the free arginine concentration and thus introduce a source of error. If contaminants are found to be present, the radiolabeled precursor should be either repurified or discarded.

2. Length of the labeling period. During pulse labeling under non-steady-state conditions, when the size and turnover rates of the intracellular precursor pools may vary, the length of the labeling period becomes important. Two antagonistic criteria must be fulfilled: (a) the labeling period must be of sufficient duration (i) to permit pool equilibration to a constant specific activity and (ii) to permit the synthesis of protein of high specific activity; (b) the labeling period must be as short as feasible (i) to minimize possible interferences from protein turnover, (ii) to minimize the catabolism of the precursor and subsequent incorporation into protein via another precursor, and (iii) to minimize complications from significant endogenous fluctuations in the precursor pools.

Although the effect of protein turnover may be assumed to be minimal, this must be considered in investigations of individual proteins that may be rapidly turned over; for example, soluble ornithine decarboxylase has a half-life reported as 10 min (Schimke and Bradley 1975).

If the cells to be labeled, are under steady-state condition, and precursor catabolism and incorporation via another route occur at an insignificant rate, then the requirement for a short labeling period may be disregarded and labeling by continuous infusion may be adopted. This method has the advantage of allowing an estimation of both the rate of protein synthesis and the rate of protein turnover. The reader is referred to review articles on protein turnover for the use of continuous infusion (Schimke and Doyle 1970; Huffaker and Peterson 1974).

3. Comparing rates of protein synthesis obtained with cells at different stages of the cell cycle. When rates of protein synthesis by cells at different stages of the cell cycle are to be compared, it must be demonstrated that differences in the measured rates actually reflect differences in the rates of protein synthesis. Variations in precursor uptake, pool sizes, or turnover rates may cause differences in the intracellular isotopic dilution of the radioactive precursor and thus result in artifactual differences in the rates of protein synthesis. Therefore, either the isotopic dilution of the label should be determined for each cell population or the precursor pools should be swamped with the labeled precursor so that the intracellular specific activity approaches that of the exogenous label (Holleman and Key 1967).

II. Materials and methods

A. Source and culture of cells

Chlamydomonas reinhardtii, strain 89[+], was grown on high-salt minimal (HSM) medium (Sueoka 1960) supplemented with a trace element and chelated iron solution (Jones 1962). Cells were cultured at 24 C in 300-ml Belco flasks on a 12-h light, 12-h dark regimen to induce synchrony of cell division (Kates and Jones 1964). The light intensity at the surface of the flasks was 800 ft-c and the cultures were aerated with 2% CO_2 in air. Cell numbers were estimated by determining the mean of 10 hemacytometer counts. All experiments involving labeling of synchronous cells were carried out at cell densities of 0.5×10^6 to 1×10^6 cells ml^{-1} (at time L_0 of the photoperiod).

B. Analytical procedures

1. Protein determination. Whole-cell protein was determined by a modification (Hartree 1972) of the Folin phenol method (Lowry et al. 1951) using bovine serum albumin as the reference standard. The absorbance measurements were taken at 750 nm to eliminate complications from chlorophyll absorbance.

2. Preparation of iodinated ferredoxin–NADP reductase and the antigen–antibody complex. (a) Radioiodination. Ferredoxin–NADP reductase (fp) was labeled with $Na^{125}I$ by the chloramine-T procedure (Greenwood et al. 1963). The reaction mixture contained 10 μmol of phosphate buffer, pH 7.0; 2 $mCiNa^{125}I$, and 20 μg fp in 75 μl. The reaction was initiated by the addition of 0.18 μmol chloramine-T (25 μl) and terminated after 1 min by the addition of 1 μmol of sodium metabisulfite. Protein-bound and free iodide was separated by exclusion chromatography on a column of Bio-Gel P-2. (b) Preparation of ^{125}I-fp–anti-fp complexes. ^{125}I-fp was incubated with the IgG fraction of anti-fp (induced in rabbits) for 0.5 h at 37 C and 24 h at 4 C. The antigen-antibody complex was separated from unbound antigen by precipitation with 50% saturated $(NH_4)_2SO_4$, pH 7.8 (Farr 1971).

C. Precursor incorporation protocols

1. Labeling cells
   a. Pulse labeling cells. Samples of 0.5×10^6 to 1×10^6 cells were withdrawn from the Belco flasks at regular intervals during the synchronous cell cycle, pelleted by centrifugation at room temperature, and resuspended at the original cell density in fresh HSM medium containing either ^{14}C-(U)-arginine or ^{14}C-(U)-phenylalanine. The concentration and specific activities of the labeled amino acids used are presented with the figures. The cells were labeled in 10×75-mm glass tubes with aeration and illumination conditions the same as for the Belco flask cultures and stirring provided with a magnetic flea. Cells were incubated for periods up to 2 h, after which the cells were killed by adding 1 volume of cold 20% trichloroacetic acid (TCA) and processed as described below (II.C.2.b).
   b. Pool swamping. Cells were harvested at regular intervals during the cell cycle and resuspended in HSM medium containing increasing concentrations of radiolabeled amino acids (^{14}C-arginine or ^{14}C-phenylalanine), but keeping the specific activity constant. After 0.5–2 h, cells were killed by the addition of an equal volume of 20% TCA,

treated as described below (II.C.2.b), and the incorporation of the radiolabeled amino acids determined.

2. Determination of precursor incorporation. The method most frequently used for processing radiolabeled samples is to collect the killed cells on filters (Mans and Novelli 1961) with hydrolysis and washing to deacylate aminoacyl-tRNA and to remove unincorporated precursor, respectively. Methods for batch washing many filters and for treating individual filters are presented.

a. Batch washing filters. Small aliquots of cells (100 μl) labeled with ^{14}C-arginine were spotted on 25-mm filter paper disks and dried for 15 sec with a hair dryer. The disks were immersed for 1 h in ice-cold 10% TCA containing 0.1 M arginine (nonradioactive). The filters were transferred to 5% TCA, at 90 C, and incubated for 30 min to hydrolyze any ^{14}C-arginyl-tRNA. After hydrolysis the filters were rewashed for 10 min in cold 10% TCA and then transferred to an ethanol-ether (1:1 v/v) bath at 37 C for 30 min. The filters were then washed twice for 15 min with anhydrous ether at room temperature and finally air-dried.

b. Washing individual filters. Much of the data presented in the results section was obtained using nitrocellulose filters that were individually washed using a modification of the procedure for batch washing filters. Radiolabeled cells that were killed with TCA, or a radiolabeled antigen-antibody complex precipitated with 10% TCA (II.B.2), were heated at 90 C for 30 min, cooled, and filtered through 25-mm Millipore nitrocellulose filters placed on a sintered glass filter holder. Glass fiber or paper filters were used as well in some experiments. The individual filters were then washed with 30 ml of cold 10% TCA and air-dried at 37 C. The filters were then washed twice for 10 min each in anhydrous ether to remove residual TCA and finally air-dried. It should be mentioned that washing with ethanol-either (1:1 v/v), as done for batch washing filters, cannot be used with nitrocellulose filters, as they slowly dissolve in ethanol.

3. Determination of radioactivity by scintillation counting. Dried filters, containing ^{14}C-labeled protein, were placed in the bottom of 24-mm diam. scintillation vials containing 10 ml of Omnifluor (New England Nuclear) and placed in the dark for 1 h. They were then counted in a Nuclear Chicago Mark III liquid scintillation counter and corrected for any quenching.

Filters containing the radioiodinated antigen-antibody complex were placed in 10 × 75-mm polystyrene tubes (Falcon Plastics) and counted in a Nuclear Chicago solid crystal gamma scintillation counter at approximately 80% efficiency.

Table 34-1. *Retention of anti-fp–^{125}I-fp complexes on different filters*

| Manufacturer | Designation | Material | Percent radioactivity found in | | |
			Filter	Tube	Filtrate
Millipore	GSA	Nitrocellulose	98.5	0.7	0.9
Reeve Angel	934AH	Glass fiber	78.2	0.6	21.4
Whatman	GF/A	Glass fiber	76.5	0.5	21.6
Whatman	GF/C	Glass fiber	79.6	0.4	19.8
Whatman	3MM	Paper fiber	27.1	0.5	71.5
Schleicher and Schüll	No. 1	Paper fiber	26.2	0.5	73.3

Note: Average of three samples. Each sample consisted of 30 ng of ^{125}I-fp in excess homologous antibody (ca. 100 μg IgG).

III. Sample data

A. Comparison of filters

The choice of the filter material used to entrap and retain the TCA-precipitated radiolabeled protein is important in quantitating the rate of protein synthesis. Significant differences can be demonstrated for paper, glass fiber, and nitrocellulose filters. As an example, ^{125}I-ferredoxin–NADP reductase was bound to the homologous antibody and precipitated with 50% saturated $(NH_4)_2SO_4$. The antibody-bound radioactivity was precipitated with 10% TCA (final concentration) and applied to various filters. The filters were washed individually (II.C.2.b). As shown in Table 34-1, only the nitrocellulose filter (Millipore) quantitatively retained the antigen-antibody complex; glass fiber and paper fibers retained 75–80% and 25–27%, respectively. It should be realized that this experiment is not meant to represent a control for other experiments presented below, but rather to point out differences in retention of protein by various filters. However, similar results are obtained with ^{14}C-labeled *Chlamydomonas* protein (Table 34-2). Three aliquots of radiolabeled cells were either killed with 10% TCA and filtered through a nitrocellulose filter or a paper filter (3MM) and washed individually (II.C.2.b), or spotted on a paper filter and batch-washed. The nitrocellulose filter retains over sevenfold more ^{14}C-labeled protein than the paper filter when either individually or batch-washed.

In addition to the superior retention of protein by nitrocellulose filters, these filters allow slightly higher counting efficiencies during liquid scintillation counting because they become transparent when

Table 34-2. *Retention of* ^{14}C-*arginine-labeled Chlamydomonas protein on filters*

Filter	Washing	Cpm retained − background
Nitrocellulose, Millipore GSA	Individual	9441
Whatman 3MM, paper	Individual	1287
Whatman 3MM, paper	Batch	1326

Note: Approximately 8 μg of protein (5 $\times$ 10^5 cells) was applied to each filter. Average of three filters.

placed in toluene- or xylene-based scintillation cocktails. This is in contrast to paper or glass fiber filters, which remain opaque or become slightly translucent. All subsequent sample data are from experiments using nitrocellulose filters.

B. Incorporation of precursors

1. Incorporation of ^{14}C-*arginine into protein at low exogenous arginine concentrations.* Usually cell suspensions are labeled at low exogenous precursor concentrations using high specific activities (Kates 1966; Jones et al. 1968; Iwanij et al. 1975). Under these conditions, significant radioactivity is incorporated into protein. In Fig. 34-1 cells from hour D_4 of the cell cycle incorporate ^{14}C-arginine into protein with zero-order kinetics when labeled with 4 μM ^{14}C-arginine (250 μCi μmol^{-1}). However, different zero-order rates of incorporation are obtained with cells at different times during synchronous cell growth. Cells supplied with either 4 or 20 μM ^{14}C-arginine exhibit a decreasing rate of incorporation of radioactivity into protein from the beginning of the photoperiod (early G_1, Fig. 34-2). During the photoperiod, protein accumulates at a zero-order rate of 2.25 pg protein h^{-1} cell^{-1}. The specific activities of the protein synthesized during the photoperiod were found to vary from 3150 cpm μg^{-1} protein at hour L_2 to 1290 cpm μg^{-1} protein at hour L_{11} when labeled with 4 μM arginine. Although a variation in the arginine content of the protein might be postulated to explain the decrease in the apparent rate of arginine incorporation into protein, a two- to three-fold decrease in the arginine content would be required to explain the data. A simpler explanation is that there is a variation in the intracellular isotopic dilution of ^{14}C-arginine by the arginine *protein precursor pool* (as defined by Holleman and Key 1967). This seems reasonable, as the *total precursor pool* of arginine expands significantly during the photoperiod, although the turnover rate and size of the protein precursor pool are not known (Jones, Iwasa, and Renke, unpublished). Consequently, when cells are labeled at low exogenous precursor concentrations, it is difficult,

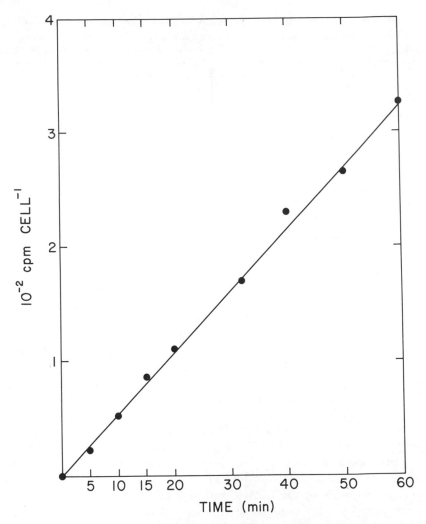

Fig. 34-1. Time course for incorporation of ^{14}C-arginine. A 10-ml sample of a cell suspension (ca. 1×10^6 mother cells per milliliter) was harvested at hour D_4 of the cell cycle and resuspended in fresh HSM medium containing 10 μCi ^{14}C-arginine (250 μCi μmol^{-1}). The final arginine concentration was 4 μM with ca. 1×10^6 mother cells per milliliter. The cells were incubated in the dark with 2% CO_2 in air and stirring. At the times indicated 1-ml samples were withdrawn from the cell suspension and mixed with an equal volume of cold 20% TCA. The samples were then processed and collected on nitrocellulose filters (II.C.2.b).

if not impossible, to state with assurance that variations in the apparent rates of precursor incorporation into protein actually reflect changes in the capacity of the cells to synthesize protein and are not attributable to changes in precursor isotope pools.

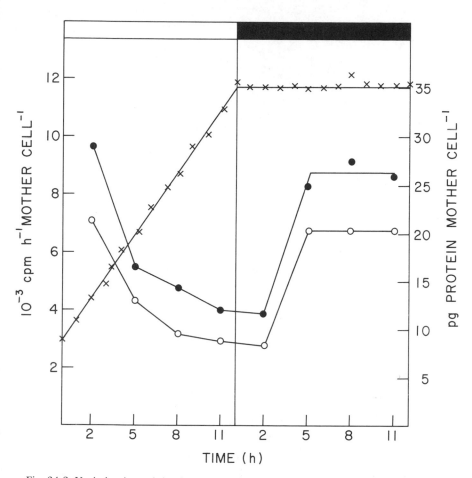

Fig. 34-2. Variation in arginine incorporation during the synchronous cell cycle at low concentrations of arginine. Cell samples were withdrawn from a master culture at the times indicated and resuspended in fresh HSM medium containing either 4 μM (white circles) or 20 (black circles) μM ^{14}C-arginine (100 μCi μmol^{-1}). The samples were incubated for 1 h in the light or dark, depending on the time in the cell cycle, and killed by the addition of an equal volume of cold 20% TCA. The samples were then processed and collected on nitrocellulose filters (II.C.2.b). Protein (crosses) was determined as described in II.B.1.

2. Swamping protein precursor pools. To minimize the effects of changing precursor pool sizes and pool turnover rates, the protein precursor pools should be swamped with precursor so that the intracellular isotopic dilution of precursor is the same for each labeling period. When protein precursor pools are completely swamped with a precursor, the specific activity of the precursor incorporated into protein approaches that of the exogenous precursor (Holleman and Key

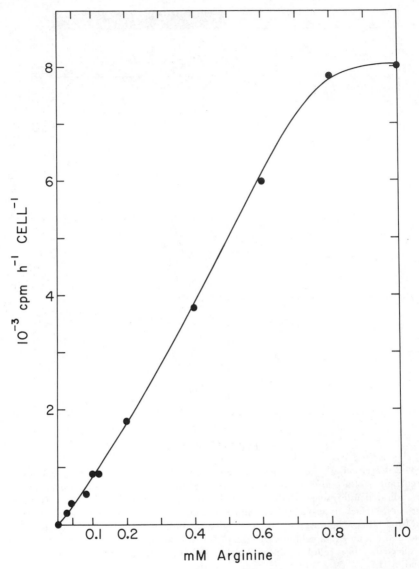

Fig. 34-3. Swamping the protein precursor pool with arginine. An aliquot of a cell suspension (ca. 2.5×10^6 cells) was harvested at hour L_{10} of the cell cycle. The cells were resuspended to 2×10^6 cells ml^{-1} in HSM medium and then 1×10^6 cells were added to an equal volume of HSM medium containing 0–2 mM ^{14}C-arginine (10 μCi μmol^{-1} in all cases). The cells were incubated for 1 h in the light with aeration, killed, and processed as described in the legend to Fig. 34-2.

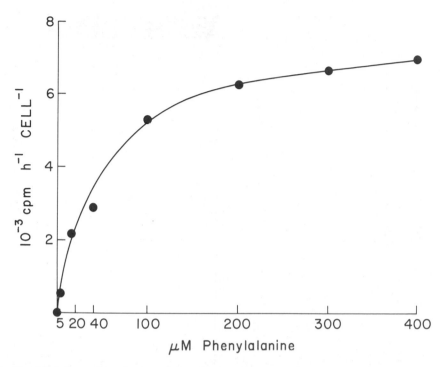

Fig. 34-4. Swamping the protein precursor pool with phenylalanine. Cells were collected and resuspended as described in the legend to Fig. 34-3. The final incubation medium contained 1×10^6 cells ml^{-1} and 0–400 μM phenylalanine (10 μCi μmol^{-1}). The cells were incubated, killed and processed as described in the legend to Fig. 34-2.

1967). To demonstrate pool swamping, cells from hour L_{10} of the photoperiod were labeled for 1 h with increasing concentrations of ^{14}C-arginine or ^{14}C-phenylalanine (Figs. 34-3 and 34-4). With either amino acid the rate of incorporation of amino acid into protein increased with increasing concentration. However, even at high concentration of exogenous arginine (0.8–1.0 mM) the protein precursor pool is not adequately swamped using strains of *Chlamydomonas* that are autotrophic with respect to arginine. This is in contrast to the phenylalanine protein precursor pool (Fig. 34-4). When the data from Fig. 34-4 are replotted in double reciprocal form (not shown), a maximum of 8.63×10^{-3} cpm ^{14}C-phenylalanine can be incorporated per hour per cell at infinite intracellular phenylalanine concentration. Thus at 300 μM and 400 μM exogenous phenylalanine, the protein precursor pool is respectively swamped with precursor to 74% and 81% of the maximum possible value at a time in the cell cycle when the endogenous total phenylalanine pool is large compared with other stages (Jones, Iwasa, and Renke, unpublished). Thus, although

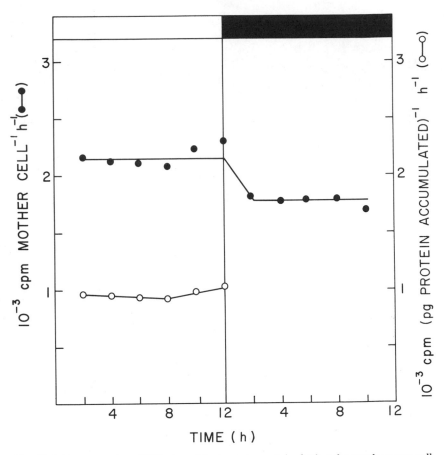

Fig. 34-5. Incorporation of ^{14}C-phenylalanine into protein during the synchronous cell cycle at high concentration of exogenous phenylalanine. Samples of 0.5×10^5 cells ml[-1] were withdrawn from a master culture 1 h before the times indicated and resuspended to the same cell density in fresh HSM medium containing 300 μM phenylalanine (4 μCi μmol[-1]). The cell samples were incubated, killed, and processed as described in the legend to Fig. 34-2.

greater specific activities of newly synthesized protein can be obtained with ^{14}C-arginine than with ^{14}C-phenylalanine (when low exogenous concentrations are used), the arginine protein precursor pool is more difficult to swamp with precursor than is the phenylalanine protein precursor pool.

3. Incorporation of ^{14}C-phenylalanine into protein during the synchronous cell cycle using swamping exogenous precursor concentrations. In Fig. 34-5 data are presented for the incorporation of ^{14}C-phenylalanine into protein for cells that were labeled for 1 h at a high concentration of exogenous phenylalanine (400 μM; 4 μCi μmol[-1]). When the rate of ^{14}C-phenylalanine incorporation is expressed as a function of either the total cells labeled or the amount of newly accumulated protein, there is little variation in rate during the photoperiod. Similarly, in the scotoperiod, the rate of protein synthesis is constant, although there is a decrease in the absolute rate of incorporation of ^{14}C-phenylalanine into protein, compared with the photoperiod, and there is no *net* protein accumulation.

These results differ from those previously published for *C. reinhardtii* labeled by a continuous infusion technique at a low exogenous phenylalanine concentration ((Jones et al. 1968) and probably represent differences in the final isotopic dilutions of the phenylalanine pools using the two different labeling protocols.

IV. Discussion

A. Changes in pool sizes during synchronous growth

The above sample data indicate that the interpretation of precursor incorporation from different labeling periods during synchronous development can be complicated by variations in pool sizes or kinetics of movement of the precursor into intracellular pools. In wild-type *C. reinhardtii*, free amino acids and low-molecular-weight, acid-soluble peptide amino acids represent two types of pools. The peptide pool might well serve as an amino acid storage pool for later utilization. During synchronous vegetative growth and gametogenesis, the total acid-soluble arginine pool fluctuates. The free amino acid pool at any given time is small compared with the low-molecular-weight peptide pool. The free arginine and total acid-soluble arginine pools expand during the photoperiod in vegetative growth and are maximal at the end of the photoperiod. However, at the end of cytokinesis (D_6), no free arginine is detectable by amino acid analysis. Similar results have been obtained with the free phenylalanine and phenylalanine peptide pools.

The effects of varying pool sizes on the intracellular isotopic dilution of arginine are suggested in Fig. 34-2. During the photoperiod, protein accumulates at a constant rate, whereas the rate of ^{14}C-arginine incorporation into protein decreases. Although the total arginine precursor pool increases during the photoperiod, this pool is capable of expanding when supplied with exogenous arginine. It is not known at this time if the total arginine precursor pool and the arginine protein precursor pool represent the same or two separate compartments.

B. Incorporation of precursors

Although labeling synchronously growing cells with high concentrations of exogenous precursor is recommended for comparing rates of protein synthesis under conditions of changing precursor pool sizes, this technique does have some disadvantages and may be unnecessary or impractical in some cases. First, in order to swamp pools with precursor, it is necessary to employ large quantities of radioactive precursor. Consequently, it is impractical to label large volumes of cells. Second, the high concentrations of precursor used to swamp pools might lead to undesirable secondary effects with some systems (Joy and Folkes 1965; Oaks and Bidwell 1970). Last, it may be unnecessary to swamp precursor pools, depending on the purpose of the experiment. For example, if cells are to be labeled for examining protein turnover (rather than protein synthesis) by following the loss of radioactivity from protein over time, then pool swamping with high exogenous precursor concentrations is undesirable.

It should be realized that the sample data presented refer to cells synchronously grown on an inorganic medium. If cells are grown on medium containing organic sources of carbon (e.g., acetate), the metabolism of the cells will be different. Interpretation of data, particularly data related to precursor pools and isotopic dilution and the concentration of precursor necessary to swamp pools, may be different under conditions of photoheterotrophic and heterotrophic nutrition.

V. Acknowledgments

This work was supported by Research Grants BMS73-06899 A01 from the National Science Foundation and Research Grant HD-07986 from the National Institute of Health and a Grant-in-Aid from the Graduate School, the State University of New York at Stony Brook.

VI. References

Farr, R. S. 1971. Determination of antigen-binding capacity. *Methods Immunol. Immunochem.* 3, 66–73.

Greenwood, F. C., Hunter, W. M., and Glover, J. S. 1963. The preparation of ^{131}I-labelled human growth hormone of high specific radioactivity. *Biochem. J.* 89, 114–23.

Hartree, E. F. 1972. Determination of protein: A modification of the Lowry method that gives a linear photometric response. *Anal. Biochem.* 48, 422–7.

Holleman, J. M., and Key, J. L. 1967. Inactive and protein precursor pools of amino acids in the soybean hypocotyl. *Plant Physiol.* 42, 29–36.

Huffaker, R. C., and Peterson, L. W. 1974. Protein turnover in plants and possible means of its regulation. *Annu. Rev. Plant Physiol.* 25, 363–92.

Iwanij, V., Chua, N.-H., and Siekevitz, P. 1975. Synthesis and turnover of ribulose biphosphate carboxylase and of its subunits during the cell cycle of *Chlamydomonas reinhardtii. J. Cell Biol.* 64, 572–85.

Jones, R. F. 1962. Extracellular mucilage of the red alga *Porphyridium cruentum. J. Cell. Comp. Physiol.* 60, 61–4.

Jones, R. F., Kates, J. R., and Keller, S. J. 1968. Protein turnover and macromolecular synthesis during growth and gametic differentiation in *Chlamydomonas reinhardtii. Biochim. Biophys. Acta* 157, 589–98.

Joy, K. W., and Folkes, B. F. 1965. The uptake of amino acids and their incorporation into the proteins of excised barley embryos. *J. Exp. Bot.* 16, 646–66.

Kates, J. R. 1966. Biochemical aspects of synchronized growth and differentiation in *Chlamydomonas reinhardtii.* Ph.D. dissertation, Princeton University, Princeton. 350 pp.

Kates, J. R., and Jones, R. F. 1964. The control of gametic differentiation in liquid cultures of *Chlamydomonas. J. Cell. Comp. Physiol.* 63, 157–64.

Lowry, O. H., Rosebrough, N. J., Farr, A. L., and Randall, R. J. 1951. Protein measurement with the Folin-phenol reagent. *J. Biol. Chem.* 193, 265–75.

Mans, R. J., and Novelli, G. D. 1961. Measurement of the incorporation of radioactive amino acids into protein by a filter-paper disk method. *Arch. Biochem. Biophys.* 94, 48–53.

Oaks, A., and Bidwell, R. G. S. 1970. Compartmentation of intermediary metabolites. *Annu. Rev. Plant Physiol.* 21, 43–66.

Schimke, R. T., and Bradley, M. O. 1975. Properties of protein turnover in animal cells and a possible role for turnover in "quality" control of proteins. In Reich, E., Rifkin, D. B., and Shaw, E. (eds.). *Proteases and Biological Control,* vol. 2, pp. 515–30. Cold Spring Harbor Conference on Cell Proliferation, Cold Spring Harbor Press, Cold Spring Harbor, N.Y.

Schimke, R. T., and Doyle, D. 1970. Control of enzyme levels in animal tissues. *Annu. Rev. Biochem.* 39, 929–76.

Sueoka, N. 1960. Mitotic replication of deoxyribonucleic acid in *Chlamydomonas reinhardi. Proc. Natl. Acad. Sci. U.S.A.* 46, 83–91.

35: Nucleic acid synthesis

ROSE ANN CATTOLICO

Botany Department AK-10,
University of Washington, Seattle, Washington 98195

CONTENTS

I	Introduction	*page* 354
II	Materials and test organisms	354
III	Methods and discussion	354
	A. Stringent maintenance of growth parameters	354
	B. Algal culture contamination	355
	C. Labeling vs. synthesis	356
IV	References	359

I. Introduction

This chapter makes no attempt to present detailed methodology for the determination of nucleic acid biosynthetic rates. Instead, an attempt has been made both to familiarize investigators with many of the technical problems that occur in this type of analysis and to cite representative studies that contain methodology useful for monitoring and/or eliminating these experimental difficulties.

II. Materials and test organisms

All chemicals were of reagent grade. Synchronous cultures of *Chlamydomonas reinhardtii* Dangeard 137 C$^+$ and *Olisthodiscus luteus* Carter were used as the test organisms in our studies.

III. Methods and discussion

A. Stringent maintenance of growth parameters

Experiments must be done with cultures maintained under strictly reproducible growth conditions. Cultures should be induced to increase in cell number at a specific rate, and experimental samples that are to be used for analysis should be taken only when a culture attains a specific cell density.

These precautions are strongly advised. Small alterations in the physiological state of an organism have been shown seriously to affect both the qualitative and quantitative program of nucleic acid production in algal cells. For example, it has been suggested (Clay et al. 1975) that the amount of cytoplasmic and organelle ribosomal RNA and the pattern of their syntheses are altered in *Chlamydomonas* cultures that have been induced to form two, four, and eight daugher cells during a single synchronous cell cycle.

Changes in cellular nucleic acid complement may also occur during different phases of a normal growth sequence. *Olisthodiscus* (Gooderham and Cattolico, unpublished), *Euglena* (Gibson and Hershberger 1975), and several species of dinoflagellates (Allen et al. 1975) shift total cellular DNA content as a culture progresses from the exponential to the stationary growth phase, and cellular RNA content (Warr and Durber 1971) increases sharply as *Chlamydomonas* cells enter the stationary phase of growth.

Modification in the growth medium used for cellular maintenance might also induce dramatic changes in cellular nucleic acid profiles. Phosphate withdrawal (frequently used to increase $^{32}PO_4^{3-}$ labeling efficiency in algal cells) results in altered timing of the cell cycle in *Chlamydomonas* and causes a depressed production of both RNA and DNA in these cells (Lien and Knutsen 1973); the DNA complement of *Euglena* is reduced by one-half (Tsubo et al. 1972) when this organism is cultured in a phosphate-depleted medium. Comparison of the nucleic acid content of algal cells maintained heterotrophically, mixotrophically, or phototrophically reveals that a significant difference in the complement of both DNA (Gibbs et al. 1974; Rawson and Boerma, 1976) and RNA (Bourque et al. 1971; Goodenough and Levine 1971; Boynton et al. 1972) occurs in response to these varied growth conditions

An algal cell may be "programmed" to produce a specific nucleic acid species during a limited portion of a cell cycle (Hopkins et al. 1972; Cattolico et al. 1973, 1976; Lee and Jones 1973). To magnify this event so that biochemical analysis may be undertaken, algal cells often are synchronized (Bernstein 1960; Cook and James 1960; Sweeney and Hastings 1962; Lewin et al. 1966) by light pulses, temperature shifts, or nutrient stress. It must be cautioned, however, that cells treated in this manner may be locked into their own specialized program of nucleic acid production, which differs from that of an unsynchronized population. For example, it has been demonstrated (Howell 1972) that the number of rDNA cistrons in cultures of *Chlamydomonas* grown in continuous light is greater than the number of rDNA cistrons found in cells maintained synchronously on a 12-h light, 12-h dark regimen.

B. *Algal culture contamination*

Extreme care must be taken to maintain algal cultures in an axenic condition. Fungal and bacterial DNA have buoyant densities (Schildkraut et al. 1962; Bauer and Vinograd 1971) which may be similar to that observed (Brawerman and Eisenstadt 1964; Kirk 1971; Cattolico et al. 1976; Siu et al. 1976) for algal DNA species. The possibility that extraneous DNA might be present in higher plant and in algal DNA preparations has caused difficulty and controversy in the identification of DNA satellite species (Kirk 1971; Bendich 1972; Green 1974).

The presence of contaminating fungal or bacterial cells in an algal culture may also seriously affect the quantitation of RNA biosynthesis. For example, bacterial ribosomal RNA has an S value similar to that of chloroplast ribosomal RNA, whereas fungal ribosomal RNA comigrates with the cytoplasmic ribosomal RNA species (Cattolico et al. 1973). Tests for contamination in our algal system are made as follows.

Bacterial. A 1.0-ml sample of the algal culture is axenically transferred to a test tube containing 15 ml of bacterial growth medium composed of 7.0 g K_2HPO_4, 3.0 g KH_2PO_4, 2.0 g $(NH_4)_2SO_4$, 0.1 g $MgSO_4 \cdot 7$ H_2O, 5.0 g glucose, and 50 g yeast extract per liter of distilled water. The inoculated medium is kept at 37 C for 24 h. Cells for nucleic acid analysis are used only if bacterial growth was negative 24 h after inoculation of the test material.

Fungal. A 1.0-ml algal culture is axenically transferred to a test tube containing 15 ml of fungal growth medium composed of 8.0 g nutrient broth, 5.0 g yeast extract, and 5.0 g glucose made in 1.0 l of distilled water. The inoculated sample is maintained at 25 C. Cells are used for nucleic acid analysis only if fungal growth did not occur in 48 h after sample inoculation.

C. Labeling vs. synthesis

It is important to determine whether the incorporation of a chosen radioactive precursor truly reflects the rate of synthesis of the specific nucleic acid under analysis. A number of factors listed below may seriously distort or even mask incorporation data.

1. Precursor transport and pool levels. Labeled nucleotides may be transported into a cell at rates that are often independent of one another (Plagemann 1971). This program of specific precursor movement may vary during different phases of the cell cycle (Stambrook and Sisken 1972) or may be induced to change by an alteration in cellular growth conditions (Cunningham and Pardee 1969). Once inside the cell, the precursor molecule may be present in a small, large, or expandable pool (Oaks and Bidwell 1970; Mitchison 1971). Pool size (type) is not a constant. Fluctuations in intracellular nucleotide pool levels are quite dependent on cell cycle parameters (Walters et al. 1973; Bersier and Braun 1974; Fink 1975) and also respond to exogenous nucleotide supply. In addition, pool size may be affected by the presence of nucleotide precursors that originate from within the cell as the degradation products of preexisting macromolecular components. The use of this nucleotide source is well demonstrated in the *Chlamydomonas* system, where precursors for DNA synthesis during gametic differentiation are obtained by the degradation of more than

90% of the vegetative RNA complement of the cell (Siersma and Chiang 1971).

In summary, the absolute rate of nucleic acid synthesis cannot be determined by the direct measurement of a radioactive precursor into a macromolecular product. It is imperative that pool size and, therefore, specific activity of precursor nucleotides be critically analyzed (Bersier and Braun 1974; Sutton and Kemp 1976).

2. Compartmentation. Differential incorporation of precursor molecules into the nucleic acids of cellular organelles will often occur in algal cells. *Chlamydomonas* (Swinton and Hanawalt 1972), *Dictyota, Padina, Bryopsis* (Steffensen and Sheridan 1965), and *Olisthodiscus* (Cattolico and Ersland, unpublished) incorporate thymidine into chloroplast DNA, but only small amounts of label appear in the nuclear DNA species. It has been suggested (Steffensen and Sheridan 1965; Swinton and Hanawalt 1972) that the enzyme thymidine kinase is present only within the chloroplast. Further examples of compartmentation in algal systems are numerous. Preferential labeling of cytoplasmic ribosomal RNA will occur if phosphate is used as the precursor molecule, whereas uracil will be predominantly incorporated into the chloroplast ribosomal RNAs (Galling and Ssymank 1970; Ssymank 1972; Cattolico 1973).

Large algal cells may present new difficulties to investigators studying RNA production. Recent studies by Naumova and co-workers (1976) indicate that an apicobasal gradient of RNA species exists in *Acetabularia,* and they suggest that this gradient is attributable to a differential rate of RNA synthesis in different portions of the organism.

3. Precursor nonspecificity. Although a cell may be labeled with a "highly specific" precursor, a false pattern of nucleic acid synthesis may be obtained if the investigator measures total counts into a product precipitable with trichloroacetic acid (TCA). Metabolic pathways exist within the cell that may funnel nucleotide precursors into nonnucleic acid products. In *Chlorella,* for example, uracil is degraded to CO_2 and β-alanine, and this conversion is followed by the immediate photoassimilation of the released CO_2 by illuminated cells (Knutsen 1972). The efficiency of the metabolic conversion of uracil varies during different phases of synchronous cell growth. Placement of a uracil degradation product into nonnucleic acid material no doubt occurs frequently in other cell systems. The specific activity of RNA isolated from *Chlamydomonas* cells that have been pulse labeled with this precursor is extremely low, even though whole cells precipitated with TCA contain a large amount of radioactivity (Cattolico, unpublished).

Although it is generally assumed that DNA alone will be labeled with a thymidine precursor, caution is advised. A degradative catabolic pathway exists in plant systems that can convert deoxythymidine to β-aminoisobutyric acid and CO_2 (Howland and Yette 1975). After pulse labeling *Euglena* with thymidine, Sagan (1965) was able to demonstrate that a significant proportion of the radioactivity present in the TCA-precipitable product was RNase sensitive. Similar data have recently (Cattolico, unpublished) been obtained for the *Olisthodiscus* system. If the thymidine-labeled nucleic acids of this organism are isolated by phenol extraction, a significant portion of the counts present in this purified macromolecular fraction is removed by RNase digestion. In animal systems, incorporation of radioactivity originating from a ^{3}H-thymidine source appears in lipid, protein, glycogen, and RNA fractions (Dobson and Cooper 1971; Schneider and Greco 1971; Goldspink and Goldberg 1973). Schneider and Greco (1971) warn that rigorous purification of DNA is necessary for estimation of DNA synthesis using thymidine as a precursor molecule.

4. Labeling artifacts. Often extraneous cellular components will contaminate nucleic acid preparations. These materials are often coextracted (Segovia et al. 1965) during nucleic acid isolation and are frequently difficult to eliminate. For example, *Chlamydomonas* RNA preparations labeled with $NaH^{14}CO_3$ contain a contaminant that appears as five reproducible radioactive peaks on polyacrylamide gel analysis (Cattolico et al. 1973). These five peaks have S values of 28–35, and each of the five peaks varies in amount in a specific pattern during different phases of synchronous cell growth. These radioactive peaks do not absorb at 260 nm and are not digested by ribonuclease, lipase, α-amylase, pronase, or deoxyribonuclease and probably represent some form of polysaccharide. The contamination of DNA preparations with polysaccharide materials has been observed in a large variety of algal cells (see Edelman et al. 1967 for discussion). It is interesting to note that label originating as a thymidine pulse appears as an α-amylase-sensitive product (Counts and Flamm 1966) that is coextracted with mouse liver DNA (see III.C.3 for discussion of nonspecific precursors).

The measurement of nucleic acid biosynthesis in algal cells is not a simple task. In summary, the following precautions are suggested:

a. Cells must be maintained axenically under specific and reproducible growth conditions.

b. Caution must be exercised in monitoring the specific activity of precursor nucleotides. Nucleotide transport mechanisms and nucleotide pools exist in algal cells, and both transport capacity and pool size may differ when whole cells and subcellular structures such as the mitochondria or the chloroplasts are compared.

c. Care must be taken to ensure that "highly specific" radioactive precursor molecules do not appear as a result of salvage pathways in other subcellular components. The investigator must also be aware of the possibility that these nonnucleic acid salvage products may be coextracted with DNA or RNA and could, therefore, interfere with analytical procedures.

Finally, it should be mentioned that this list of precautions is not meant to frighten potential investigators away from the field of nucleic acid analysis, but is meant to stress the importance of careful, complete control of all experimental parameters when undertaking this type of macromolecular quantitation.

IV. References

Allen, J. R., Roberts, T. M., Loeblich, A. R., III, and Klotz, L. C. 1975. Characterization of the DNA from the dinoflagellate *Crypthecodinium cohnii* and implications for nuclear organization. *Cell* 6, 161–9.

Bauer, W. R., and Vinograd, J. 1971. Sedimentation velocity experiments in the analytical ultracentrifuge. *Proc. Nucleic Acid. Res.* 2, 297–354.

Bendich, A. J. 1972. Effect of contaminating bacteria on the radiolabeling of nucleic acids from seedlings: False DNA "satellites." *Biochim. Biophys. Acta* 272, 494–503.

Bernstein, E. 1960. Synchronous division in *Chlamydomonas moewusii. Science* 131, 1528–9.

Bersier, D., and Braun, R. 1974. Pools of deoxyribonucleoside triphosphates in the mitotic cycle of *Physarum. Biochim. Biophys. Acta* 340, 463–71.

Bourque, D. P., Boynton, J. E., and Gillham, N. W. 1971. Studies on the structure and cellular location of various ribosome and ribosomal RNA species in the green alga *Chlamydomonas reinhardi. J. Cell Sci.* 8, 153–83.

Boynton, J. E., Gillham, N. W., and Chabot, J. F. 1972. Chloroplast ribosome deficient mutants in the green alga *Chlamydomonas reinhardi* and the question of chloroplast ribosome function. *J. Cell Sci.* 10, 267–305.

Brawerman, G., and Eisenstadt, J. M. 1964. Deoxyribonucleic acid from the chloroplasts of *Euglena gracilis. Biochim. Biophys. Acta* 91, 477–85.

Cattolico, R. A. 1973. Changes in cytoplasmic and chloroplast ribosomal RNA during synchronous growth in *Chlamydomonas reinhardtii.* Ph.D. thesis, State University of New York, Stony Brook, N.Y. 250 pp.

Cattolico, R. A., Gooderham, K., and Ersland, D. 1976. Regulation of organelle biogenesis in *Olisthodiscus luteus. Plant Physiol.* 57 (Suppl.), 100.

Cattolico, R. A., Senner, J. W., and Jones, R. F. 1973. Changes in cytoplasmic and chloroplast ribosomal ribonucleic acid during the cell cycle of *Chlamydomonas reinhardtii. Arch. Biochem. Biophys.* 156, 58–65.

Clay, W. F., Matsuda, K., Hoshaw, R. W., and Rhodes, P. R. 1975. Nucleic acid metabolism in *Chlamydomonas moewusii* undergoing daily cell doubling. *Ann. Bot. (London)* 39, 525–33.

Cook, J. R., and James, T. W. 1960. Light-induced division synchrony in *Euglena gracilis* var. *bacillaris. Exp. Cell Res.* 21, 583–9.

Counts, W. B., and Flamm, W. G. 1966. An artifact associated with the incorporation of thymine into DNA preparations. *Biochim. Biophys. Acta* 114, 628–30.

Cunningham, D. D., and Pardee, A. B. 1969. Transport changes rapidly initiated by serum addition to "contact inhibited" 3T3 cells. *Proc. Natl. Acad. Sci. U.S.A.* 64, 1049–56.

Dobson, R. L., and Cooper, M. F. 1971. Incorporation of radioactivity from thymidine into mammalian glucose and glycogen. *Biochim. Biophys. Acta* 254, 393–401.

Edelman, M., Swinton, D., Schiff, J. A., Epstein, H. T., and Zeldin, B. 1967. Deoxyribonucleic acid of the blue-green algae (Cyanophyta). *Bacteriol. Rev.* 31, 315–31.

Fink, K. 1975. Fluctuations in deoxyribo- and ribonucleoside triphosphate pools during the mitotic cycle of *Physarum polycephalum. Biochim. Biophys. Acta* 414, 85–9.

Galling, G., and Ssymank, V. 1970. Bevorzugter Einbau markierten Uridins in die Vorläüfer von Chloroplasten-Ribosomen in Algenzellen. *Planta* 94, 203–12.

Gibbs, S. P., Mak, R., Ng, R., and Slankis, T. 1974. The chloroplast nucleoid in *Ochromonas danica.* II. Evidence for an increase in plastid DNA during greening. *J. Cell Sci.* 16, 579–91.

Gibson, W. H., and Hershberger, C. L. 1975. Changing proportions of DNA components in *Euglena gracilis. Arch. Biochem. Biophys.* 168, 8–14.

Goldspink, D. F., and Goldberg, A. L. 1973. Problems in the use of (Me-^{3}H) thymidine for the measurement of DNA synthesis. *Biochim. Biophys. Acta* 299, 521–32.

Goodenough, U. W., and Levine, R. P. 1971. The effects of inhibitors of RNA and protein synthesis on the recovery of chloroplast ribosomes, membrane organization, and photosynthetic electron transport in the ac-20 strain of *Chlamydomonas reinhardi. J. Cell Biol.* 50, 50–62.

Green, B. R. 1974. Nucleic acids and their metabolism. In Stewart, W. D. P. (ed.). *Algal Physiology and Biochemistry,* pp. 281–313. Blackwell, Oxford.

Hopkins, H. A., Flora, J. B., and Schmidt, R. R. 1972. Periodic DNA accumulation during the cell cycle of a thermophilic strain of *Chlorella pyrenoidosa. Arch. Biochem. Biophys.* 153, 845–9.

Howell, S. H. 1972. The differential synthesis and degradation of ribosomal DNA during the vegetative cell cycle in *Chlamydomonas reinhardi. Nature (London) New Biol.* 240, 264–7.

Howland, G. P., and Yette, M. L. 1975. Simultaneous inhibition of thymidine degradation and stimulation of incorporation into DNA by 5-fluorodeoxyuridine. *Plant Sci. Lett.* 5, 157–62.

Kirk, J. T. O. 1971. Will the real chloroplast DNA please stand up? In Boardman, N. K., Linnane, A. W., and Smillie, R. M. (eds.). *Autonomy and Biogenesis of Mitochondria and Chloroplasts,* pp. 267–76. North Holland, Amsterdam.

Knutsen, G. 1972. Degradation of uracil by synchronous cultures of *Chlorella fusca. Biochim. Biophys. Acta* 269, 333–43.

Lee, R. W., and Jones, R. F. 1973. Induction of Mendelian and non-Mendelian streptomycin resistant mutants during the synchronous cell cycle of *Chlamydomonas reinhardtii. Mol. Gen. Genet.* 121, 99–108.

Lewin, J. C., Reimann, B. E., Busby, W. F., and Volcani, B. E. 1966. Silica shell formation in synchronously dividing diatoms. In Cameron, I. L., and Padilla, G. M. (eds.). *Cell Synchrony,* pp. 169–88. Academic Press, New York.

Lien, T., and Knutsen, G. 1973. Phosphate as a control factor in cell division of *Chlamydomonas reinhardti,* studied in synchronous culture. *Exp. Cell Res.* 78, 79–88.

Mitchison, J. M. 1971. *The Biology of the Cell Cycle,* pp. 11–16. Cambridge University Press, London.

Naumova, L. P., Pressman, E. K., and Sandakchiev, L. S. 1976. Gradient of RNA distribution in the cytoplasm of *Acetabularia mediterranea. Plant Sci. Lett.* 6, 231–5.

Oaks, A., and Bidwell, R. G. S. 1970. Compartmentation of intermediary metabolites. *Annu. Rev. Plant Physiol.* 21, 43–66.

Plagemann, P. G. W. 1971. Nucleoside transport by Novikoff rat hepatoma cells growing in suspension culture: Specificity and mechanism of transport reactions and relationship to nucleoside incorporation into nucleic acids. *Biochim. Biophys. Acta* 233, 688–701.

Rawson, J. R. Y., and Boerma, C. 1976. Influence of growth conditions upon the number of chloroplast DNA molecules in *Euglena gracilis. Proc. Natl. Acad. Sci. U.S.A.* 73, 2401–4.

Sagan, L. 1965. An unusual pattern of tritiated thymidine incorporation in *Euglena. J. Protozool.* 12, 105–9.

Schildkraut, C. L., Marmur, J., and Doty, P. 1962. Determination of the base composition of deoxyribonucleic acid from its buoyant density in CsCl. *J. Mol. Biol.* 4, 430–43.

Schneider, W. C., and Greco, A. E. 1971. Incorporation of pyrimidine deoxyribonucleosides into liver lipids and other components. *Biochim. Biophys. Acta* 228, 610–26.

Segovia, Z. M. M., Sokol, F., Graves, I. L., and Ackermann, W. W. 1965. Some properties of nucleic acids extracted with phenol. *Biochim. Biophys. Acta* 95, 329–40.

Siersma, P. W., and Chiang, K.-S. 1971. Conservation and degradation of cytoplasmic and chloroplast ribosomes in *Chlamydomonas reinhardtii. J. Mol. Biol.* 58, 167–85.

Siu, C.-H, Swift, H., and Chiang, K.-S. 1976. Characterization of cytoplasmic and nuclear genomes in the colorless alga *Polytoma.* II. General characterization of organelle nucleic acids. *J. Cell Biol.* 69, 371–82.

Ssymank, V. 1972. Influence of nitrogen deficiency on uridine incorporation into ribosomes in the green alga *Chlorella. Arch. Mikrobiol.* 82, 311–24.

Stambrook, P. J., and Sisken, J. E. 1972. The relationship between rates of [^{3}H]uridine and [^{3}H]adenine incorporation into RNA and the measured rates of RNA synthesis during the cell cycle. *Biochim. Biophys. Acta* 281, 45–54.

Steffensen, D. M., and Sheridan, W. F. 1965. Incorporation of H^3-thymidine into chloroplast DNA of marine algae. *J. Cell Biol.* 25, 619–26.

Sutton, D. W., and Kemp, J. D. 1976. Calculation of absolute rates of RNA synthesis, accumulation, and degradation in tobacco callus *in vivo*. *Biochemistry* 15, 3153–7.

Sweeney, B. M., and Hastings, J. W. 1962. Rhythms. In Lewin, R. A. (ed.). *Physiology and Biochemistry of Algae*, pp. 687–700. Academic Press, New York.

Swinton, D. C., and Hanawalt, P. C. 1972. *In vivo* specific labeling of *Chlamydomonas* chloroplast DNA. *J. Cell Biol.* 54, 592–7.

Tsubo, Y., Ueda, T., and Yokomura, E. 1972. Chloroplast autonomy revealed by nutritionally controlled *Euglena gracilis*. *Proc. Int. Seaweed Symp.* 7, 336–8.

Walters, R. A., Tobey, R. A., and Ratliff, R. L. 1973. Cell-cycle-dependent variations of deoxyribonucleoside triphosphate pools in Chinese hamster cells. *Biochim. Biophys. Acta* 319, 336–47.

Warr, J. R., and Durber, S. 1971. Studies on the expression of a mutant with abnormal cell division in *Chlamydomonas reinhardi*. *Exp. Cell Res.* 64, 463–9.

36: Nitrogen fixation

TIMOTHY H. MAGUE

*Bigelow Laboratory for Ocean Sciences,
West Boothbay Harbor, Maine 04575*

CONTENTS

I	Introduction	*page* **364**
II	Equipment and supplies	**365**
III	Assay methods suitable for whole cells	**365**
	A. Growth on N_2	365
	B. Fixation of $^{15}N_2$	365
	C. C_2H_2 reduction	369
IV	Assay methods suitable for cell-free preparations	**372**
	A. Isolation of heterocysts and preparation of nitrogenase	372
	B. Reaction mixture	374
	C. Production of NH_3	374
	D. H_2 evolution	375
	E. Utilization of ATP	375
	F. Oxidation of dithionite	375
	G. Alternative substrates	376
V	Acknowledgment	**376**
VI	References	**376**

I. Introduction

Nitrogen fixation by blue-green algae has been extensively studied in the field and in pure culture in the laboratory, but relatively little attention has been paid to the enzymology of the process. This chapter describes methods suitable for both ecological and laboratory work and outlines procedures for isolation of the N_2-fixing enzyme complex for purification and study in vitro. Recent work (Gallon et al. 1972; Haystead and Stewart 1972) has demonstrated that algal nitrogenase is similar to that of bacteria and exhibits the following characteristic properties of the enzyme:

1. Multiple substrate affinity. Nitrogenase is capable of binding and reducing the compounds listed in Table 36-1. Several binding sites appear to be involved, and alternate substrate interactions are complex.

2. O_2 sensivity. Purified nitrogenase is irreversibly inactivated when exposed to air.

3. Cold lability. The enzyme from most species is considerably less stable at -20 to 0 C than at 20 C.

4. Requirement for energy and reductant. ATP plus Mg^{2+} and an electron donor, such as $Na_2S_2O_4$, are required for activity.

5. H_2 evolution. In the presence of an electron donor and the absence of a reducible substrate, nitrogenase exhibits an ATP-dependent H_2 evolution.

6. Sensitivity to CO. The reducing capability of nitrogenase will be virtually eliminated by 0.05 atm. of CO.

Table 36-1. *Reactions catalyzed by purified nitrogenase*

(1) $N_2 \rightarrow NH_3$	(2) $C_2H_2 \rightarrow C_2H_4$
(3) $N_2O \rightarrow N_2 + H_2O$	(4) $N_3^- \rightarrow N_2 + NH_3$
(5) $CN^- \rightarrow CH_4 + NH_3 + CH_3NH_2 +$ traces of C_2H_4 and C_2H_6	
(6) $CH_3NC \rightarrow CH_4 + C_2H_6 + C_2H_4 + C_3H_6 + C_3H_8 + CH_3NH_2$	
(7) $CH_3CCH \rightarrow CH_3CHCH_2$	(8) $CH_3CH_2CCH \rightarrow CH_3CH_2CHCH_2$

Source: Burns and Hardy (1972).

[364]

II. Equipment and supplies

(1) Compressed gases and gas mixtures (Matheson). (2) Biochemicals (Sigma). (3) Gas chromatographs (Varian) and Porapak column packing (Waters Associates). (4) $^{15}N_2$ gas (Bio-Rad). (5) Sephadex (Pharmacia). (6) Serum bottles, stoppers, syringes, and needles (A. H. Thomas). (7) Ion-exchange cellulose (Whatman). (8) Rittenberg tubes (Eck and Krebs).

III. Assay methods suitable for whole cells

A. Growth on N_2

Growth and accumulation of nitrogen in initially nitrogen-free medium are good evidence for the ability of blue-green algae to fix N_2. Suitable media are those of Allen and Arnon (1955) and Kratz and Myers (1955, N-free modifications). Cultures should be inoculated with a minimum of material to prevent carryover of combined N and exposed to a suitable light source ("daylight" fluorescent or tungsten lamps at 2–10 klux). The culture vessels are continuously bubbled with air passed first through 1 N solutions of NaOH and then H_2SO_4 to remove traces of nitrogen oxides and ammonia. After a suitable growth period, total nitrogen content of the cultures is measured by Kjeldahl digestion and ammonium analysis, then compared with controls consisting of uninoculated medium sampled before and after the growth period as well as medium inoculated with killed algae. Increases in nitrogen content of the growing cultures can be attributed to N_2 fixation by the algae.

Although this method is helpful in screening new isolates or communities for the ability to fix N_2, there are several caveats. Errors inherent in the Kjeldahl nitrogen analysis make the technique suitable only for systems that fix substantial quantities of N_2 (i.e., generally more than 5% of the total N initially present). Additionally, some newly transferred organisms may not grow in N-free medium, but require a limited supply of combined N to initiate growth.

Direct manometric measurement of the uptake of N_2 in a closed incubation vessel may be possible when examining vigorous N_2-fixation rates, but the method is laborious and insensitive and has largely been replaced by newer methods.

B. Fixation of $^{15}N_2$

Unequivocal evidence for N_2 fixation by an algal species is provided by demonstrating the incorporation of the stable isotope, ^{15}N, from ^{15}N-enriched N_2 gas (hereafter referred to as $^{15}N_2$) into cellular material. This isotope tracer technique is considerably more sensitive than either chemical or manometric measurement of N_2 utilization and affords the additional advantage of allowing fixed N to be traced to other organisms.

1. Preparation and purification of $^{15}N_2$. When the only available source of ^{15}N is an ammonium salt ($^{15}NH_4NO_3$ or $^{15}NH_4Cl$), $^{15}N_2$ must be prepared by oxidation with CuO at 650 C of the $^{15}NH_3$ generated from the salt by addition of 13 N NaOH. A suitable vacuum apparatus for this procedure and detailed instructions are given in Burris and Wilson (1957).

Commercially prepared $^{15}N_2$ is readily available in enrichments of up to 95 at. % ^{15}N, but should be cleaned as follows before use to remove possible contaminants such as ammonia or nitrogen oxides. Attach the stock $^{15}N_2$ container to a vacuum manifold equipped with a Toepler pump (SGA Scientific Inc.) and transfer the gas to a washing chamber containing 25 ml of an aqueous solution containing 50 g $KMnO_4$ and 25 g KOH l^{-1}. Any NO will be oxidized to higher oxides of nitrogen and absorbed by the KOH. Exchange the stock container for a storage reservoir such as that shown in Fig. 36-1, and transfer the $^{15}N_2$ back from the washing chamber. The acidic displacing fluid used in the reservoir will absorb any residual NH_3.

2. Assay. Samples from laboratory cultures or field specimens containing at least 1 mg N are conveniently exposed to $^{15}N_2$ by placing them in 7-ml glass serum bottles fitted with rubber septum-type vaccine or serum stoppers. $^{14}N_2$ must be removed from the gas phase by flushing with a mixture of 80% Ar, 20% O_2, and 0.03% CO_2 (for aerobic samples). Introduce the flushing mixture through a hypodermic needle (22- to 25-gauge) inserted into the serum stopper, and insert a second needle to serve as a vent. Flushing is preferred to evacuation and replacement in order to avoid damage to fragile cells or those containing gas vacuoles. Remove the needles and withdraw approximately 30% of the gas-phase volume with a disposable hypodermic syringe (these disposable syringes, having a glass or polyethylene barrel and a rubber-tipped plunger, are sufficiently gas-tight and accurate and are a great economy). Using a syringe, add sufficient $^{15}N_2$ from the reservoir to give a final pN_2 of approximately 0.3 atm. Incubate the samples under appropriate conditions for 30–120 min depending on the rates of N_2 fixation expected. Long incubation times should be avoided unless time-course experiments demonstrate linear uptake rates throughout the incubation period. Inactivate samples by injecting 0.5 ml of 5 N H_2SO_4 (a convenient reagent because of the subsequent Kjeldahl digestion), and seal the stopper against leakage with a

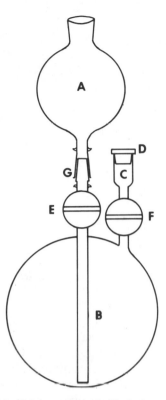

Fig. 36-1. Reservoir for $^{15}N_2$. $^{15}N_2$ is stored in *B*. Flask *A* contains displacing fluid (20% Na_2SO_4 in 5% v/v H_2SO_4), which can be introduced into *B* through *E*. With *F* closed, *C* is evacuated through a hypodermic needle piercing vaccine stopper *D*. After *C* is evacuated, *F* and *E* are opened to fill *C* with $^{15}N_2$. $^{15}N_2$ is withdrawn through *D* with a hypodermic syringe and needle. Detach *A* at 10/30 tapered joint *G* to attach reservoir to a vacuum manifold for filling with $^{15}N_2$. (After Burris 1972 with permission.)

self-curing silicone rubber compound (General Electric, type 102 RTV, or Dow-Corning marine silicone sealant).

3. Analysis of ^{15}N. The gas phase of each sample should be analyzed for ^{15}N content as soon as possible after incubation, as incomplete flushing and leakage may have diluted the $^{15}N_2$ with $^{14}N_2$. Attach a 25-gauge hypodermic needle to a short piece of heavy-wall tubing (preferably butyl rubber, to minimize gaseous diffusion) and connect the other end to a liquid N_2 cold trap on the mass spectrometer inlet manifold. Push the needle part way into the silicone sealant on the sample bottle stopper to plug the needle opening and evacuate the hose and cold trap. Then, push the needle through the sample bottle septum and allow the gas mixture to fill the vacuum manifold. Proceed to ana-

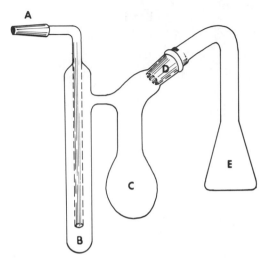

Fig. 36-2. Apparatus for converting NH_4^+ to N_2. Standard taper joint A attaches to mass spectrometer manifold. Trap B is 20×150 mm, bulb C holds 15 ml, joint D is 19/38 standard taper, and chamber E is constructed from a 50-ml Erlenmeyer flask.

lyze the sample on the mass spectrometer, allowing for the partial pressures of the accompanying O_2, Ar, and CO_2. All three ion peaks corresponding to masses 28, 29, and 30 must be measured separately and corrected for background, because any leakage during incubation will have resulted in a nonequilibrium mixture of ^{14}N and ^{15}N. Calculations are as follows:

$$\text{at. } \% \text{ } ^{15}N \text{ in gas phase} = \frac{V_{30} + \frac{1}{2}V_{29}}{V_{28} + V_{29} + V_{30}} \times 100 \qquad (1)$$

where V_{28}, V_{29}, V_{30} are the corrected voltages of the peaks corresponding to ion masses 28, 29, and 30, respectively.

Digest the sample by standard Kjeldahl techniques. Remove NH_4^+ by basification and steam distillation (see Burris and Wilson 1957 for discussion and precautions) and collect it in 10 ml of 0.1 N H_2SO_4. If total nitrogen content of the sample is desired, remove an aliquot of the distillate for NH_4^+ analysis by Nesslerization or other means; otherwise, the mass spectrometric analysis alone will yield fixation rate per unit nitrogen of the sample. Alternatively, the nitrogen content may be measured manometrically in the mass spectrometer manifold (if so equipped) after conversion to N_2.

Ensure that the distillate is acidic and evaporate it to 1–2 ml. Convert the sample to N_2 in the apparatus shown in Fig. 36-2 (Rittenberg tubes are commercially available). Attach joint A to the mass spectrometer manifold and immerse trap B in a dry-ice bath (a liquid N_2

bath may cause too rapid evaporation of sample and subsequent freezing). Add sample to chamber C through opening at D. Place approximately 2 ml of alkaline hypobromite solution in E and attach at D. (Prepare NaOBr solution by adding slowly, with swirling, 8 ml of Br_2 to 40 ml of 13 N NaOH in 60 ml of 0.18% (w/v) KI cooled in ice; store at 4 C.) Evacuate the system rapidly, allowing liquid in C and E to boil and de-gas; otherwise, residual dissolved N_2 in the solution can be released during conversion and ruin the analysis. As an extra precaution when converting small amounts of NH_4^+, samples and the hypobromite stock can be sparged with argon prior to introduction into the apparatus. After evacuation, rotate E around D and dump NaOBr into C to generate N_2. Because this conversion results in an equilibrium mixture of the N isotopes, only the mass 28 and 29 peaks need be measured. Calculations are as follows:

$$\text{at. \% } {}^{15}N = \frac{R}{2 + R} \times 100 \qquad (2)$$

where $R = V_{29}/V_{28}$ (see equation 1). The natural abundance of ${}^{15}N$ is approximately 0.368 at. %, but this should be determined in unexposed controls of the sample material and with the instrument to be used. With sufficient replication, an enrichment of 0.015% of ${}^{15}N$ is a conservative demonstration of N_2 fixation. To determine the rate of N_2 fixation, calculate the at. % ${}^{15}N$ in excess of natural abundance. Then:

$$\frac{\text{at. \% } {}^{15}N \text{ excess in sample}}{\text{at. \% } {}^{15}N \text{ excess in gas phase}} = \frac{N \text{ fixed}}{\text{unit N of sample}} \qquad (3)$$

C. C_2H_2 reduction

By far the most useful and widely applied method for estimation of N_2 fixation is the reduction of C_2H_2 to C_2H_4 (Hardy et al. 1973). The gases are separated by gas chromatography and measured by hydrogen flame ionization. This method is at least 10^3 times more sensitive than ${}^{15}N_2$ uptake and is considerably faster and cheaper than mass spectrometric analysis.

1. Analytical system for C_2H_4. A column of aluminum or stainless-steel tubing, 2 mm i.d. × 2 m, packed with Porapak N, 80- to 100-mesh, gives good separation of C_2 hydrocarbons when operated isothermally at 80–100 C with N_2 at 25 ml min^{-1} as the carrier gas. In a properly constructed chromatographic system, C_2H_4 should elute in 45–60 sec, followed by C_2H_2 in 60–85 sec. There will generally be several orders of magnitude more C_2H_2 than C_2H_4 in samples, so the C_2H_4 peak must be separated well enough that baseline can be discerned on each side of the peak. Calibrate the flame detector by injecting known amounts of high-purity C_2H_4 diluted in the gas mixture to be used for sample exposure (air is a satisfactory diluent). Serial dilutions of C_2H_4 can be prepared in large (50-ml) disposable syringes capped with serum stoppers; a useful concentration range is $10^{-6}-10^{-12}$ mol C_2H_4 per sample injected (0.1–0.2 ml volume). At the same time, prepare a large volume of one of the intermediate dilutions to serve as a daily reference standard, against which the original calibration can be compared. (Commercial gas vendors will supply cylinders of C_2H_4-in-air mixtures that will last for years.)

LaRue and Kurz (1973) have described a less sensitive colorimetric assay for C_2H_4.

2. Preparation of C_2H_2. C_2H_2 can be generated from CaC_2 and water and stored in a reservoir, as shown in Fig. 36-1. For field work, C_2H_2 can be stored in a rubber basketball or football bladder and withdrawn through an adapter constructed from the blunt inflating needle. Commercial grade C_2H_2 is convenient to use in the laboratory, but it is stored in cylinders containing acetone. These solvent vapors interfere with the C_2H_4 analysis and may be inhibitory to certain enzyme systems, so it is wise first to pass the C_2H_2 through a washing chamber of distilled water, concentrated H_2SO_4, or a dry-ice trap. Some batches of gas contain unacceptably high levels of C_2H_4, and several cylinders may have to be analyzed in order to obtain one low in background C_2H_4.

3. Assay. The serum bottles, stoppers, and plastic syringes described in III.B.2 are quite suitable for C_2H_2 reduction assays. Introduce the sample, stopper the bottle, flush with the desired gas phase, and initiate the reaction by injecting C_2H_2. If the sample medium does not contain a sufficient carbonate-buffer system, 0.001 atm of CO_2 should be added during gassing. Include controls with algae but without C_2H_2 and with C_2H_2 but not algae. For routine testing of N_2 fixation by algae under aerobic conditions, it is not necessary to replace the air in the bottle if 0.1–0.2 atm of C_2H_2 is employed, and this is a great convenience in the field (Stewart et al. 1971). Because of the sensitivity of nitrogenase to O_2, it may be desirable to test for activity under anaerobic conditions. However, it should be borne in mind that strict anaerobiosis may be unfavorable for maximum expression of C_2H_2 reduction activity, and a low pO_2 (0.05 atm) may be necessary (Millbank 1970).

If possible, samples should be incubated with shaking in a water bath; however, with samples in suspension or on the surface of media, there is no measurable lag in C_2H_4 production, indicating rapid equilibration of the gaseous phase. Large clumps of algal material or soil samples may show considerable lag periods, and the linearity of C_2H_4 production over time should be checked with the system to be used.

Drozd and Postgate (1970) have shown reversible inhibitory effects of O_2 on C_2H_2 reduction by bacteria during shaking, as well as adaptation to various ambient levels of O_2 during incubation. The possibility that these effects may interfere with the assay of blue-green algae should be checked.

A 30-min incubation is normally sufficient to produce substantial amounts of C_2H_4 when active cultures of algae are used; for example, *Anabaena* and *Nostoc* species could produce 0.5–5 nmol C_2H_4 min^{-1} mg^{-1} protein or 15–150 nmol C_2H_4 per assay bottle in 30 min (Stewart et al. 1968). A 0.1-ml gas sample would contain 0.3–3 nmol C_2H_4, compared with the lower limit of detection for the gas chromatograph of 10 pmol C_2H_4. In practice, sensitivity is usually limited by the background level of C_2H_4 in the C_2H_2.

After incubation, the samples may be inactivated by injecting 0.5 ml of 5 N H_2SO_4 (convenient if a subsequent total N determination is to be made) and sealed against leakage with silicone rubber compound. Samples may be stored for several days before analysis. Solutions of 50% (w/v) trichloroacetic acid, sodium azide, $HgCl_2$, or saturated $CuSO_4$ may be used to stop the reaction. Because of possible artifacts noted below, the desirable procedure is to add 1 ml of distilled water to each bottle at the end of the incubation to provide a positive internal pressure, immediately withdraw a gas sample with a gas-tight syringe, and inject it into the chromatograph. Gas samples may be stored briefly in the syringe by pushing the needle into the side of a stopper to seal it. Schell and Alexander (1970) have withdrawn and stored gas samples in evacuated Vacutainer blood-collection tubes. This "nondestructive" sampling is advantageous if subsequent measurements of labile compounds (e.g., chlorophyll) or enzymes are to be made on the sample. Leaving a slight positive pressure in the incubation bottle after injection of C_2H_2 allows several samples to be withdrawn serially for monitoring changes in C_2H_4 production rates.

Although evidence to date indicates that C_2H_2 reduction is a measure of nitrogenase activity, it is at best an indirect method for assaying N_2 fixation and should be compared with the uptake of $^{15}N_2$ for purposes of calibration. Rivera-Ortiz and Burris (1975) have shown how N_2 and C_2H_2 may behave differently on the enzyme surface depending upon the supply of reductant available. In addition, C_2H_4 may be liberated from organic material, especially in soil samples, by the action of strong acid, and may be absorbed or released by the rubber serum stoppers or by polyethylene syringes. Spurious C_2H_4 production may also occur if rubber materials are autoclaved. C_2H_4 may be oxidized by methane-utilizing bacteria (DeBont and Mulder 1974). The C_2H_2 reduction assay is simple and rapid enough that sufficient controls and checks can be included to account for artifacts.

Because of reports that acetylene and ethylene inhibit the growth and metabolism of certain anaerobic microorganisms (Brouzes and Knowles 1971; Ormeland and Taylor 1975), one may wish to avoid the use of C_2H_2 when studying mixed populations or natural communities with anaerobic zones. 1-Propyne, 1-butyne, and allene are reduced to the corresponding alkenes (Hardy and Jackson 1967) and may be suitable in special cases.

IV. Assay methods suitable for cell-free preparations

A. Isolation of heterocysts and preparation of nitrogenase

The preparation of purified nitrogenase from blue-green algae has not been perfected to the same degree as for certain bacteria. The organism that has received the most attention is *Anabaena cylindrica* Lemm. (Cambridge Culture Collection No. 1403/2) (Bothe 1970; Haystead et al. 1970; Haystead and Stewart 1972), and the following procedure, adopted for this alga, may not be satisfactory with other species. Specifically, *Gloeocapsa* requires modified handling (Gallon et al. 1972). All steps require strict anaerobiosis, and all solutions must be purged with Ar or N_2 to remove O_2, to which nitrogenase is very sensitive.

To prepare extracts of nitrogenase it is not necessary to isolate heterocysts. The following growth procedure will maximize nitrogenase yield from heterocystous algae as well as from nonheterocystous species by allowing N_2-fixing activity to develop in vegetative cells. It may be desirable to grow the algae in air, under natural conditions, and to separate heterocysts from vegetative cells; however, heterocyst isolation must be done anaerobically.

Grow the alga under continuous illumination in 20-l carboys in N-free medium purged with 5% CO_2 in 95% N_2 at 200 ml h^{-1}. During logarithmic growth, harvest the cells from 40 l of medium by low-speed centrifugation (300 × g for 10 min) while under N_2, and resuspend in one-fifth the original volume of new, sterile medium previously purged with Ar. Change the gas mixture to 5% CO_2 in 95% Ar and continue to purge for 36 h to increase nitrogenase yield prior to harvesting of cells by centrifugation or by a continuous-flow cell harvester.

1. Isolation of heterocysts

a. Sonication. Wash the cells twice with Ar-purged, N-free medium and finally resuspend in a minimum volume of 20 mM Na-HEPES buffer, pH 7.55. Add $Na_2S_2O_4$ to a final concentration of 0.5 mM. Place the algal suspension in the sample cup of a probe-type ultrasonic cell disintegrator and place the cup in crushed ice. Operate the probe at an output of approximately 75–100 W, for five 1-min intervals spaced 1 min apart (to minimize sample heating). Centrifuge the sample at 1300 × g for 10 min at 10 C and discard the supernatant containing soluble material (see IV.A.2) and algal mucilage. Resuspend the pellet in the buffer, centrifuge at 100 × g for 10 min to sediment heterocysts, and discard the supernatant containing vegetative cellular debris. Repeat this final step twice to obtain a relatively pure heterocyst preparation.

b. French pressure cell. Wash the algal cells by centrifugation at 300 × g as above and suspend the pellet in an equal volume of the HEPES buffer. Transfer the suspension to a French pressure cell precooled to 4 C, and force through the needle valve at a pressure of 760 atm. Dilute the suspension in 10 times its volume of buffer and purify by centrifugation as in IV.A.1.a.

c. Enzymatic digestion. Fay and Lang (1971) have shown by electron microscopy that heterocysts isolated by the above methods, although intact, are damaged to an extent that may allow leakage of cellular contents. They describe a method that uses lysozyme to remove vegetative cells, leaving heterocysts undamaged.

2. *Preparation of crude cell-free extracts.* The 1300 × g × 10 min supernatant from sonicated cells contains soluble enzymes from vegetative cells and includes most of the recoverable nitrogenase activity from cells grown in the absence of O_2. Heterocyst preparations from the above separation procedures are most conveniently broken in the French pressure cell at maximum attainable pressures (in excess of 2100 atm). Centrifuge either of these cell-free extracts at 40 000 × g for 20 min at 10 C. Nitrogenase from *Anabaena cylindrica* and *Plectonema boryanum* remains in the supernatant (Haystead et al. 1970), although Gallon et al. (1972) found that nitrogenase from *Gloeocapsa* is sedimented at 10 000 × g for 20 min. Specific activity at this point should be approximately 4 nmol C_2H_4 min^{-1} mg^{-1} protein and recovery, compared with whole filaments, 20%.

3. *Purification of nitrogenase.* Mix the supernatant of the 40 000 × g centrifugation (containing 2–5 g protein in 100 ml) with an equal volume of a 50% (w/v) slurry of DEAE-cellulose (Whatman DE 52) in 25 mM HEPES buffer, pH 7.55, which has been purged with Ar. (To prepare the DEAE-cellulose for use, suspend in 0.5 M HCl, decant, wash with water, resuspend in 0.5 M KOH, and wash again with water. Place in a Büchner funnel and leach with the HEPES buffer until the pH of the eluate is 7.0 or above.) Stir under Ar for 20 min at 10 C and pour into a column approximately 5 × 15 cm fitted with a valve and hypodermic needle at the bottom. (Abundant membranous material in the 40 000 × g supernatant necessitates this batch-mixing technique.) Allow the suspension to settle, draw down the excess liquid, and wash the column at 1–2 ml min^{-1} with 5 volumes of N_2-purged 25 mM HEPES buffer, pH 7.55, containing 0.1 mM $Na_2S_2O_4$

and 10 mg l⁻¹ dithiothreitol (Cleland's reagent). Follow with 5 volumes of the same buffer, containing 50 mM $MgCl_2$, to remove phycocyanin. Elute the nitrogenase at 2 ml min⁻¹ with this same buffer containing 150 mM $MgCl_2$, and collect 10- to 15-ml fractions in stoppered serum bottles that are continuously flushed with N_2 through hypodermic needles. Elution of the protein components can be monitored by the relative absorbancy at 280 nm or by analysis of individual fractions for protein content. The partially purified nitrogenase-containing fraction is greenish-brown. Centrifugation of this fraction at 144 000 × g for 3 h at 10 C should leave all nitrogenase activity in the supernatant. This should have a specific activity (activity per minute per milligram of protein) 25- to 50-fold greater than that of the initial crude extract. Recovery of activity at this step should be 75% of that in the crude extract. The nitrogenase complex can be preserved for at least 2 weeks by dropwise addition to plastic bottles containing liquid N_2 and storage under liquid N_2.

Further purification by gel filtration on Sephadex G-100 or G-200 should be possible, but has not been well documented. The methods for purification of bacterial nitrogenase described in San Pietro (1972) should be consulted.

B. Reaction mixture

To assay cell-free extracts for nitrogenase, a supply of energy and electrons must be provided. In a 7-ml serum bottle, prepare a reaction mixture of final volume 1.5 ml containing Na-HEPES buffer, pH 7.55, 30 μmol; $MgCl_2$, 5 μmol; ATP, 5 μmol; creatine phosphate, 20 μmol; creatine phosphokinase, 50 μg; $Na_2S_2O_4$, 2.5 μmol. Purge the bottle well with Ar or N_2 and add a sample of crude extract containing 10 mg protein or 1–2 mg of protein as purified enzyme. (The ATP-generating system is necessary because ATP concentrations greater than 2 mM are inhibitory.) Physiological electron donor systems other than $S_2O_4^{2-}$ have been described (Smith et al. 1971; Haystead and Stewart 1972).

The most convenient assay for nitrogenase is the acetylene reduction technique (III.C). The gas phase must be anaerobic, and Ar should be used to prevent competition from N_2. For partially purified extracts, 0.04 atm of C_2H_2 in the assay bottle is sufficient for maximal activity, and a 30-min incubation produces relatively large amounts of C_2H_4. One should expect activities of 20–50 nmol C_2H_4 min⁻¹ mg⁻¹ protein. ¹⁵N_2 fixation (III.B) could also be used.

C. Production of NH_3

When purified enzyme preparations are assayed under N_2, the ammonia produced can be measured by microdiffusion and Nesslerization. After incubation, inactivate the samples with 1 ml of saturated K_2CO_3 solution. Immediately replace the serum stopper with a solid stopper supporting a roughened glass rod moistened with 1 N H_2SO_4. Do not let the rod touch the bottle or enzyme mixture. Place the bottle on a rotating shaker for 3 h at 25 C to facilitate diffusion of NH_3 to the acid. Remove the stopper assembly carefully and use the rod to mix together the components of Nessler's reagent (Umbreit et al. 1964) or of the indophenol reaction (Chaykin 1969).

D. H_2 evolution

Nitrogenase-catalyzed H_2 evolution is ATP-dependent (unlike hydrogenase) and is not inhibited by CO. The reaction is carried out in respirometer-type vessels using standard manometric techniques. Place the $Na_2S_2O_4$ solution in the flask side arm and the rest of the reaction mixture in the main vessel. Flush with Ar and add 0.05 atm CO. Mix the contents of the side arm at zero time and measure production of H_2. A 40 000 $\times$ g $\times$ 20 min supernatant fraction should yield 1–5 nmol H_2 min^{-1} mg^{-1} protein.

If a mass spectrometer is available, the enzyme-catalyzed exchange between H_2 and D_2 can be followed (Burris 1972).

E. Utilization of ATP

Because ATP and ADP are continually recycled in the reaction mixture, their turnover cannot be measured directly. Instead, the formation of creatine from creatine phosphate during regeneration of ATP serves as a measure of ATP utilization.

The method of Eggleton et al. (1943, cf. Burris 1972) is satisfactory. To a small amount of reaction mixture containing less than 0.5 μmol creatine, add 2 ml of an aqueous solution containing 10 g of α-naphthol, 60 g of NaOH, and 160 g of Na_2CO_3 per liter (add the correct proportion of α-naphthol to the desired volume of alkaline carbonate just before assay). Add 1 ml of 0.05% (v/v) aqueous diacetyl (prepared from a 1% stock) and dilute the sample to 10 ml. Allow 20 min at 25 C for color development and determine the absorbance at 530 nm.

Include some controls without $Na_2S_2O_4$ (or other electron donor for nitrogenase) to account for ATP hydrolyzed by enzymes other than nitrogenase, and others with 0.05 atm CO in the gas phase to inhibit ATP-dependent N_2 reduction while leaving ATP-dependent H_2 evolution active.

F. Oxidation of dithionite

Nitrogenase activity and kinetics of the reaction can be monitored by observing the decrease in absorbance at 315 nm caused by the oxidation of $S_2O_4^{2-}$ during transfer of electrons to nitrogenase (Ljones and Burris 1972). The oxidation products do not interfere at this wavelength. However, $S_2O_4^{2-}$ disappearance owing to decomposition and reactions not responsible for N_2 reduction must be accounted for in controls without enzyme (decomposition), without ATP (ATP-independent hydrogenase), or with 0.05 atm $\dot{C}O$ added (ATP-dependent H_2 evolution).

Select 1-cm pathlength quartz spectrophotometer cells with stoppers and fit with small serum stoppers. Evacuate and flush the cuvettes with N_2 and add reaction mixture components with a syringe. Add the nitrogenase last to initiate the reaction and monitor absorbance change in a spectrophotometer.

G. Alternative substrates

The ability of the nitrogenase complex to reduce several substrates (Table 36-1) may provide useful methods for measuring activity; however, these are usually used only to elucidate properties of the enzyme or to provide additional assurance that one has, in fact, isolated the nitrogenase complex. These substrates all exhibit various competitive interactions with one another (Rivera-Ortiz and Burris 1975), and some are generally toxic to cellular metabolism, so caution should be exercised in interpreting results from unpurified preparations.

Azide reduction under Ar can be measured by ammonia production and microdiffusion. Add NaN_3 to a final concentration of 5 mM in the reaction mixture after flushing the gas phase; then add enzyme to initiate the reaction. Cyanide and isocyanide reduction can be measured by determination of CH_4 formation by gas chromatography. Add $NaCN$ or CH_3CN to a final concentration of 2 mM (concentrations greater than 4 mM are inhibitory) after flushing assay bottles with Ar; then add enzymes. CH_3CN must be stored well sealed, under N_2 preferably on dry ice. Withdraw gas samples and analyze CH_4 as for C_2H_2 reduction. CH_4 elutes before C_2H_4 on the Porapak N column. Include controls to account for nonenzymatic formation of CH_4.

V. Acknowledgment

This contribution was supported by NSF grant DES 75-20956.

VI. References

Allen, M. B., and Arnon, D. I. 1955. Studies on nitrogen-fixing blue-green algae. I. Growth and nitrogen fixation by *Anabaena cylindrica* Lemm. *Plant Physiol.* 30, 366–72.

Bothe, H. 1970. Photosynthetische Stickstoffixierung mit einem zellfreien Extrakt aus der Blaualge *Anabaena cylindrica*. *Ber. Dtsch. Bot. Ges.* 83, 421–32.

Brouzes, R., and Knowles, R. 1971. Inhibition of growth of *Clostridium pasteurianum* by acetylene: Implication for nitrogen fixation assay. *Can. J. Microbiol.* 17, 1483–90.

Burns, R. C., and Hardy, R. W. F. 1972. Purification of nitrogenase and crystallization of its Mo-Fe protein. *Methods Enzymol.* 24B, 480–96.

Burris, R. H. 1972. Nitrogen fixation-assay methods and techniques. *Methods Enzymol.* 24B, 415–31.

Burris, R. H., and Wilson, P. W. 1957. Methods for measurement of nitrogen fixation. *Methods Enzymol.* 4, 355–66.

Chaykin, S. 1969. Assay of nicotinamide deamidase: Determination of ammonia by the indophenol reaction. *Anal. Biochem.* 31, 375–82.

DeBont, J. A. M., and Mulder, E. G. 1974. Nitrogen fixation and co-oxidation of ethylene by a methane-utilizing bacterium. *J. Gen. Microbiol.* 83, 113–21.

Drozd, J., and Postgate, J. R. 1970. Interference by oxygen in the acetylene-reduction test for aerobic nitrogen-fixing bacteria. *J. Gen. Microbiol.* 60, 427–9.

Eggleton, P., Elsden, S. R., and Gough, N. 1943. The estimation of creatine and of diacetyl. *Biochem. J.* 37, 526–9.

Fay, P., and Lang, N. J. 1971. The heterocysts of blue-green algae. I. Ultrastructural integrity after isolation. *Proc. R. Soc. London, Ser. B.* 178, 185–92.

Gallon, J. R., LaRue, T. A., and Kurz, W. G. W. 1972. Characteristics of nitrogenase activity in broken cell preparations of the blue-green alga *Gloecapsa* sp. LB 795. *Can. J. Microbiol.* 18, 327–32.

Hardy, R. W. F., Burns, R. C., and Holsten, R. D. 1973. Applications of the acetylene-ethylene assay for measurement of nitrogen fixation. *Soil Biol. Biochem.* 5, 47–81.

Hardy, R. W. F., and Jackson, E. K. 1967. Reduction of model substrates–nitriles and acetylenes–by nitrogenase (N_2ase). *Fed. Proc.* 26, 725.

Haystead, A., Robinson, R., and Stewart, W. D. P. 1970. Nitrogenase activity in extracts of heterocystous and non-heterocystous blue-green algae. *Arch. Mikrobiol.* 74, 235–43.

Haystead, A., and Stewart, W. D. P. 1972. Characteristics of the nitrogenase system of the blue-green alga *Anabaena cylindrica*. *Arch. Mikrobiol.* 82, 325–36.

Kratz, W. A., and Myers, J. 1955. Nutrition and growth of several blue-green algae. *Am. J. Bot.* 42, 282–7.

LaRue, T. A., and Kurz, W. G. W. 1973. Estimation of nitrogenase using a colorimetric determination for ethylene. *Plant Physiol.* 51, 1074–5.

Ljones, T., and Burris, R. H. 1972. Continuous spectrophotometric assay for nitrogenase. *Anal. Biochem.* 45, 448–52.

Millbank, J. W. 1970. The effect of conditions of low oxygen tension on the assay of nitrogenase in moulds and yeasts using the acetylene reduction technique. *Arch. Mikrobiol.* 72, 375–7.

Oremland, R. S., and Taylor, B. F. 1975. Inhibition of methanogenesis in marine sediments by acetylene and ethylene: Validity of the acetylene reduction assay for anaerobic microcosms. *Appl. Microbiol.* 30, 707–9.

Rivera-Ortiz, J. M., and Burris, R. H. 1975. Interactions among substrates and inhibitors of nitrogenase. *J. Bacteriol.* 123, 537–45.

San Pietro, A. (ed.). 1972. Nitrogen fixation. *Methods Enzymol.* 24B, 413–504.

Schell, D. M., and Alexander, V. 1970. Improved incubation and gas sampling techniques for nitrogen fixation studies. *Limnol. Oceanogr.* 15, 961–2.

Smith, R. V., Noy, R. J., and Evans, M. C. W. 1971. Physiological electron donor systems to the nitrogenase of the blue-green alga *Anabaena cylindrica*. *Biochim. Biophys. Acta* 253, 104–9.

Stewart, W. D. P., Fitzgerald, G. P., and Burris, R. H. 1968. Acetylene reduction by nitrogen-fixing blue-green algae. *Arch. Mikrobiol.* 62, 336–48.

Stewart, W. D. P., Mague, T. H., Fitzgerald, G. P., and Burris, R. H. 1971. Nitrogenase activity in Wisconsin lakes of differing degrees of eutrophication. *New Phytol.* 70, 497–509.

Umbreit, W. W., Burris, R. H., and Stauffer, J. F. 1964. *Manometric Techniques*, 4th ed. Burgess, Minneapolis. 305 pp.

37: Uptake of organic substrates

JOHAN A. HELLEBUST AND
YI-HUNG LIN

*Department of Botany,
University of Toronto, Toronto, Ontario M5S 1A1, Canada*

CONTENTS

I Introduction   *page* **380**
II Materials and test organism   **380**
  A. Materials   380
  B. Test organism   381
III Methods   **381**
  A. Time course for uptake   381
  B. Properties of the uptake system   382
IV Sample data   **384**
  A. Time course for glucose uptake
    in light and dark   384
  B. Time course for glucose uptake:
    rapid sampling procedure   385
V Discussion   **386**
VI References   **387**

I. Introduction

Many algae are able to utilize organic substrates for heterotrophic growth or as additional carbon, nitrogen, or energy sources for growth in the presence of light (Danforth 1962; Neilson and Lewin 1974; Hellebust and Lewin 1977). We present methods for determining uptake rates of microalgae and stress the importance of environmental conditions in regulating both transport capacities and activities of algal cells. Methods for determining uptake and assimilation of specific compounds for investigating respiratory metabolism and protein and nucleic acid synthesis are presented in Chps. 32, 34, and 35.

II. Materials and test organisms

A. Materials

1. Radioactive substrates. ^{14}C-(U)-D-glucose, 150–250 mCi mmol^{-1} (New England Nuclear). Radioactive substrates used in transport studies should be of high specific activity to permit sufficient uptake of isotope at low substrate concentrations and after short incubation times for accurate determination of uptake rates. It is important to check the purity of radioactive substrates used. Chromatographic methods should be used, as suggested in Chps. 32 and 35.

2. Equipment. Millipore filtration apparatus and 2.5-cm diam. HA 0.45 μm pore size membrane filters (Millipore). Screw-cap glass vials (5–10 ml) for uptake experiments. Aqueous solutions of glucose or other unlabeled substrates at appropriate concentrations. These should be prepared in sterile f/2 medium and kept frozen when not in use to prevent microbial growth. Aluminum planchets for mounting Millipore filters with radioactive algal cells (e.g., Hruden plain aluminum planchets, $1\frac{1}{4} \times \frac{3}{32}$ in., Canadian Scientific Products). Scintillation vials and scintillation cocktail (e.g., Aquasol, New England Nuclear). Isotope-counting facilities for gas-flow counting (e.g., model 470 Gas Flow Detector, model 186 Decade Scaler, model C-110B Automatic

[380]

Sample Changer, Nuclear-Chicago) or scintillation counting (e.g., model LS-230 Liquid Scintillation System, Beckman Instruments).

B. Test organism

Axenic cultures of *Cyclotella cryptica* (clone T-13-L) from the culture collection of Dr. R. R. L. Guillard (Woods Hole Oceanographic Institution, Woods Hole, Mass.) were used to produce the sample data presented. This marine centric diatom was grown in a seawater medium f/2 (Guillard and Ryther 1962; see McLachlan 1973) at 20 C and 6 klux of continuous light (cool-white fluorescent tubes). The cells should be transferred from light to dark conditions while still in their exponential growth phase, but should at the same time be sufficiently dense (1×10^5 to 2×10^5 cells ml^{-1}) to allow accurate determinations of uptake rates. The cells must be kept in the dark for 24 h to allow induction of transport capacity for glucose. Cells growth at light-saturating intensities for photosynthesis have negligible uptake capacities for glucose (Hellebust 1971) and also relatively low amino acid uptake capacities (Liu and Hellebust 1976; Hellebust, unpublished data). It is essential that cultures be strictly axenic, and frequent tests for the presence of fungi or bacteria should be made by inoculation into seawater nutrient broth (Hoshaw and Rosowski 1973).

III. Methods

A. Time course for uptake

For kinetic studies of how the rate of uptake varies as a function of substrate concentration or environmental conditions, it is essential that initial uptake rates be determined. It is, therefore, necessary to determine the extent of linear uptake with time, so that the experimental period chosen will not exceed this initial period of constant uptake rate.

Uptake system. 1 ml algal culture (1×10^5 to 2×10^5 cells ml^{-1}) incubated in the dark at 20 C for 24 h prior to the uptake experiment; 100 μl of unlabeled substrate (e.g., 1.1 mM glucose if a final concentration of glucose of 0.1 mM in the uptake system is desired); 10 μl of high-specific-activity (> 50 mCi mmol^{-1}), uniformly labeled ^{14}C substrate of known total radioactivity (e.g., 0.1 μCi ^{14}C-(U)-D-glucose). Make up triplicate vials for each sampling time.

The components are added to 5- or 10-ml vials in the above order. As soon as the radioactive substrate is added to the system, shake the vial quickly and immediately pour the contents into the filtration apparatus, which already contains a few milliliters of unlabeled culture medium and is under low suction. The cells on the filter are then washed with 2 ml medium. The washing is repeated once, and the filter removed with forceps and placed on aluminum planchets finely coated with Vaseline or silicone grease to make the filters adhere tightly. These filters are considered zero-time samples. The contents of vials for other sampling times are treated in the same way at the appropriate times (e.g., 10, 20, 40, 60, and 120 min). The filters on the aluminum planchets are dried at room temperature for at least 4 h and are then counted in a gas-flow counter. If determinations of radioactivity are made using a scintillation counter, the filters should be transferred directly after filtration to scintillation vials containing 10 ml scintillation cocktail and counted. The average radioactivity of the samples taken at different times should then be plotted to determine the duration of the initial constant uptake rates.

If it is necessary to take samples at brief time intervals, it is advisable to prepare one uptake system with sufficient volume for the required replicate time samples. These are then removed with an automatic pipettor, e.g., 1-ml samples with a 1.0-ml Biopette (Schwarz/Mann), and are squirted into ca. 5 ml ice-cold medium in the filtration apparatus. These conditions will stop the uptake almost immediately and allow rapid and accurate timing of the samples.

B. Properties of the uptake system

1. Transport rate as a function of substrate concentration. It is important to determine the linearity of uptake for the lowest substrate concentrations employed. In the case of glucose uptake by 1×10^5 to 2×10^5 cell ml^{-1} suspensions of *C. cryptica,* the rate of uptake is linear for at least 30 min for substrate concentrations down to 1×10^{-5} M. The uptake system outlined above can be used in cases where the substrate concentration is varied by adding 100 μl of unlabeled glucose of different concentrations. The incubation period should be 10 min for linear kinetics down to 1×10^{-6} M, and sufficient high-specific-activity ^{14}C-glucose should be added (ca. 0.2 μCi per sample) to yield accurate uptake rates for these short uptake periods. The amount of radioactivity added should be determined by plating on planchets and counting in a gas-flow system or by adding to a scintillation vial and counting in a scintillation counter, depending on the method used for determining the radioactivity incorporated by the cells.

The rate of uptake, V, is calculated as follows: $V = (A \times B)/(C \times D \times E)$, where A is the average radioactivity (counts per minute) in cells on the filter, B is the amount of substrate in the uptake system (micromoles), C is the radioactivity (counts per minute) added to the uptake system, D is the number of cells in the uptake system, and E is the time (minutes) of incubation. The units for the uptake rate will then be micromoles of substrate per cell per minute.

For accurate determination of uptake kinetic constants, plot S/V against S (Dowd and Riggs 1965), where V is the uptake rate and S the substrate concentration. This plot will yield a straight line if simple Michaelis-Menten kinetics occur. K_s is obtained as the negative intercept of the straight line with the abscissa, and the slope of the line is equal to $1/V_{max}$, where V_{max} is the maximum uptake rate, and K_s the substrate concentration that supports an uptake rate equal to $\frac{1}{2} V_{max}$.

2. Specificity of uptake system. This can be determined by addition to the uptake system of unlabeled substrates of chemical structures similar to the radioactive substrate investigated. The concentration of these analogues may be similar or considerably higher than that of the radioactive substrate tested. In the case of glucose uptake, interference by different hexoses and pentoses with the uptake of radioactive glucose may be determined by adding 10 μl of 0.11 mM of unlabeled sugar to the uptake system (III.A) containing 100 μl of 1.1 mM unlabeled glucose and 0.05 μCi ^{14}C-(U)-D-glucose, and incubating the algae for 30 min.

3. Test for active uptake. One of the most clear-cut criteria for active uptake of a neutral organic substrate is that it is taken up against a concentration gradient. The easiest way to demonstrate this is by using radioactive substrate analogues that cannot be metabolized by the cells. In the case of glucose uptake by *C. cryptica*, ^{14}C-(U)-D-deoxyglucose, which appears to be taken up by the same transport system as glucose, may be used. Incubate the cells for 60 min in the uptake system described above with 1 μM final ^{14}C-(U)-D-deoxyglucose concentration. From the radioactivity taken up by the cells, and the known specific activity of the ^{14}C-(U)-D-deoxyglucose, the amount of deoxyglucose in the cells can be calculated. By estimating the cell volume from light microscope determinations (ocular micrometer) of the average dimensions of the cells, one can determine the concentration of the substrate in the cells and obtain the accumulation ratio (concentration in cells/that in the medium). Accumulation ratios of several hundred have been obtained in this way for *C. cryptica* (unpublished data).

Evidence for active uptake can also be obtained by demonstrating that uncoupling agents of energy metabolism (Chp. 46), such as 2,4-dinitrophenol (1 mM final concentration in uptake system), tetrachlorosalicylaniline (5 μM), or carbonylcyanide *m*-chlorophenylhydrazone (25 μM), inhibit ^{14}C-glucose uptake. As the solubilities in water of some of these inhibitors are low, they may be added to the uptake system in a small amount of ethanol (10–50 μl). The effect of ethanol on substrate uptake should be determined separately.

4. Regulation of uptake capacity by environmental conditions. The capacity for uptake of organic substrates may vary considerably depending on the environmental conditions to which an alga is exposed. In particular, it has been demonstrated that light conditions (Hellebust and Lewin 1977), presence or absence of substrate (Tanner et al. 1970), and nitrogen deficiency (North and Stephens 1972) are determining factors for the transport capacities for several substrates.

When testing for the ability of a cell to take up a particular substrate, one should measure potential uptake rates for cultures raised under different conditions (e.g., continuous light, light-dark periods, preincubation for different periods of darkness before the uptake experiment) and also in the presence of different concentrations of the unlabeled substrate during dark or light preincubation conditions. To test the potential transport capacities for nitrogen-containing substances, the cells should be investigated after growth under nitrogen-deficient as well as sufficient conditions (North and Stephens 1972). *C. cryptica,* when grown at high light intensities (> 5 klux) possesses very low transport capacities for glucose (Hellebust 1971). One can demonstrate the almost exponential increase in transport capacity of this alga during 24 h incubation of cells previously grown in continuous light (10 klux) by removing samples at regular time intervals and determining glucose uptake using the uptake system outlined above (III.A; 1 mM final glucose concentration, 30-min incubation period). The induction of the transport system can be abolished by cycloheximide, an inhibitor of protein synthesis, added to the culture at 100 μg ml^{-1} before dark incubation (Hellebust 1971). The inactivation of the transport system by high light intensities can be prevented almost completely by adding to the culture 1×10^{-5} M DCMU, an inhibitor of photosynthesis (Jolley et al. 1976; Hellebust, unpublished data).

IV. Sample data

A. Time course for glucose uptake in light and dark

A 50-ml culture of *C. cryptica* (initial cell density 2×10^4 cells ml^{-1}) was incubated for 3 days in continuous light (4 klux), and then incubated in the dark for 24 h before the experiment. Uptake rates were determined using the standard uptake system: 1 ml cell suspension (1.9×10^5 cells ml^{-1}), 100 μl 1.1 mM glucose, and 10 μl ^{14}C-(U)-D-glucose (0.1 μCi), yielding a final glucose concentration of 0.1 mM. Quadruplicate samples were taken each time, and the radioactivity was determined by gas-flow counting. Cells killed by formalin (50 μl added to the uptake system before addition of isotope) or heat (cell suspension heated to 80 C for 5 min before addition of isotope) resulted in approximately the same uptake of ^{14}C-glucose (2–4% of the 30-min uptake values). The average radioactivity for the zero-time

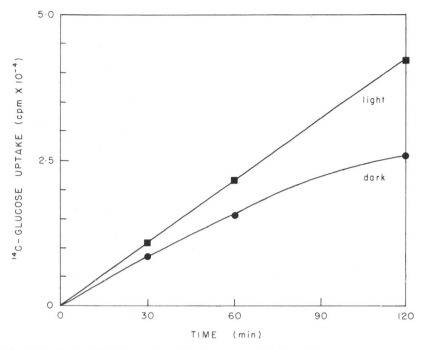

Fig. 37-1. Uptake of ^{14}C-glucose by *Cyclotella cryptica* in light (4 klux) and dark. Glucose concentration, 0.1 mM; cell density, 1.9×10^5 cells ml^{-1}; 0.1 μCi ^{14}C-(U)-D-glucose ml^{-1}.

samples, assuming it was attributable to physical adsorption, was subtracted from all the subsequent experimental values. Figure 37-1 shows that the uptake of glucose was linear for 2 h when the cells were incubated in light (10 klux), whereas linear uptake lasted less than 1 h for dark-incubated cells. Light increased glucose uptake considerably at all sampling times. Standard deviations for each point were less than ±4%.

B. Time course for glucose uptake: rapid sampling procedure

When uptake studies have to be made with experimental systems of high cell densities and low substrate concentrations, a rapid sampling procedure must be employed because the uptake rate will decrease as the substrate is rapidly removed from the external medium. The cells used here were grown on a 3-day light-dark regimen, then incubated for 24 h in the dark prior to the experiment.

Uptake system. A 3-ml sample of cell suspension (4×10^6 cells ml^{-1}), 30 μl 5 mM glucose (final concentration 5×10^{-5} M), 10 μl ^{14}C-(U)-D-

Table 37-1. *Uptake of ^{14}C-glucose (cpm/0.1 ml sample) by Cyclotella cryptica incubated in the dark*[a]

	Time (min)					
	0	1	2	5	10	30
Sample						
1	167	649	1094	2508	3857	6116
2	208	581	1123	2748	4339	5468
3	224	774	1110	1929	4291	4744
4	278	497	1092	2718	4774	5471
Average uptake	219	625	1105	2476	4314	5449
Zero-time corrected uptake	0	406	886	2257	4095	5230
Rate of uptake (μmol cell^{-1} min$^{-1} \times 10^{-10}$)	—	5.3	5.8	5.9	5.3	2.3
Percent of total glucose in cells	0	4	9	23	41	53

[a] Glucose concentration, 5×10^{-5} M; cell density, 4×10^{6} cells ml^{-1}; 0.1 μCi ^{14}C-(U)-D-glucose ml^{-1} yielding an initial radioisotope concentration by scintillation counting of 9960 cpm/0.1 ml.

glucose (0.3 μCi). The cells were incubated in the dark following addition of the ^{14}C-glucose and shaken intermittently. The uptake was stopped as described above (III.A) by removing 0.1-ml samples and adding these to a filtration apparatus containing 5 ml of ice-cold medium. After filtration to dryness, the cells were washed twice with 5-ml portions of ice-cold medium. The filters were counted first by gas-flow counting (Nuclear Chicago model 470 Gas Flow Detector), then the filters were removed from the counting planchets, transferred to scintillation vials containing 10 ml of Aquasol, and counted in a scintillation system (Beckman model LS-230). Only the scintillation counting data are presented (Table 37-1). The gas-flow data yielded similar results (less than 15% difference between same sets of values with a scintillation/gas-flow counting efficiency ratio of 1.7).

The rate of uptake was linear for ca. 10 min, then decreased rapidly owing to the decrease in glucose concentration in the medium. At 30 min about 50% of the available glucose had accumulated in the cells. Most of the glucose had probably been taken up by the cells by this time, as about 50% of the total glucose can be expected to have been respired as ^{14}CO$_2$.

V. Discussion

The methodology for determining uptake of organic substances by microalgae is relatively simple. However, because recent studies have shown that transport systems in algae vary considerably depending on environmental conditions and the physiological state of the cells, it is essential to determine uptake rates for a given substance under different conditions of pretreatment prior to, as well as during, the uptake period. Light appears to affect glucose and amino acid transport systems in two different ways: (1) there is an inverse relationship between the light intensity at which the cells have been grown prior to the experiment and the transport capacity, and (2) the light intensity during the uptake period positively affects the uptake rate (Hellebust 1971; Liu and Hellebust 1976). Different uptake rates for organic substrates may be obtained for cells at different stages of their life cycle (Knutsen 1972) or at different growth phases in a batch culture (Hellebust, unpublished data). Methods for studying the mechanisms of energy coupling to the transport of sugars by algae can be found in publications by Tanner and co-workers (Komor et al. 1973a, 1973b).

VI. References

Danforth, W. F. 1962. Substrate assimilation and heterotrophy. In Lewin, R. A. (ed.). *Physiology and Biochemistry of Algae,* pp. 99–123. Academic Press, New York.

Dowd, J. W., and Riggs, D. S. 1965. A comparison of Michaelis-Menten kinetic constants from various linear transformations. *J. Biol. Chem.* 240, 863–9.

Guillard, R. R. L., and Ryther, J. H. 1962. Studies of marine planktonic diatoms. I. *Cyclotella nana* Hustedt and *Detonula confervacea* (Cleve) Gran. *Can. J. Microbiol.* 8, 229–39.

Hellebust, J. A. 1971. Glucose uptake by *Cyclotella cryptica:* Dark induction and light inactivation of transport system. *J. Phycol.* 7. 345–9.

Hellebust, J. A., and Lewin, J. C. 1977. Heterotrophy. In Werner, D. (ed.). *The Biology of Diatoms,* pp. 169–97. Blackwell, Oxford.

Hoshaw, R. W., and Rosowski, J. R. 1973. Methods for microscopic algae. In Stein, J. R. (ed.). *Handbook of Phycological Methods: Culture Methods and Growth Measurements,* pp. 53–68. Cambridge University Press, London.

Jolley, E. T., Jones, A. K., and Hellebust, J. A. 1976. A description of glucose uptake in *Navicula pelliculosa* (Bréb.) Hilse including a brief comparison with an associated *Flabovacterium* sp. *Arch. Microbiol.* 109, 127–33.

Knutsen, G. 1972. Uptake of uracil by synchronous *Chlorella fusca. Physiol. Plant.* 27, 300–9.

Komor, E., Haass, D., Komor, B., and Tanner, W. 1973a. The active hexose-uptake system of *Chlorella vulgaris. Eur. J. Biochem.* 39, 193–200.

Komor, E., Loos, E., and Tanner, W. 1973b. A confirmation of the proposed model for the hexose uptake system of *Chlorella vulgaris:* Anaerobic studies in the light and in the dark. *J. Membr. Biol.* 12, 89–99.

Liu, M. S., and Hellebust, J. A. 1976. Regulation of proline metabolism in the marine centric diatom *Cyclotella cryptica. Can. J. Bot.* 54, 949–59.

McLachlan, J. 1973. Growth media – marine. In Stein, J. R. (ed.). *Handbook of*

Phycological Methods: Culture Methods and Growth Measurements, pp. 25–51. Cambridge University Press, London.

Neilson, A. H., and Lewin, R. A. 1974. The uptake and utilization of organic carbon by algae: An essay in comparative biochemistry. *Phycologia* 13, 227–64.

North, B. B., and Stephens, G. C. 1972. Amino acid transport in *Nitzschia ovalis* Arnott. *J. Phycol.* 8, 64–8.

Tanner, W., Grünes, R., and Kandler, O. 1970. Spezifität und Turnover des induzierbaren Hexose-Aufnahmesystems von *Chlorella*. *Z. Pflanzenphysiol.* 62, 376–86.

38: Release of organic substances

C. NALEWAJKO

Scarborough College,
University of Toronto, West Hill, Ontario M1C 1A4, Canada

CONTENTS

I	Introduction	*page* **390**
II	Materials and test organisms	**390**
	A. Special materials and equipment	390
	B. Test organisms	391
III	Methods	**391**
	A. Excretion of extracellular products labeled with ^{14}C	391
	B. Molecular size estimation of extracellular products labeled with ^{14}C	392
	C. Dissolved organic carbon	393
	D. Chemical methods	393
IV	Sample data	**394**
V	Discussion	**396**
VI	References	**397**

I. Introduction

Extracellular release or excretion is the liberation of soluble organic substances by living organisms. This definition attempts to separate "true" extracellular substances from organic compounds released by moribund organisms or liberated during cell lysis, although in practice a clear-cut distinction may not be possible. Fogg (1966) distinguishes two types of extracellular products: type I, small-molecular-weight metabolic intermediates; type II, larger molecular weight end products of metabolic pathways. Capsules and mucilage sheaths may be sources of certain type-II compounds.

Two main approaches have been used in studies of extracellular release (Fogg 1971; Hellebust 1974; Nalewajko 1978). In one, ^{14}C-labeled extracellular products are measured during photosynthesis in $^{14}CO_2$ and expressed as a percentage of cellular ^{14}C or as an excretion rate (micrograms of carbon per time period). In the other, molecular weights of individual substances are determined and/or the substances are identified chemically and their concentrations measured. Very few attempts have been made to measure extracellular products directly as dissolved organic carbon (DOC) accumulating in the medium during short-term experiments or during growth.

The main objective of this chapter is to describe methods for these two approaches and, because separation of algae from the medium is an essential step in measuring extracellular release, to describe precautions needed to prevent cell damage and to avoid other artifacts that may arise during filtration (e.g., retention of organic substances in membrane filters by processes such as adsorption).

II. Materials and test organisms

A. Special materials and equipment

(1) High-specific-activity $NaH^{14}CO_3$ (>40 mCi/mmol). Membrane filters, Millipore type HA, 0.45-μm pore size. Scintillation cocktails; per liter of dioxane: 60 g naphthalene, 4 g 2,5-diphenyloxazole (PPO), 0.2 g 1,4-bis-(5-phenyloxazol-2-yl)-benzene (POPOP). (2) Seph-

adex column with flow adapter. Sephadex G-15 or G-25 beads. Pump: Buchler Polystáltic 4 channel; Isco Golden Retriever fraction collector. (3) Whatman GF/C filters baked at 550 C for 2 h. (4) Gas chromatograph. (5) Aluminum oxide active, Brockman grade 1; 2,7-dihydroxynaphthalene.

B. Test organisms

Chlorella pyrenoidosa Chick UTEX 26. *Anabaena flos-aquae* (Lyngbye) UTEX 1444. *Navicula pelliculosa* (Bréb.) Hilse UTEX 645. *Scenedesmus basiliensis* Vischer UTEX 79.

III. Methods

A. Excretion of extracellular products labeled with ^{14}C

The cultures may be used directly, after dilution with new medium, or after centrifugation and resuspension in new medium. The following are important points in choosing one of the above: (1) population density affects extracellular release (Watt 1966); (2) dense cultures may clog filters and result in cell breakage (III.A.I); (3) pH and especially CO_2 concentration may affect extracellular release (Watt and Fogg 1966); (4) extracellular substances in the old medium may affect release rates of ^{14}C-labeled extracellular substances during the experiment; and (5) centrigugation may be a shock that affects excretion.

The cultures, preferably in closed containers (stoppered Erlenmeyer flasks or Pyrex bottles), should be equilibrated for about 1 h under the experimental conditions of light and temperature and agitated gently to prevent settling of the cells. $NaH^{14}CO_3$ is added at zero time to give 1–5 $\mu Ci/100$ ml of culture, depending on population density and expected photosynthetic rates. Immediately after tracer addition, and subsequently at suitable (15–20 min) intervals, appropriate volumes (III.A.I) of the culture are filtered on 0.45-μm membrane filters, using gentle suction (75–100 mm Hg vacuum). The suction should be turned off when about 1 ml of culture remains on the filter, and this last aliquot should be allowed to drain very slowly (ca. 10 sec). As soon as this liquid has filtered through, the filter is removed (air should not be sucked through the filter) and placed in a scintillation vial. Then 1 ml of distilled water is added followed by scintillation cocktail.

The filtrate is acidified to pH 3–3.5 with HCl, transferred to an aeration unit (Schindler et al. 1972), and bubbled with air vigorously, but without overflow, for about 20 min. The exact time needed to expel all $^{14}CO_2$ should be determined for each batch of ^{14}C tracer and culture medium (III.A.2). Aliquots of 4–5 ml of the filtrate are transferred into scintillation vials, and 13 to 15-ml aliquots of scintillation cocktail are added. Radioactivity of these and of the filters is measured in a Beckman model 230 scintillation counter. Radioactivity values should be converted to carbon using data on dissolved inorganic carbon (DIC) concentrations. A suitable method for DIC is given in Stainton et al. (1974). Percentage extracellular release is calculated as: (disintegrations per minute in filtrate)/(disintegrations per minute on filter + filtrate).

1. Selection of appropriate volume of the culture for filtration. It is advisable to select a volume such that the filtration is completed within 2–3 min when gentle suction is applied and such that the rate of filtration does not slow (i.e., no clogging of the filter occurs). However, it is best to avoid filtering very small volumes (0.5–3 ml) because adsorption of any ^{14}C-labeled dissolved substances by the filters may cause a significant error by overestimating the radioactivity in the particulate fraction and at times underestimating that in the extracellular fraction. To select the volume, a range of volumes should be filtered and the duration of filtration timed. Calculations of radioactivity per milliliter on filter or in filtrate for each volume filtered indicate the acceptable range of volumes that may be filtered. Filter clogging is indicated by a slowing of filtration rate (Nalewajko and Lean 1972).

2. Blanks. To test NaH^{14}CO$_3$ tracer purity and completeness of ^{14}CO$_2$ removal during the aeration, it is essential to carry out some preliminary experiments. The tracer is added to filtered lake water (seawater) or medium without the algae. Aliquots are acidified and bubbled for varying times to select an appropriate aeration period. If expulsion of ^{14}C is incomplete in spite of prolonged aeration, the ^{14}C tracer may be impure (i.e., contaminated with ^{14}C organic compounds).

B. Molecular size estimation of extracellular products labeled with ^{14}C

Approximate molecular weights of ^{14}C-labeled extracellular products in filtrates may be estimated using Sephadex gel fractionation. After aeration, an exactly measured aliquot of about 4 ml is injected into a column 2.5 × 45 cm containing Sephadex G-15 or G-25 (exclusion MW of 2500 and 5000, respectively) in a bed of about 200 ml. The eluant is 0.3% NaCl with 0.02% NaN$_3$; the rate of flow is about 8 ml min^{-1}. Successive 3- to 4-ml portions of the effluent should be collected to a total of ca. 200 ml. The radioactivity of the injected aliquot should be known and should be no less than 1000 dpm ml^{-1}. To check for possible retention of ^{14}C-labeled materials on the column, the percentage recovery of the injected radioactivity should always be determined. With a suitable measuring device such as the drop counter or timer on the fraction collector, all the effluent portions are exactly of equal volume and may be collected directly in scintillation vials. The radioactivity of these is determined as described previously. Background radioactivity of blanks (eluant plus scintillation cocktail) should be determined and subtracted from each sample.

C. Dissolved organic carbon

DOC in filtrates of short-term excretion or long-term growth experiments may be determined by the method of Stainton et al. (1974). Preignited Whatman GF/C filters must be used for filtration. The filtrates, after removal of CO_2 by vigorous bubbling with helium at pH below 4, are treated with potassium persulfate at elevated temperature and pressure in sealed Pyrex glass ampules. During this procedure, organic carbon compounds are oxidized to CO_2, which is measured in a gas chromatograph.

D. Chemical methods

1. Determination of glycolic acid. Glycolic acid may be determined directly in filtrates (freshwater or seawater media) using the Calkins (1943) method provided certain interfering substances such as nitrate are absent (Watt and Fogg 1966; Smith 1974; Al-Hasan et al. 1975).

Natural samples may need concentrating before glycolate levels are within the sensitivity range of the Calkins method. An alumina extraction method described by Shah and Fogg (1973) and Shah and Wright (1974) achieves approximately a 20-fold concentration and has proved reliable in marine and freshwater work. We have found some modifications of the method necessary (K. Lee, unpublished data). It proved very important to avoid storage of lake water samples in polyethylene. Stainless steel was the best material for freezing filtered lake water. After extraction of alumina with 18 N H_2SO_4, it proved more satisfactory to centrifuge the extract and use the clear supernatant for analysis with 2,7-dihydroxynaphthalene rather than filtering on Gelman type-A glass fiber filters. Also the purple color with the above reagent was not stable. It was essential to treat all samples and blanks in an identical manner and to time and standardize the interval between boiling (to develop the color) and absorbance measurement. The 2,7-dihydroxynaphthalene reagent was dissolved in Analar grade conc. H_2SO_4 (BDH) and used when 16–18 h old.

2. Other substances. Besides glycolate, several other substances have been reported among algal extracellular products (e.g., various amino acids, amides, peptides, polysaccharides, enzymes, organic acids, lipids, and vitamins) (Fogg 1966; Fogg 1971; Hellebust 1974). Reliable quantitative methods are available for some, but not all, of the

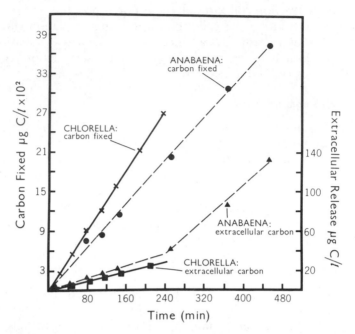

Fig. 38-1. Time course of ^{14}C fixation and extracellular release in *Chlorella pyrenoidosa* and *Anabaena flos-aquae*. Radioactivity values were converted to micrograms of carbon using DIC values in the medium (Chu No. 10) of 8.87 mg l^{-1}.

above substances. The limitation appears to be the low concentration of some substances. Hellebust (1974) presents tables of the substances reported as algal extracellular products. These should be consulted for references giving analytical details for specific substances.

IV. Sample data

In tests on NaH^{14}CO$_3$ tracer purity, 20-min aeration of pH 3.2 reduced the initial radioactivity from 47 460 dpm ml^{-1} to 14 dpm ml^{-1} in Chu No. 10 medium and from 52 160 dpm ml^{-1} to 4 dpm ml^{-1} in artificial seawater medium. The stock tracer had been diluted to the required radioactivity using filtered distilled water at pH 8.5, dispensed in glass ampules, and autoclaved. Repeated freezing and thawing of the ^{14}C tracer should be avoided, as it may result in the appearance of particulate radioactivity.

A time course of ^{14}C fixation and extracellular release, expressed as micrograms of carbon per liter, in axenic *Chlorella pyrenoidosa* and *Anabaena flos-aquae* cultures is shown in Fig. 38-1.

Figure 38-2 shows a Sephadex gel fractionation of extracellular

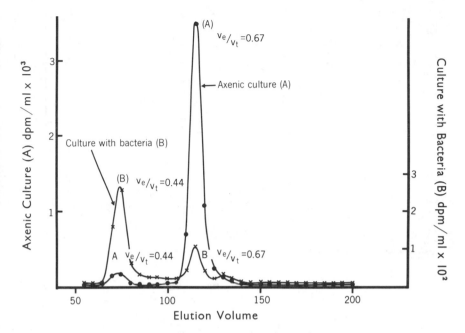

Fig. 38-2. Sephadex gel fractionation (G-15) of axenic and contaminated *Chlorella pyrenoidosa* filtrate. The cultures were 11 days old; bacteria were added to the second culture 5 days prior to this analysis.

products from axenic and contaminated *Chlorella* cultures. Peaks at V_e/V_t of 0.67 correspond to glycolate or other organic compounds of low molecular weight, but the chemical identity of substances eluting in the void volume ($V_e/V_t = 0.44$) has not been determined. Percentage recovery of ^{14}C from the column was 97.2%.

Figure 38-3 shows accumulation of DOC in filtrates of axenic *Chlorella pyrenoidosa*, *Anabaena flos-aquae*, *Navicula pelliculosa*, and *Scenedesmus basiliensis* cultures. Because the medium (Chu No. 10) contains iron as Fe-EDTA, initial DOC values were high. Presumably utilization by algae accounted for the decrease of DOC from day zero to about day 4. Subsequently, DOC values increased and seemed to parallel the increase in population density. In *Anabaena* cultures growing on $FeCl_3$ as an iron source DOC increased directly from day zero.

Fig. 38-4 shows a time course of glycolate accumulation during growth of axenic *C. pyrenoidosa* in Chu No. 10 in which nitrate was replaced by urea. A final glycolate concentration of 14 mg l^{-1} was reached after 12 days. No glycolate was detected in the medium of axenic *N. pelliculosa* growing under identical conditions.

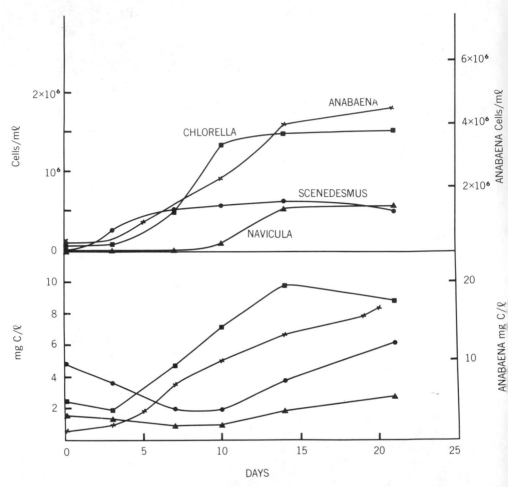

Fig. 38-3. Growth and accumulation of DOC in filtrates of four species of axenic algae.

V. Discussion

Extracellular release values are viewed by some with suspicion, primarily because of the possibility of cell breakage during filtration and because of other artifacts that may occur during the filtration process. Particular concern has been expressed with natural phytoplankton populations, rather than cultures. Tests outlined above (III.A.2 and 3) should eliminate sources of error associated with cell breakage, filter retention of $H^{14}CO_3^-$ and ^{14}C-DOC, and contaminated $H^{14}CO_3^-$ tracer. Following the time course of extracellular release, as opposed to once-only determination of extracellular substances, is also highly desirable, as excretion does not always follow simple linear kinetics.

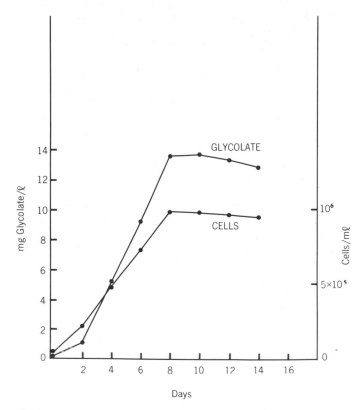

Fig. 38-4. Growth and glycolate excretion by axenic *Chlorella pyrenoidosa*.

In cultures or natural populations with low population density the amounts of extracellular products released may be very tiny. If ^{14}C-labeled extracellular products of photosynthesis are being analyzed, use of higher levels of radioactivity initially may overcome this problem. If chemical identification is to be attempted, some concentration procedure must be employed, but with due precautions so that breakdown of compounds does not occur. Sephadex gel fractionation may be a useful first step in separating organic compounds in freshwater as well as marine culture media.

VI. References

Al-Hasan, R. H., Coughlan, S. J., Pant A., and Fogg, G. E. 1975. Seasonal variations in phytoplankton and glycollate concentrations in the Menai Straits, Anglesey. *J. Mar. Biol. Assoc. U.K.* 55, 557–65.

Calkins, V. P. 1943. Microdetermination of glycollic and oxalic acids. *Ind. Eng. Chem. Anal. Ed.* 15, 762–3.

Fogg, G. E. 1966. The extracellular products of algae. *Oceanogr. Mar. Biol. Annu. Rev.* 4, 195–212.

Fogg, G. E. 1971. Extracellular products of algae in freshwater. *Arch. Hydrobiol. Beih. Ergeb. Limnol.* 5, 1–25.

Hellebust, J. A. 1974. Extracellular products. In Stewart, W. D. P. (ed.). *Algal Physiology and Biochemistry,* pp. 838–63. Blackwell, Oxford.

Nalewajko, C. 1978. Extracellular release in freshwater algae and bacteria: Extracellular products of algae as a source of carbon for heterotrophs. In Cairns, J., Jr. (ed.). *The Aquatic Microbial Communities.* Dekker. New York. (In press.)

Nalewajko, C., and Lean, D. R. S., 1972. Retention of dissolved compounds by membrane filters as an error in the ^{14}C method of primary production measurement. *J. Phycol.* 8, 37–43.

Schindler, D. W., Schmidt, R. V., and Reid, R. A. 1972. Acidification and bubbling as an alternative to filtration in determining phytoplankton production by the ^{14}C method. *J. Fish. Res. Board Can.* 29, 1627–31.

Shah, N. M., and Fogg, G. E. 1973. The determination of glycollic acid in sea water. *J. Mar. Biol. Assoc. U.K.* 53, 321–4.

Shah, N. M., and Wright, R. T. 1974. The occurrence of glycollic acid in coastal sea water. *Marine Biol.* 24, 121–4.

Smith, W. O., Jr. 1974. The extracellular release of glycolic acid by a marine diatom. *J. Phycol.* 10, 30–3.

Stainton, M. P., Capel, M. J., and Armstrong, F. A. J. 1974. *The Chemical Analysis of Fresh Water.* Misc. Special Publ. 25, Fishery and Marine Service of Canada, Otawa. 119 pp.

Watt, W. D. 1966. Release of dissolved organic material from the cells of phytoplankton populations. *Proc. R. Soc. London, Ser. B* 164, 521–51.

Watt, W. D., and Fogg, G. E. 1966. The kinetics of extracellular glycollate production by *Chlorella pyrenoidosa. J. Exp. Bot.* 17, 117–34.

Section V

Nutrients

39: Nitrate uptake

RICHARD W. EPPLEY

*Institute of Marine Resources, A-018, Scripps Institute of Oceanography,
University of California, San Diego, La Jolla, California 92093*

CONTENTS

I	**Introduction**	*page*	**402**
II	**Equipment**		**402**
	A. Nitrate reduction columns		402
	B. Spectrophotometer		403
	C. Filtration equipment		404
	D. Pipettes		404
	E. Special glassware		404
	F. Special reagents		404
	G. Software		404
III	**Methods**		**405**
	A. Culturing and incubation procedures		405
	B. Measurement of nitrate		405
IV	**Capabilities**		**407**
V	**Limitations and troubleshooting**		**408**
VI	**Acknowledgments**		**408**
VII	**References**		**408**

I. Introduction

The community of aquatic botanists has benefited significantly from the development of an adequate chemical method for determining submicromolar concentrations of nitrate. This development was achieved by chemical oceanographers in the years 1962–7 (see Strickland and Parsons 1972, p. 71). Their contribution was the development of a convenient method of reducing nitrate to nitrite. The subsequent determination of nitrite is via formation of a highly colored azo dye based on the classic Griess reaction. Given this new reduction method, it was easy to measure nitrate uptake by cultures of unicellular algae or pieces of macroalgal tissue. For perspective it is interesting to note that before this method was available, Hattori (1962) performed nitrate-to-nitrite reduction using *Escherichia coli* nitrate reductase in his study of nitrate assimilation by *Anabaena cylindrica*. Nevertheless, measurement of nitrate uptake in planktonic situations cannot usually be accomplished with this chemical method unless the standing stock is very high; with low crops, ^{15}N-labeled nitrate and mass spectroscopy may also be employed (Dugdale and Goering 1967). The radioactive isotope of nitrogen, ^{13}N, has an inconveniently short half-life (ca. 10 min) and is only rarely used in this context. What follows is a description of the spectrophotometric nitrate method, taken largely from Strickland and Parsons (1972), and its use in measuring nitrate uptake by cultures of unicellular algae (Eppley and Coatsworth 1968; Eppley et al. 1969; Carpenter and Guillard 1970).

II. Equipment

A. Nitrate reduction columns

The columns currently in use in this laboratory are shown in Fig. 39-1. Coarse cadmium metal filings, plated with copper, are used as the reductant in the columns. The columns are basically those described by Strickland and Parsons (1972), but scaled down in size to handle samples 25–30 ml in volume.

[402]

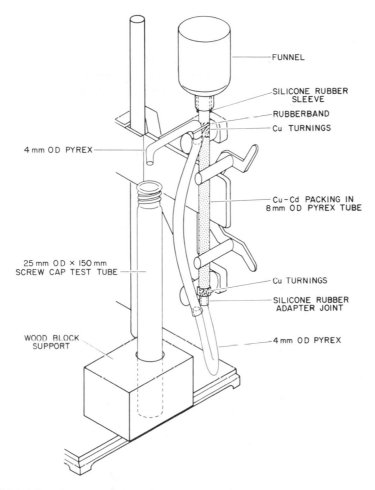

FUNNEL

SILICONE RUBBER SLEEVE

RUBBERBAND

Cu TURNINGS

4 mm OD PYREX

Cu-Cd PACKING IN 8 mm OD PYREX TUBE

25 mm OD × 150 mm SCREW CAP TEST TUBE

Cu TURNINGS

SILICONE RUBBER ADAPTER JOINT

WOOD BLOCK SUPPORT

4 mm OD PYREX

Fig. 39-1. Nitrate reduction column. Use of the column is described in the text; some details on construction are noted here. The length of the column is 15 cm overall, with an 11-cm length of Cu-Cd packing. The funnel is a 50-ml plastic bottle cut off near the base and inverted. A short length of silicone rubber tubing serves as a sleeve to effect a good seal between the neck of the funnel and the column. The liquid level in the column (it must not be allowed to go dry) is maintained by the position of the 4-mm Pyrex outlet tube. A length of flexible tubing such as tygon, attaching the outlet tube to the column, allows easy positioning.

B. *Spectrophotometer*

A prism or grating instrument operable at 543 nm and equipped to accommodate both 1- and 10-cm pathlength absorption cells as required. The capacity of the 10-cm cells should be no more than 10 ml.

C. Filtration equipment

A small filter flask (100- to 250-ml capacity) fitted with a Millipore or Gelman-type filter base and funnel is convenient. The filters should be selected with a view to retaining the smallest microalgae to be studied, but fast enough to allow quick filtering, avoiding undue nitrate uptake or loss during filtration. Reeve Angel 984H glass fiber filters and Whatman GF/C have functioned well. Centrifugation is to be avoided, as the procedure is too slow and nitrate losses will occur with large, mechanically fragile vacuolated cells.

D. Pipettes

An automatic pipette of 10-ml volume, and one of 0.2-ml volume with disposable tips.

E. Special glassware

Screw-capped test tubes of 25-ml volume are needed for nitrate samples.

F. Special reagents

1. Conc. NH_4Cl solution. Dissolve 125 g reagent grade NH_4Cl in 500 ml distilled water and store in a glass or plastic bottle.

2. Dilute NH_4Cl solution. Dilute 50 ml of the conc. NH_4Cl solution to 2000 ml with distilled water.

3. Sulfanilamide solution. Dissolve 5 g of sulfanilamide in a mixture of 50 ml conc. HCl (reagent grade) and about 300 ml distilled water. Dilute to a final volume of 500 ml with distilled water. The solution is stable at room temperature for months in either glass or plastic bottles.

4. N-(1-naphthyl)-ethylenediamine dihydrochloride (NEDA) solution. Dissolve 0.50 g NEDA in 500 ml distilled water. Store the solution in a brown bottle in the refrigerator. Prepare a fresh solution monthly or as soon as the solution develops a brown color.

G. Software

If kinetic analysis of nitrate uptake is contemplated, computer programs for calculating confidence limits for the maximum uptake rate and half-saturation constants will be useful. Kilham (1975) suggests use of procedures in Bliss and James (1966) and Hanson et al. (1967).

III. Methods

A. Culturing and incubation procedures

Culture media are prepared with nitrate as the nitrogen source at a concentration of 20–50 μM. It should be present at one-fifth or less than the usual amount added to culture media to ensure that, during growth, nitrate will be depleted before other nutrients. The depletion of nitrate in the medium during growth of the culture is followed daily at first, then more frequently in order to assess the time, within an hour, of nitrate depletion (to less than a few tenths of a micromole per liter). When the nitrate is depleted, the culture is either (1) spiked with nitrate to 10–20 μM and the depletion of nitrate followed over time (Caperon and Meyer 1972), or (2) divided into aliquots, different concentrations of nitrate being added to each, and the nitrate of each measured after a carefully timed incubation period (Eppley et al. 1969; Carpenter and Guillard 1970). Agitation should be provided during the incubation periods to minimize the depletion of nitrate in the "unstirred layer" about each cell (Pasciak and Gavis 1974). It is conveninent to carry out the nitrate uptake measurements in the same incubator, with the same light and temperature conditions that prevailed during growth of the cultures. Of course, the light and temperature dependence of nitrate uptake can be the subject of special studies.

A volume of 25 ml is taken for nitrate analysis. Filter this as quickly as possible. The filtrates for nitrate analysis can then be set aside, if bacterial contamination is negligible, until the incubation phase of the experiment is completed and all the samples are ready to be passed through the columns. A few 25-ml volumes of distilled water should be passed through the filters and set aside for nitrate analysis in order to assess the quantity of nitrate washed out of the filters. This is usually quite small, but not negligible. Prewashing the filters, by passing some distilled water through them before filtering the experimental aliquots, may be required if the nitrate blank is either appreciable or variable.

B. Measurement of nitrate

1. Preparing the nitrate reaction columns for use. Sticks of metallic cadmium are commercially available. Coarse filings, retained on a sieve with 0.5-mm openings, but passing a 2-mm sieve, are prepared using a coarse wood rasp. Stir several grams of the filings in a 2% w/v solution of copper sulfate ($CuSO_4 \cdot 5\ H_2O$) for several minutes. Prepare small plugs of fine copper turnings for the bottom and top of the columns. Insert the bottom plug into the column and fill the column with dilute NH_4Cl solution. Now slowly pour in sufficient cadmium-copper mixture to nearly fill the column, tapping the column at each addition for proper packing, and insert the upper copper plug. Avoid trapping air in the columns (although this is easily corrected by removing the upper plug and backwashing with dilute NH_4Cl solution). The void volume of the column should be < 10 ml. When not in use, the columns should be kept filled with the dilute NH_4Cl solution.

The columns can be regenerated by emptying the cadmium-copper into a beaker, washing it with 5% v/v HCl solution, then with water until the pH exceeds 5, and replating the cadmium with copper sulfate solution. This is usually required after about 100 determinations.

2. Determination of nitrate. Add 0.5 ml of the conc. NH_4Cl solution to each 25-ml sample. With the 10-ml automatic pipette, well rinsed with distilled water, place 10 ml of sample for nitrate analysis on the column. Allow this to pass through the column, collecting it in a test tube, and discard it as a column rinse. Pipette a second 10-ml aliquot onto the column, allow it to pass through the column into a clean test tube and save this sample for nitrate analysis. Although there is little "memory" effect associated with these columns, it is good practice to pass a 10-ml volume of the dilute NH_4Cl solution through the column as an additional rinse between samples if the first sample contains a nitrate concentration much higher than the second (e.g., 10 μM followed by 1 μM).

Color development of the nitrite formed from the reduction of nitrate on passage through the column is accomplished by adding 0.2 ml sulfanilamide solution, shaking the contents of the tube to ensure complete mixing, then adding 0.2 ml NEDA solution and mixing again. Color development is complete within 10 min. Determine the absorbance of the sample at 543 nm with the spectrophotometer. For nitrate concentrations less than about 2.5 μM, 10-cm absorption cells will be required. For nitrate concentrations > 25 μM, use 0.5-cm absorption cells or dilute the samples with nitrite-free distilled water. Correct the observed absorbance for the reagent blank determined by treating 25 ml distilled water exactly as the samples; add conc. NH_4Cl solution, pass two 10-ml aliquots through the column, saving the second, add color reagents, etc. Calculate the nitrate concentration as follows: micromoles of NO_3^- per liter = corrected absorbance $\times$ F (see III.B.4).

It is well to note that any nitrite in the samples will be measured as nitrate. This is readily checked by adding 0.2 ml sulfanilamide and 0.2 ml NEDA to 10 ml of the filtered sample and determining the absorbance at 543 nm. The latter value, corrected for the dilution of nitrate samples with the NH_4Cl (i.e., multiplied by 10/10.2), should be subtracted from the result above.

3. Reagent blanks. The blank is significant when 10-cm cells are used and should also be checked with 1-cm cells. Nitrate and nitrite in the distilled water are the usual source of the blank values. Ordinary distilled water has been satisfactory in our experience. Strickland and Parsons (1972) suggest distilling the water from a little alkaline permanganate, rejecting the first few milliliters of distillate, if blanks are high.

4. Standardization. Dissolve 1.02 g reagent grade KNO_3 in 1 l of distilled water; 1 ml of this contains 10.0 μmol NO_3^-. Store this primary standard in the refrigerator. Working standards are prepared by diluting this with distilled water or synthetic seawater, depending on whether the study concerns freshwater or marine algae. There is a small salt effect with this method. Discard the working standards each day. Working standards should cover the range of nitrate values analyzed, as standard curves may not be precisely linear. Add 0.5 ml conc. NH_4Cl to 25-ml volumes of the standard solutions, pass them through the columns, and add color reagents exactly as with unknown samples. Compute the value of F for each standard concentration (F = standard concentration/absorbance at 543 nm). If the value of F varies with the nitrate concentration, prepare a graph of F vs. absorbance in order that appropriate values of F may be interpolated. With 1-cm absorption cells F is usually about 24 for nitrate = 20 μM, and may fall to as low as 21.5 for nitrate = 1.0 μM. F values with 10-cm cells are one-tenth of these.

5. Synthetic seawater. Dissolve 310 g reagent grade NaCl, 10 g reagent grade $MgSO_4 \cdot 7H_2O$, and 0.5 g $NaHCO_3$ in 10 l of distilled water.

IV. Capabilities

The range of the nitrate method is 0.05–45 μM nitrate. At a concentration of 20 μM the precision is $\pm 0.05/n^{1/2}$, with 1-cm absorbance cells, where n is the number of replicates. At a concentration of 1 μM the precision is $\pm 0.05/n^{1/2}$ using 10-cm cells. The limit of detection is about 0.01 absorbance units with a 10-cm cell, or about 0.02 μM.

Nitrate uptake rates can be conveniently measured for ambient nitrate concentrations in the range 1–45 μM. The upper end of the range is easily extended by diluting samples with distilled water.

The 95% confidence limits for kinetic parameters, the maximum uptake rate and half-saturation constant, are usually of the order $\pm 10\%$ of the mean for the former and ± 10–20% of the mean for the latter, although even less precision is common if the half-saturation constant is $\geqslant 1$ μM.

V. Limitations and troubleshooting

With many algae ammonium in the medium inhibits nitrate assimilation at concentrations as low as 1 μM ammonium. It is recommended that a method for measuring ammonium (Solórzano 1969; see Strickland and Parsons 1972) be developed and kept at hand for checking ammonium levels during these experiments. Other organic N sources assimilated by the algae in use may produce similar, although less spectacular, effects.

Solutions other than dilute NH_4Cl have been proposed for rinsing the nitrate columns. Dilute EDTA solutions have not worked here. The efficacy of the nitrite color reagents sulfanilamide and NEDA is readily checked with a standard $NaNO_2$ solution. But it is the nitrate columns that have usually given trouble to the inexperienced, although the copper-treated cadmium method is nearly foolproof compared with the cadmium-mercury amalgam used earlier.

The organisms may lose the ability to assimilate nitrate if deprived too long of a nitrogen source as well as if ammonium is present. The nitrate uptake ability can usually be restored by adding 2–5 μM nitrate to the culture. There should, of course, be a quantity of cells available in the culture which, when the cells are in fact assimilating nitrate, will result in a measurable change in the added nitrate concentration. The recommended nitrate levels, 20–50 μM, present in the culture initially will result in that same quantity of cell nitrogen when the nitrate is depleted and the experiment commenced. Rates of nitrate uptake can then be expected of the order 0.01–0.1 μmol per micromole of cell N per hour.

VI. Acknowledgments

J. L. Coatsworth, J. N. Rogers, and E. H. Renger provided technical assistance during the development of the method and its application to marine phytoplankton. The small-volume nitrate columns (Fig. 39-1) and accompanying technology were E. H. Renger's invention. Several former colleagues have used the method and improved it. These include J. J. McCarthy, M. J. Perry, and W. G. Harrison.

VII. References

Bliss, C. I., and James, A. T. 1966. Fitting the rectangular hyperbola. *Biometrics* 22, 573–602.

Caperon, J., and Meyer, J. 1972. Nitrogen-limited growth of marine phytoplankton. II. Uptake kinetics and their role in nutrient-limited growth of phytoplankton. *Deep-Sea Res.* 19, 619–32.

Carpenter, E. J., and Guillard, R. R. L. 1970. Interspecific differences in nitrate half-saturation constants for three species of marine phytoplankton. *Ecology* 52, 183–5.

Dugdale, R. C., and Goering, J. J. 1967. Uptake of new and regenerated forms of nitrogen in primary productivity. *Limnol. Oceanogr.* 12, 196–206.

Eppley, R. W., and Coatsworth, J. L. 1968. Uptake of nitrate and nitrite by *Ditylum brightwellii:* Kinetics and mechanisms. *J. Phycol.* 4, 151–6.

Eppley, R. W., Rogers, J. N., and McCarthy, J. J. 1969. Half-saturation constants for uptake of nitrate and ammonium by marine phytoplankton. *Limnol. Oceanogr.* 14, 912–20.

Hanson, K. R., Ling, R., and Havir, E. 1967. A computer program for fitting data to the Michaelis-Menten equation. *Biochem. Biophys. Res. Commun.* 29, 194–7.

Hattori, A. 1962. Light-induced reduction of nitrate, nitrite and hydroxylamine in a blue-green alga, *Anabaena cylindrica. Plant Cell Physiol.* 3, 355–69.

Kilham, S. S. 1975. Kinetics of silicon-limited growth in the freshwater diatom *Asterionella formosa. J. Phycol.* 11, 396–9.

Pasciak, W. J., and Gavis, J. 1974. Transport limitation of nutrient uptake in phytoplankton. *Limnol. Oceanogr.* 19, 881–8.

Solórzano, L. 1969. Determination of ammonia in natural waters by the phenolhypochlorite method. *Limnol Oceanogr.* 14, 799–801.

Strickland, J. D. H., and Parsons, T. R. 1972. *A Practical Handbook of Seawater Analysis.* Bull. Fish. Res. Board Can. 167, 310 pp.

40: Phosphate uptake

F. P. HEALEY

Department of the Environment, Fisheries and Marine Service, Freshwater Institute, 501 University Crescent, Winnipeg, Manitoba R3T 2N6, Canada

CONTENTS

I	Introduction	page 412
II	Materials and test organisms	412
	A. Materials	412
	B. Test organisms	413
III	Methods	413
	A. Preparation of cell suspensions	413
	B. Incubation and sampling	414
	C. Calculations	415
IV	Sample data	415
V	Discussion	416
VI	Acknowledgments	416
VII	References	416

I. Introduction

The purpose of this chapter is to present a simple method for determining the sustained rate of phosphate uptake by unicellular algae. The procedure involves the measurement of the net rate of loss of dissolved orthophosphate from the medium in which the experimental population is suspended. It does not give information about the metabolism of P subsequent to uptake, exchange or release of phosphate, or short-term kinetics such as initial adsorption of phosphate prior to uptake. However, it should be useful for testing the effect of various environmental factors on the sustained (0.5 to several hours) rate of phosphate uptake by laboratory populations.

This method has the advantages of being rapid, simple, inexpensive, and applicable to most laboratory and some denser natural populations. However, it is not useful where phosphate uptake is low (below 0.3 μmol l^{-1} h^{-1}) because of either low populations or low intrinsic rates (as in P-sufficient cells), and is of limited usefulness at low phosphate concentrations (< 1 μM). In these situations, use of radioactive P (^{32}P or ^{33}P) is required. The same basic procedure as described here can be used in measuring the uptake of labeled P, with a counting procedure replacing the colorimetric measurement of phosphate. The use of radioactive P will be considered briefly in the discussion.

II. Materials and test organisms

A. Materials

1. Constant-temperature bath. Because metabolic rates are temperature-dependent, some way of maintaining a constant temperature is essential. An aquarium with an inexpensive circulating thermoregulator is adequate for measurements above room temperature. Below room temperature, a unit circulating coolant through coils, preferably attached to the thermoregulator, is required. I have used a KR-70 cooler with a Haake model E51 thermoregulator (both PolyScience). An all glass or plastic aquarium can be lighted from any direction. I have used fluorescent lights on two sides of a glass aquarium.

2. Filtration equipment. Either a small vacuum pump or an aspirator on a water line can be used. The pump allows better control of the vacuum. For most algae either membrane or glass fiber filters can be used. Some of the smallest algae will pass through glass fiber filters, but where retention is adequate, these have the advantage of more rapid filtration.

3. Glassware. (a) Filtration flask or bell jar and filter holder. (b) Test tubes: 18 × 150 mm (should be soaked in 5% v/v HCl, rinsed, and dried before use); 29 × 300 mm (some darkened with black electrical tape if rates in darkness are to be measured). (c) Pipettes: 1.0 ml, 10 ml.

4. Chemicals. (a) Stock phosphate solution (100 mM). Dissolve 1.361 g KH_2PO_4 in distilled water to a total volume of 100 ml solution. Dilute as required with distilled water to obtain substrate and standard solutions. (b) Ammonium molybdate solution. Dissolve 3.0 g $(NH_4)_6Mo_7O_{24} \cdot 4H_2O$ in 100 ml distilled water. Store in darkness. (c) Dilute sulfuric acid. Add 28.0 ml conc. H_2SO_4 to 180 ml distilled water. (d) Antimony potassium tartrate solution. Dissolve 0.068 g $KSbO \cdot C_4H_4O_6$ in 50 ml distilled water. Use within 4 months. (e) Ascorbic acid solution. Make immediately before use by dissolving 0.54 g ascorbic acid in 10.0 ml distilled water. (f) Mixed reagent. Use within 6 h of preparation. Combine ammonium molybdate solution (6.0 ml), dilute sulfuric acid solution (15.0 ml), ascorbic acid solution (6.0 ml), and antimony potassium tartrate solution (3.0 ml). The reagents and procedure for phosphate measurement are based on the directions given by Strickland and Parsons (1965).

5. Miscellaneous. (a) Vortex mixer. (b) Spectrophotometer or colorimeter. (c) Air pump (or source of compressed air).

B. Test organisms

This procedure has been used successfully with several planktonic microalgae, and probably can be used with any alga that forms a uniform suspension.

III. Methods

A. Preparation of cell suspensions

The purpose of this step is to obtain the desired cell concentration in the desired medium, preferably initially free of phosphate. When these conditions are met in the experimental population, it can be used directly. If uptake from a standard medium is to be measured, the algae can be harvested by centrifugation and resuspended in the desired medium. One or more washes in the desired medium should be included if some component in the growth medium must be removed. Larger algae or net plankton from a natural community can be caught and washed with fresh medium or filtered lake water before resuspension in the final medium. For laboratory cultures, resuspension in the growth medium lacking phosphate is suggested, whereas natural net plankton would usually be resuspended in filtered (0.45 μm pore size membrane) lake water. A cell concentration of about 5 mg organic dry weight per liter (150 000 cells ml^{-1}) of P-deficient algae should give an easily measurable rate; up to 10 times this concentration of P-sufficient algae may be required. These figures should be used only as a rough guide to be revised in accordance with initial results.

B. *Incubation and sampling*

1. Incubate 50-ml samples of the cell suspension in 29 × 300-mm test tubes held vertically in a water bath. A slow stream of air bubbles introduced by tubing extending to the bottom of the tube ensures rapid mixing of additions and homogeneous cell suspensions.

2. After the cell suspensions have come to the desired temperature (usually the temperature at which the algae were grown or from which they were collected), add 0.5 ml of 1.0 mM KH$_2$PO$_4$ to obtain a final concentration of 10 μM.

3. Sample 1 min after addition and at 5- to 10-min intervals by pipetting 10-ml samples. These can be treated in either of two ways.

a. Filter and collect the filtrate in acid-cleaned 18 × 150-mm test tubes. Either immediately or when convenient, add 1.0 ml of mixed reagent and immediately mix well on a vortex mixer.

b. Pipette the sample into acid-cleaned 18 × 150-mm test tubes, add 1.0 ml of mixed reagent, and immediately mix well on a vortex mixer. If only the rate of phosphorus uptake is desired, the algae need not be removed before measuring the absorbancies of the solutions, but the tubes should be vortexed just before measurement so that the algae are evenly suspended to provide a constant background. If the absolute phosphate concentration is needed, the particulate material must be filtered out before the absorbancies of the solutions are measured. For each alga used, this procedure should be checked against the preceding one to ensure that treatment with the acid does not result in the release of phosphate from the cells.

In preliminary work it is often advantageous to assay for phosphate immediately. Color development reaches a maximum within 5 min and is stable for at least 2 h. Visual inspection of the samples indicates whether subsequent samples should be taken at longer or shorter time intervals.

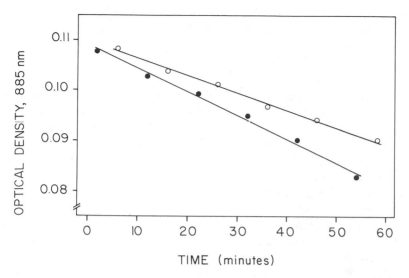

Fig. 40-1. Phosphate uptake by *Scenedesmus quadricauda* when P-sufficient (white circles, 38 mg dry weight per liter) and when P-deficient (black circles, 4.5 mg dry weight per liter).

C. Calculations

Read absorbancies of the samples and standards (0,2,5,10 μM) at 885 nm in a spectrophotometer. Plot the absorbancies against time. Calculate the change in absorbance per unit time and convert this to a decrease in phosphate concentration using a standard curve relating absorbance to phosphate concentration. Rate measurements should be accompanied by one or more measurements of biomass so that the rate can be expressed per unit of biomass. Where comparisons are to be made between species, this should include rate per unit carbon or dry weight.

IV. Sample data

Figure 40-1 compares phosphate uptake by *Scenedesmus quadricauda* from P-sufficient and P-deficient cultures. Both cultures were grown in medium WC (Guillard and Lorenzen 1972) with KH_2PO_4 replaced by 10 μM KH_2PO_4 for the low-P culture and 20 μM KH_2PO_4 for the high-P culture. Both cultures were harvested by centrifugation 1 day after the low-P culture had depleted its phosphate, and were resuspended in medium WC, with KH_2PO_4 replaced with equimolar KCl, at pH 7.5. Phosphate was added and samples removed, filtered, and assayed for phosphate as above. The P-deficient algae had a phosphate uptake rate of 0.30 μmol h^{-1} mg^{-1} dry weight; the P-sufficient suspension showed an uptake rate of 0.026 μmol h^{-1} mg^{-1} dry weight.

V. Discussion

By introducing variations in the growth of the culture, the medium composition, or the incubation conditions, this procedure permits the measurement of the effects of several variables on phosphate uptake by algae. In planning experiments, the considerably higher rates of phosphate uptake by P-deficient cells as compared with P-sufficient ones should be kept in mind (Healey 1973). Where low rates are encountered, such as with low cell or substrate concentrations, use of radioactive P provides greater sensitivity. Uptake of ^{33}P followed by autoradiography can be used to associate phosphate uptake with particular cells in mixed populations (Fuhs and Canelli 1970). Labeled substrate is also needed where short-term reactions, such as initial binding (Medveczky and Rosenberg 1971) or identification of subsequent metabolic products (Ullrich 1972), is needed. Use of radioactive P involves the addition to the cell suspension of ^{32}P- or ^{33}P-labeled phosphate alone or with unlabeled phosphate as needed, usually in 1- to 10 μCi amounts. Sampling is by automatic or Cornwall syringe, followed by filtration. Activity in either or both the filter and filtrate is followed using a scintillation counter. Measurement of both filter and filtrate activities and expression of the results as a percentage of the activity taken up with time eliminates errors caused by variation in sample volumes. At some point filter adsorption should be measured. The short half-lives of the P isotopes (14.3 days for ^{32}P and 25 days for ^{33}P) should be kept in mind. Further details on using ^{32}P for measuring phosphate uptake by natural populations are provided by Rigler (1966).

VI. Acknowledgments

I acknowledge with thanks the comments of Miss S. Levine and Dr. D. Planas on the manuscript and particularly on the use of radioactive P.

VII. References

Fuhs, G. W., and Canelli, E. 1970. Phosphorus-33 autoradiography used to measure phosphate uptake by individual algae. *Limnol. Oceanogr.* 15, 962–7.
Guillard, R. R. L., and Lorenzen, C. J. 1972. Yellow-green algae with chlorophyllide *c*. *J. Phycol.* 8, 10–14.
Healey, F. P. 1973. Characteristics of phosphorus deficiency in *Anabaena*. *J. Phycol.* 9, 383–94.

Medveczky, N., and Rosenberg, H. 1971. Phosphate transport in *Escherichia coli. Biochim. Biophys. Acta* 241, 494–506.

Rigler, F. 1966. Radiobiological analysis of inorganic phosphorus in lake water. *Int. Assoc. Theor. Appl. Limnol. Proc.* 16, 465–70.

Strickland, J. D. H., and Parsons, T. R. 1965. *A Manual of Sea Water Analysis.* Bull. Fish. Res. Board Can. 125, 203 pp.

Ullrich, W. R. 1972. Der Einfluss von CO_2 und pH auf die ^{32}P-Markierung von Polyphosphaten und organischen Phosphaten bei *Ankistrodesmus braunii* im Licht. *Planta* 102, 37–54.

41: Sulfate uptake

MONICA LIK-SHING TSANG

*Institute for Photobiology of Cells and Organelles Brandeis University,
Waltham, Massachusetts 02154*

ROBERT C. HODSON

*Department of Biology, University of Delaware,
Newark, Delaware 19711.*

JEROME A. SCHIFF

*Institute for Photobiology of Cells and Organelles,
Brandeis University, Waltham, Massachusetts 02154*

CONTENTS

I	Introduction	*page* **420**
II	Materials and organisms	**420**
III	Methods	**420**
	A. Sulfate uptake by *Chlorella pyrenoidosa*	421
	B. Sulfate uptake by *Chondrus crispus*	421
IV	Sample data	**422**
V	Discussion	**422**
VI	Acknowledgment	**424**
VII	References	**424**

I. Introduction

Sulfate is an important macronutrient for all algae and is used to form the ester sulfates of polysaccharides such as agar, carrageenan, and fucoidan; for partial reduction to the level of the sulfolipid and other sulfonic acids; and for complete reduction to form the thiol groups of the protein amino acids cysteine and methionine as well as other important molecules such as coenzyme A, thiamine, and biotin. Sulfate uptake is the first step in sulfate utilization and is an active process with a K_m for sulfate of about $10^{-5}-10^{-4}$ M (Schiff and Hodson 1973; Schmidt et al. 1974; Tsang and Schiff 1976).

II. Materials and organisms

Carrier-free ^{35}S-sulfuric acid (New England Nuclear). Equipment used included Sorvall RC-2B refrigerated automatic centrifuge; Thrombocytocrit, Van Allen (Arthur H. Thomas); Metabolyte Water Bath Shaker (New Brunswick Scientific); Nuclear-Chicago gas-flow counter (Searle Inc.); and Beckman Scintillation Spectrometer. Etched-glass planchets (Belton's); plastic planchet holders (Unipec Inc.).

Test organisms used included *Chlorella pyrenoidosa* chick Emerson strain 3 and *Chondrus crispus*.

III. Methods

No single detailed method can be given that will suit all organisms. The basic procedure, however, is quite uniform; to provide examples, the methods used for one unicellular alga (*Chlorella*) (Wedding and Black 1960; Hodson et al. 1968; Vallee and Jeanjean 1968a, 1968b) and for one multicellular alga, *Chondrus* (Loewus et al. 1971) will be given. For the details applicable to other organisms, the reader is referred to the original papers dealing with sulfate uptake in bacteria (Dreyfuss and Pardee 1966; Jones-Mortimer 1968; Kredich 1971); fungi (Metzenberg and Parson 1966; Yamamoto and Segel 1966;

Bradfield et al. 1970); *Porphyridium* (Ramus and Groves 1972); *Fucus* (Quatrano and Crayton 1973); *Euglena* (Goodman and Schiff 1964); higher plants (Nissen 1971; Ferrari and Renosto 1972); and cell culture suspensions from higher plants (Hart and Filner 1969; Smith 1975).

A. Sulfate uptake by Chlorella pyrenoidosa

Sulfate uptake by this unicellular green alga was studied according to the method described by Hodson et al. (1968). Cultures in the exponential phase of growth were harvested by centrifugation for 15 min at $2000 \times g$. The pellets were resuspended in fresh sterile sulfur-free medium, combined, and centrifuged for 15 min at $6000 \times g$. This pellet was resuspended in S-free medium to give a cell suspension of about 10% (v/v). Cell volume was determined by centrifugation in a Van Allen Thrombocytocrit tube at 3300 rpm (20-cm radius) for 1 h (Hodson et al. 1968). The concentrated cell suspension was dispensed in small volumes (10 ml) into 50-ml Erlenmeyer flasks and was equilibrated for 1 h with shaking at 26 C under normal laboratory lighting conditions. After equilibration of the cells, solutions containing radioactive sulfate ($1.0 \ \mu\text{mol} \ SO_4^{2-}$, 17.2×10^6 cpm) were added to the cell suspensions, and at various times thereafter samples were removed to conical centrifuge tubes, rapidly chilled in ice, and centrifuged. The clear supernatant fluid was removed and frozen until assayed for radioactivity. All determinations were made in triplicate. Aliquots were dried onto etched-glass planchets and counted in a Nuclear-Chicago counter. A zero-time control served to define the 100% value for radioactivity in the medium. The percent uptake was calculated as the percent of radioactivity that was missing from the medium after various times of incubation.

B. Sulfate uptake by Chondrus crispus

Sulfate uptake by the multicellular red alga *C. crispus* was studied according to the method described by Loewus et al. (1971). *C. crispus* was harvested during August 1970 in the vicinity of Nobska Point, Cape Cod, Mass., and was held in flowing seawater until used, in no instance over 18 h. Selected fronds free of epiphytes were cut into 0.1-g pieces, rinsed in sterile ASP_{12} medium in which $MgSO_4$ had been replaced with $MgCl_2$, and transferred to a sterile flask containing sterile medium of the same composition supplemented with trace elements in the form of salts other than sulfate. The chemical and radioactive level of sulfate in each flask was determined by appropriate additions of $Na_2{}^{32}SO_4$ and carrier-free $H_2{}^{35}SO_4$. Traces of sulfate present in the medium prior to addition of carrier-free $H_2{}^{35}SO_4$ were estimated at $0.1 \ \mu\text{M}$.

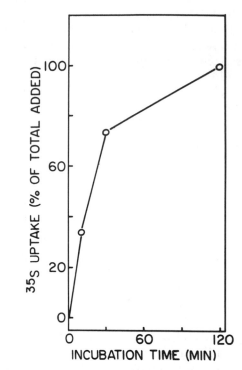

Fig. 41-1. Uptake of radioactive sulfate by *Chlorella pyrenoidosa* chick Emerson strain 3. To 10 ml of a 10% (v/v) suspension of cells 1.0 μmol (17.2 × 10^6 cpm) $^{35}SO_4^{2-}$ was added.

In a typical experiment, 10 pieces of seaweed were suspended in 20 ml of medium containing $^{35}SO_4^{2-}$ (5 μCi) in a 50-ml Erlenmeyer flask. The flask was shaken in a thermostated gyratory shaker (18 C) under cool-white fluorescent light (500 ft-c). At intervals, aliquots of medium were removed and counted in Bray's solution in a liquid scintillation spectrometer. Sulfate uptake was calculated from disappearance of radioactivity from the medium, as in the case of *Chlorella* (III.A).

IV. Sample data

Figure 41-1 shows a typical sulfate uptake curve for *Chlorella* measured as described in III.A (Hodson et al. 1968). Figure 41-2 shows typical sulfate uptake curves for *Chondrus* at various sulfate concentrations as described in III.B (Loewus et al. 1971).

V. Discussion

If the cells are isolated by centrifugation, the lowest g-forces for the shortest time should be used that will adequately sediment the cells.

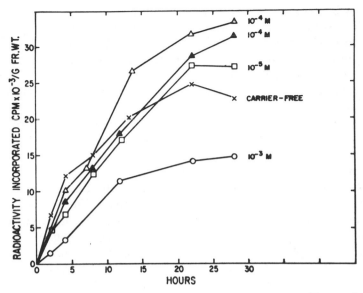

Fig. 41-2. Incorporation of $^{35}SO_4^{2-}$ into *Chondrus crispus* as measured by uptake from the medium. The concentration of sulfate used in each experiment is recorded alongside the appropriate symbol. Data reported at 0.1 mM SO_4^{2-} represent two experiments run 7 days apart.

Some procedures involve centrifugation at room temperature; others require centrifugation in the cold. The proper temperature to be used has to be determined for each organism. It is frequently necessary to allow a recovery period after centrifugation to allow the cells to return to a condition where they will take up sulfate normally. Good indicators of recovery from the centrifugation procedures are (depending on the organism) metaboly, swimming, intracellular movement, etc. For example, cells of *Euglena* that do not recover metaboly or ability to swim after centrifugation generally will not take up sulfate (Goodman and Schiff 1964).

The proper concentration of sulfate to be used depends on the organism as well. Radioactive sulfate is obtained as carrier-free $^{35}SO_4^{2-}$ from the suppliers. Two variables that operate in opposite directions influence the amount of radioactivity taken up by the cells and the rate of uptake. Because the sulfate is carrier-free, every sulfur atom is radioactive. The effective concentration of sulfate, therefore, is extremely low. And because sulfate uptake follows Michaelis-Menten kinetics, this means that the rate of sulfate uptake will be very low. If nonradioactive carrier sulfate is added, however, as the concentration is increased the rate of sulfate uptake will increase until maximum velocity is achieved. However, the more nonradioactive sulfate is added, the more the radioactive sulfate will be diluted, resulting in a lower uptake of radioactivity, the parameter being measured. As these influences act in opposite directions, an optimal concentration of nonradioactive sulfate can be found which gives a good compromise between the rate of uptake and dilution of radioactivity. Frequently this optimum is around 10^{-4} M total sulfate concentration. The simplest means of determining sulfate uptake is to incubate the cells with an appropriate specific activity of sulfate and to measure with time the disappearance of radioactivity from the medium by centrifuging out the cells and counting the supernatant fluid. This eliminates problems of self-absorption, which are encountered with counting the cells themselves, and does not require cell fractionation. It would be appropriate only in cases where no radiosulfur compounds are excreted into the medium by the cells. This is usually the case for algae, but should be checked for each organism. This method works well where uptakes are appreciable, as they usually are when conditions are optimized. However, if uptakes are very low, the large background of radioactivity in the medium may defeat the precise measurement of the sulfate removed by the cells. In these cases, the cells must be recovered and the radioactivity taken up by them measured. This usually involves corrections for self-absorption and/or cell fractionation; the reader is referred to the original references cited above for detailed methods.

VI. Acknowledgment

Supported in part by a grant from the National Science Foundation, 01-PCM-76-21486.

VII. References

Bradfield, G., Somerfield, P., Meyn, T., Holby, M., Babcock, D., Bradley, D., and Segel, I. H. 1970. Regulation of sulfate transport in filamentous fungi. *Plant Physiol.* 46, 720–7.

Dreyfuss, J., and Pardee, A. B. 1966. Regulation of sulfate transport in *Salmonella typhimurium*. *J. Bacteriol.* 91, 2275–80.

Ferrari, G., and Renosto, F. 1972. Regulation of sulfate uptake by excised barley roots in the presence of selenate. *Plant Physiol.* 49, 114–6.

Goodman, N. S., and Schiff, J. A. 1964. Studies of sulfate utilization by algae. 3. Products formed from sulfate by *Euglena. J. Protozool.* 11, 120–7.

Hart, J. W., and Filner, P. 1969. Regulation of sulfate uptake by amino acids in cultured tobacco cells. *Plant Physiol.* 44, 1253–9.

Hodson, R. C., Schiff, J. A., and Scarsella, A. J. 1968. Studies of sulfate utilization by algae. 7. *In vivo* metabolism of thiosulfate by *Chlorella. Plant Physiol.* 43, 570–7.

Jones-Mortimer, M. C. 1968. Positive control of sulfate reduction in *Escherichia coli. Biochem. J.* 110, 597–602.

Kredich, N. M. 1971. Regulation of L-cysteine biosynthesis in *Salmonella typhimurium*. I. Effect of growth on varying sulfur sources and O-acetyl-l-serine. *J. Biol. Chem.* 246, 3474–84.

Loewus, F., Wagner, G., Schiff, J. A., and Weistrop, J. 1971. The incorporation of ^{35}S-labeled sulfate into carrageenan in *Chondrus crispus. Plant Physiol.* 48, 373–5.

Metzenberg, R. L., and Parson, J. W. 1966. Altered repression of some enzymes of sulfur utilization in a temperature-conditional lethal mutant of *Neurospora. Proc. Natl. Acad. Sci. U.S.A.* 55, 629–35.

Nissen, P. 1971. Uptake of sulfate by roots and leaf slices of barley: Mediated by single, multiphasic mechanisms. *Physiol. Plant.* 24, 315–24.

Quatrano, R. S., and Crayton, M. A. 1973. Sulfation of fucoidan in *Fucus* embryos. *Dev. Biol.* 30, 29–41.

Ramus, J., and Groves, S. T. 1972. Incorporation of sulfate into the capsular polysaccharide of the red alga *Porphyridium. J. Cell Biol.* 54, 399–407.

Schiff, J. A., and Hodson, R. C. 1973. The metabolism of sulfate. *Annu. Rev. Plant Physiol.* 24, 381–414.

Schmidt, A., Abrams, W. R., and Schiff, J. A. 1974. Reduction of adenosine 5′-phosphosulfate to cysteine in extracts from *Chlorella* and mutants blocked for sulfate reduction. *Eur. J. Biochem.* 47, 423–34.

Smith, I. K. 1975. Sulfate transport in cultured tobacco cells. *Plant Physiol.* 55, 303–7.

Tsang, M. L.-S., and Schiff, J. A. 1976. Sulfate-reducing pathway in *Escherichia coli* involving bound intermediates. *J. Bacteriol.* 25, 923–33.

Vallée, M., and Jeanjean, R. 1968a. Le système de transport de SO_4^{2-} chez *Chlorella pyrenoidosa* ct sa régulation. I. Etude cinétique dc la perméation. *Biochim. Biophys. Acta* 150, 599–606.

Vallée, M., and Jeanjean, R. 1968b. Le système de transport de SO_4^{2-} chez *Chlorella pyrenoidosa* et sa régulation. II. Recherches sur la régulation de l'entrée. *Biochim. Biophys. Acta* 150, 607–17.

Wedding, R. T., and Black, M. K. 1960. Uptake and metabolism of sulfate by *Chlorella*. I. Sulfate accumulation and active sulfate. *Plant Physiol.* 35, 72–80.

Yamamoto, L. A., and Segel, I. H. 1966. The inorganic sulfate transport system of *Penicillium chrysogenum. Arch. Biochem. Biophys.* 114, 523–38.

42: Uptake of silicic acid

FAROOQ AZAM AND SALLIE W. CHISHOLM*

Institute of Marine Resources, A-018, Scripps Institution of Oceanography, University of California, San Diego, La Jolla, California 92093

CONTENTS

I Introduction — *page* 428
II Materials and test organisms — 428
  A. Radiochemicals — 428
  B. Reagents — 429
  C. Special glassware and plasticware — 429
  D. Equipment — 429
III Methods — 429
  A. Culturing — 429
  B. Harvesting and silicon starvation — 430
  C. Uptake of $Si(OH)_4$ by diatom cultures — 430
  D. Uptake of $Si(OH)_4$ by natural planktonic populations — 431
IV Capabilities and sample data — 432
V Discussion — 432
  A. Problems and limitations — 432
  B. Alternative techniques — 433
VI Acknowledgments — 433
VII References — 433

* Present address: Ralph M. Parsons Laboratory– Building 48, Department of Civil Engineering, Massachusetts Institute of Technology, Cambridge, Massachusetts 02139.

I. Introduction

Studies on silicic acid, Si(OH)$_4$, uptake and metabolism by diatoms have been hampered by the fact that the only available radioisotope of silicon, ^{31}Si, is short-lived (half-life, 156 min). The spectrophotometric method for molybdate-reactive Si(OH)$_4$ (Strickland and Parsons 1972) does not provide the necessary sensitivity for measuring short-term uptake. This is particularly true when working with natural planktonic populations.

Two tracer methods have been developed in recent years. One method (Goering et al. 1973) uses the stable isotopes of silicon, ^{29}Si or ^{30}Si, followed by measurement of the ratio of natural (^{28}Si) to heavy isotope by solid-state mass spectrometry. The second method (Azam 1974; Azam and Chisholm 1976) is based on the use of radioisotopically labeled ^{68}Ge-germanic acid, ^{68}Ge(OH)$_4$, as a tracer for Si(OH)$_4$ uptake. ^{68}Ge(OH)$_4$, which is chemically similar to Si(OH)$_4$, acts as a metabolic analogue of Si(OH)$_4$ in *Nitzschia alba* and *Navicula pelliculosa* (Azam 1974) and in other diatoms examined (Chishom, Azam, and Eppley, unpublished). When ^{68}Ge(OH)$_4$ is added in trace amounts relative to Si(OH)$_4$ to a silicifying diatom culture, the tracer uptake parallels the uptake of molybdate-reactive Si(OH)$_4$ without significant "isotope discrimination." Kinetics of Si(OH)$_4$ uptake measured with ^{68}Ge(OH)$_4$ as a tracer compare favorably with those measured using ^{31}Si(OH)$_4$ (Azam 1974). This method is simple and sensitive, and as it uses a radioactive tracer, any laboratory equipped with a liquid scintillation counter, a well scintillation crystal counter, or a beta counter should be able to use it. ^{68}Ge has a convenient half-life of 282 days.

II. Materials and test organisms

A. Radiochemicals

Carrier-free ^{68}GeCl$_4$ (99.9% radionuclide purity, in 0.5 M HCl solution, New England Nuclear). Neutralize the HCl solution with a slight excess of NaOH, to convert ^{68}GeCl$_4$ to ^{68}Ge(OH)$_4$, and dilute with distilled water to a convenient radiochemical concentration, such as 10

μCi ml^{-1}. Store frozen in a thick lead container. *Caution:* ^{68}GeCl$_4$ is somewhat volatile (b.p. 84 C). Therefore, open the vial only after NaOH has been injected into the solution with a hypodermic needle through the rubber septum. This should be done in a fume hood. Also, take necessary precautions to ensure protection against radiation. ^{68}Ge is a high-energy beta and gamma emitter ($\beta^+ = 1.89$ meV; $\gamma = 1.07$ meV; annihilation radiation 0.51 meV).

B. Reagents

(1) Neutralized Si(OH)$_4$ solution. Make a 1×10^{-3} M stock solution by dissolving 284.2 mg Na$_2$SiO$_3$·9H$_2$O in about 800 ml distilled H$_2$O. Adjust the pH of this solution to 8 with 1 N HCl using a pH meter and make the volume to 1 l. Store in a plastic container. (2) KNO$_3$, analytical reagent grade. (3) H$_2$SO$_4$, conc. analytical reagent. (4) Aquasol liquid scintillation cocktail (New England Nuclear).

C. Special glassware and plasticware

(1) Polypropylene Erlenmeyer flasks, 250 ml. (2) Polypropylene centrifuge tubes, 12 ml, tapered. (3) Pyrex glass centrifuge tubes, 12 ml, tapered. (4) Pipettes, 5 and 10 ml, plastic. (5) Pyrex glass tube, 20 ml. (6) Pyrex glass Erlenmeyer flasks. (7) Scintillation vials, glass or plastic.

D. Equipment

(1) Boiling-water bath. (2) Magnetic stirrer with a Teflon-coated magnetic stir bar. (3) Filtration equipment, a side-arm filter flask (1–4 l) and a Millipore or Gelman filter base and funnel for 25-mm diam. filters. (4) A vacuum pump, an aspirator or laboratory vacuum line. (5) Membrane filters, 25-mm diam., 0.8 μm pore size (Millipore or Nuclepore). (6) Clinical centrifuge. (7) Liquid scintillation or Geiger end-window counter.

III. Methods

A. Culturing

Culture media and conditions will vary for different diatoms. *Nitzschia alba* grows well in a synthetic seawater medium, SSW (Hemmingsen 1971), with 0.2% D-glucose as the carbon and energy source, at 30 C on a gyratory shaker. *Navicula pelliculosa* is cultured in a synthetic salts medium (Darley and Volcani 1971) at 20–22 C on a reciprocal shaker under continuous illumination of 5 klux from cool-white and warm-white fluorescent lamps (Sylvania).

B. Harvesting and silicon starvation

Late-exponential-phase *N. alba* cultures are harvested by centrifugation at $1600 \times g$ for 1 min in a clinical centrifuge. The cells are washed by gently resuspending the pellet in SSW containing no $Si(OH)_4$ and centrifuging again for 1 min at $1600 \times g$. The washed cell pellet is resuspended in $Si(OH)_4$-free SSW in a polypropylene flask and stirred with a magnetic stirrer at 30 C for 2 h. If Si starvation is not required, the stirring step is omitted.

C. Uptake of Si(OH)₄ by diatom cultures

1. Incubation. A cell suspension is placed in a polypropylene Erlenmeyer flask under the desired temperature and light regimen. Cell density should not be so great as to cause light or oxygen limitation. For *N. alba* a cell density of 1×10^6 ml^{-1} may be used. To initiate $Si(OH)_4$ uptake, place in a polypropylene tube an amount of neutralized $Si(OH)_4$ solution needed to achieve the desired final $Si(OH)_4$ concentration in the incubation mixture and add to it 0.2 μCi $^{68}Ge(OH)_4$. Add this mixture to the cell suspension. To determine the time course of uptake, withdraw 1- to 5-ml aliquots periodically.

2. Termination of incubation. Place the sample in a 12-ml polypropylene centrifuge tube and centrifuge at $1600 \times g$ for 1 min. Decant the supernatant. Wash the pellet twice by resuspension in 10 ml ice-cold Si-free SSW.

   When short-term uptake is to be measured, rapid filtration is necessary. Millipore membrane filters with a pore size of 0.8 μm are suitable. Apply gentle suction to facilitate filtration ($\Delta P = 10$ cm Hg). Wash the cell cake with two aliquots of 2 ml ice-cold Si-free SSW.

3. Radioassay. Add 10 ml Aquasol and measure radioactivity in a liquid scintillation counter. Wait at least 12 h before radioassaying to allow establishment of radiochemical equilibrium between ^{68}Ge and ^{68}Ga.

   ^{68}Ge decays according to the following scheme:

$$^{68}\text{Ge} \xrightarrow[\substack{(t_{\frac{1}{2}} = 282 \text{ days})}]{\text{electron capture}} {}^{68}\text{Ga} \xrightarrow[\substack{(t_{\frac{1}{2}} = 68 \text{ min})}]{\alpha, \beta^+} {}^{68}\text{Zn}$$

(β^+ results in annihilation radiation of 0.51 meV.)

   If available, a liquid scintillation counter should be used. It will measure β^+ with very high efficiency, and γ with low efficiency. The total counting efficiency will be greater than 100%. However, a Geiger end-window counter is quite adequate if a liquid scintillation counter is not available. If samples are highly quenched, a well scintillation crystal counter may be used, which will measure γ radiation and annihilation radiation.

4. *Adsorption blank.* Use incubation conditions identical to those for uptake (III.C), but add 1% final concentration of formaldehyde to the cell suspension before adding $Si(OH)_4 + {}^{68}Ge(OH)_4$ mixture.

5. *Incorporation of Si(OH)₄.* Incorporation of silica into diatom frustules can be determined after removing the soluble silicon pool within the cells by acid digestion. After incubation with $Si(OH)_4 + {}^{68}Ge(OH)_4$, mixture, and termination of incubation (III.C.2), 2 ml conc. H_2SO_4 is added to the cell pellet/filter cake in 20-ml Pyrex glass test tubes. The tubes are placed in a boiling-water bath for 5 min. The contents of the tubes will become black at this point. Add about 50 mg KNO_3 crystals and return the tubes to the boiling-water bath for 5 min. The contents should have a pale yellow color. If the color is still dark, return to the water bath for a few minutes. Remove the tubes from the water bottle and allow them to cool to the room temperature. Then pour the contents of each tube into a 125-ml Erlenmeyer flask containing about 50 ml distilled water. *Caution:* do not add water to the test tubes, except to rinse them after most of the mixture has been transferred to the flask. Filter the contents of the flask on a 0.8-μm pore-size filter. Wash the filter cake with three aliquots of about 10 ml distilled water. Transfer the filter to a scintillation vial, add 10 ml Aquasol, and assay for radioactivity. Acid digestion should be done in a fume hood. Formaldehyde-fixed blanks (III.C.4) are taken through the same procedure and their radioactivity subtracted from the samples.

6. *Method of calculation.* The rate of $Si(OH)_4$ uptake, ρ, is computed by the following equation: $\rho = (f \times Si_0)/(t \times n)$, where f is cpm ${}^{68}Ge$ taken up/cpm ${}^{68}Ge$ added; Si_0 is $Si(OH)_4$ concentration at the beginning of the incubation (moles per liter); t is time of incubation in hours; n is number of cells per sample; ρ has the dimensions of moles per hour per cell.

D. *Uptake of Si(OH)₄ by natural planktonic populations*

1. *Procedure.* Seawater (or lake water) is collected in clean polyvinyl-chloride or other plastic samplers. Place 100-ml aliquots into 125-ml plastic bottles. Add 1 μCi ${}^{68}Ge(OH)_4$ and incubate under the desired light and temperature regimen. Terminate incubation by filtering the particulate matter on a 0.8-μm pore size filter. Wash twice with filtered seawater. Prepare formaldehyde-fixed blanks by adding formaldehyde to 1% final concentration before adding ${}^{68}Ge(OH)_4$. From here on, treat the samples as described under III.C.5. A 50-ml sample of seawater is filtered through a 0.8-μm pore-size filter and the filtrate assayed for Si(OH)$_4$ by the method of Strickland and Parsons (1972).

2. *Calculations.* The rate of Si(OH)$_4$ incorporation, ρ, is computed by the following equation: $\rho = (f \times Si_0)/t$. The terms of the equation are defined under III.C.6, but ρ has the units of moles per liter per hour. It can be converted to the specific rate, v, (per hour) by dividing by the concentration of particulate silicon (silicified organisms) in the sample volume. An approximation would be to use total particulate silicon concentration (moles per liter) in the sample for this purpose (Azam and Chisholm 1976).

IV. Capabilities and sample data

The method is useful in measuring Si(OH)$_4$ uptake as low as 0.5 nmol, this limit being set by the adsorption blank. In determination of kinetic parameters K_m and V_{max} of a diatom culture the sensitivity of the method should not be limiting even for short-term "initial uptake rate" measurements. For example, in a typical experiment (Azam 1974, Figs. 3 and 4), Si(OH)$_4$ uptake by *N. alba* was measured at Si(OH)$_4$ concentration range of 0.89–89 μM; 0.1 μCi ^{68}Ge(OH)$_4$ was added to each flask. ^{68}Ge taken up in the first minute ranged from 6×10^2 cpm at 89 μM Si(OH)$_4$ to 1.2×10^4 cpm at 0.89 μM Si(OH)$_4$. This represented uptake rates in the range of about 1–10.5 μmol g^{-1} (wet weight) min^{-1}. If it is necessary to use a smaller amount of cells per sample, the concentration of ^{68}Ge(OH)$_4$ may be increased to increase the sensitivity.

When measuring Si(OH)$_4$ incorporation by natural planktonic populations, where the diatom density may be quite low, larger samples and longer incubation periods are necessary. Thus, in a typical experiment (Chisholm et al., 1977) designed to follow diel rhythms in Si(OH)$_4$ uptake by a natural planktonic population collected 3 miles off La Jolla, California, 50 μCi ^{68}Ge(OH)$_4$ was added to a 2.5-l seawater sample and 50-ml subsamples were needed to determine uptake and incorporation rates at 2-h intervals. The rates varied from 5 to 54 nmol h^{-1}l^{-1}.

V. Discussion

A. Problems and limitations

1. *Variability of the blank.* With pure cultures this is not a problem. However, with natural planktonic populations the blank varies unpredictably, in that ^{68}Ge(OH)$_4$ adsorption is not a simple function of Si(OH)$_4$ concentration. In a study of Si(OH)$_4$ incorporation by natural populations (Azam and Chisholm 1976) the average adsorption blank

was 0.08% of the ^{68}Ge added to the samples (range was 0.01–2.1% in different experiments), representing 2.1–28.8% of the total tracer uptake in different experiments. For 11 pairs of replicates, the average standard error of the blank was ±13.9% with a range of 5.4–33%.

2. *Validity of ^{68}Ge(OH)$_4$ as a tracer.* There is no a priori reason to believe that ^{68}Ge(OH)$_4$ will be a tracer of Si(OH)$_4$ in all diatom species. Validity of its use as a tracer is based mainly on the work with *Nitzschia alba* and *Navicula pelliculosa* (Azam 1974) and is supported by the recent work of Chisholm et al. (in press) on *Thalassiosira pseudonana* (FCRG-66A), *Thalassiosira fluviatilis* (FCRG-T Fluv.), *Chaetoceros gracilis* (FCRG-TC), *Ditylum brightwellii* (FCRG-64), *Lithodesmium undulatum* (FCRG-25), and *Skeletonema costatum* (FCRG-Skele.). Before using the method for any other species, it would be necessary to establish that ^{68}Ge(OH)$_4$ uptake does in fact follow Si(OH)$_4$ uptake. With pure cultures this can be done by measuring the disappearance from the medium of molybdate-reactive silicon and of ^{68}Ge(OH)$_4$ added in trace amounts, using dense suspensions of silicifying cells.

B. Alternative techniques

The radiotracer ^{31}Si(OH)$_4$ has an inconvenient half-life of 156 min. However, for laboratories located close to a nuclear reactor facility, it is a feasible method for short-term experiments (Azam et al. 1974).

The heavy isotope method of Goering et al. (1973) is excellent and reliable. If the necessary equipment is available it is a good alternative.

VI. Acknowledgments

We thank Dr. Richard W. Eppley for encouragement and valuable discussions during the development of this method. We are grateful to Dr. Theodore Enns of Scripps Institution of Oceanography for valuable discussion. We thank Ms. Terry Boyce Tracey for typing the manuscript.

This work was supported by ERDA Contract E(11-1)GEN 10, PA 20.

VII. References

Azam, F. 1974. Silicic acid uptake in diatoms studied with [^{68}Ge]germanic acid as tracer. *Planta* 121, 205–12.

Azam, F., and Chisholm, S. W. 1976. Silicic acid uptake and incorporation by natural marine phytoplankton populations. *Limnol. Oceanogr.* 21, 427–35.

Azam, F., Hemmingsen, B. B., and Volcani, B. E. 1974. Role of silicon in diatom metabolism. V. Silicic acid transport and metabolism in the heterotrophic diatom *Nitzschia alba*. *Arch. Microbiol.* 97, 103–14.

Chisholm, S. W., Azam, F., and Eppley, R. W. 1977. Silicic acid incorporation in marine diatoms on light/dark cycles: Use as an assay for phased cell division. *Limnol. Oceanogr.* (submitted).

Darley, W. M., and Volcani, B. E. 1971. Synchronized cultures: Diatoms. *Methods Enzymol.* 23A, 85–96.

Goering, J. J., Nelson, D. M., and Carter, J. A. 1973. Silicic acid uptake by natural populations of marine phytoplankton. *Deep-Sea Res.* 20, 777–89.

Hemmingsen, B. B. 1971. A mono-silicic acid stimulated adenosinetriphosphatase from protoplasts of the apochlorotic diatom *Nitzschia alba*. Ph.D. dissertation, University of California, San Diego. 132 pp.

Strickland, J. D. H., and Parsons, T. R. 1972. *A Practical Handbook of Seawater Analysis*. Bull. Fish. Res. Board Can. 167, 310 pp.

43: Uptake of micronutrients and toxic metals

ANTHONY G. DAVIES

Marine Biological Association, Plymouth PL1 2PB, England

CONTENTS

I Introduction   *page* **436**
II Materials   **436**
    A. Glassware   436
    B. Culture media   437
    C. Trace components and
        radioactive labeling   439
    D. Culture apparatus   441
    E. Other equipment   441
III Methods   **441**
    A. Test organism   441
    B. Procedure for studying the
        uptake of mercury and its effect
        on growth   441
IV Sample data   **443**
V Discussion   **445**
VI References   **445**

I. Introduction

Batch cultures provide by far the most straightforward method for studying how the growth rate of an alga is related to the quantity of a rate-limiting trace component – vitamin or metal – taken up by the cells. The basic experimental data are obtained by setting up a group of test cultures that contain a range of concentrations of the trace component under examination but that are, otherwise, initially identical. Samples are then periodically removed from each culture to follow changes in the algal population, to measure the total concentration of the trace component present in the culture (medium plus cells), and to determine the cellular content (or "cell quota," as it is often termed) of the trace component by separating some of the algae from the culture medium.

Because the preparation of the culture medium depends on the nature of the trace component being studied, it is not possible to describe a general all-purpose technique. Thus, a specific example – the effect of mercuric ions on the growth of *Isochrysis galbana* Parke – is dealt with in some detail. Variations in the method necessary for working with different types of trace components are also discussed in general.

The techniques described are based on experience gained with unicellular organisms and would need certain modifications for the study of multicellular algae.

II. Materials

A. Glassware

Borosilicate glassware should be used throughout.

1. General items. A wide range of glassware is required including large-volume (10 l) containers for storing basic culture medium or filtered seawater, small containers for stock solutions, measuring flasks and cylinders, assorted pipettes, test tubes, vials for liquid scintillation counting where appropriate, a 25-mm diam. membrane filtration apparatus, and an apparatus for filtering the culture medium or seawater (see Stein 1973, Chp. 2, II.D.2).

2. Culture vessels. These should be all-glass. Suitable containers are provided by round-bottomed flasks (250 or 500 ml), each fitted with a ground-in stopper giving access for sampling purposes, and a bubbler for aerating the culture medium with purified, sterile air (see Stein 1973, Chp. 14, VI.B.4, for details of air purification). The inlet and the outlet for the air should be protected with cotton-wool filters.

3. Cleaning of glassware. All glassware must be scrupulously cleaned before use and between experiments. New glassware should be degreased; warm dilute alkali is most suitable for this purpose. Washing with conc. HCl removes most metallic contamination, and it is recommended that all glassware be stored in dilute acid when not in use to keep it metal-free. More rigorous cleaning, e.g., with a HNO_3-ethanol solution (*caution*), may be necessary to remove organic residues, but chromic acid treatment should be avoided. After the cleaning process, the glassware should be thoroughly rinsed with distilled water and, if possible, finally steamed.

The treatment of glassware with silicone compounds in order to make the surfaces hydrophobic and thus prevent losses of trace components by adsorption can be temporarily effective. These coatings are, however, slowly leached from the glass by the culture medium and could affect the growth of the algae. It is better not to use them, but to determine the extent of any losses that take place.

B. Culture media

1. Metal uptake studies. Most of the media used for the routine culturing of algae contain chelating agents (Stein 1973, Chps. 1 and 2). By forming soluble complexes with the micronutrient metals, the chelating agents prevent the metals from being lost from the medium by precipitation or adsorption onto the container surface.

The presence of these complexes affects the uptake of the metals by the algae, however, and can also moderate the toxicity of heavy metal ions. As natural waters contain very much lower levels of chelating materials and the aim of the study is usually to discover how metals control algal growth under environmental conditions, chelating agents should not be used as constituents of the culture medium unless, of course, their own specific role is being investigated. Organic pH buffers may similarly interact with metal ions and should also be excluded from the medium; it is then necessary to aerate the medium and rely on the carbonate-bicarbonate buffering system for pH control. The absence of the chelating agents will inevitably lead to losses of some metals from the medium, but these can be taken into account by routinely measuring the total concentration present at each sampling.

Either synthetic media or nutrient-enriched natural waters are suitable for this work, but it must be ensured that the concentration of the metal to be studied in the medium being used is low relative to the levels of the metal that will be introduced for the measurements. If this is not the case, it is first necessary to remove the metal. Most soluble metals can be extracted by passing the medium through a chelating ion-exchange resin such as Chelex-100 (Bio-Rad), though care must be taken with the purification of the resin before it is used (Davey et al. 1970). The concentrations of metals present in a particulate form (e.g., iron) can be considerably reduced by filtration through membranes or filter cartridges having as small a pore size as possible (≤ 0.3 μm).

These treatments also remove most of the other metals, and those essential for growth must be replaced by adding to the medium a suitable trace metal mixture (not containing the one being studied) at the time of subculturing.

2. Vitamin uptake studies. Carefully prepared synthetic culture media are unlikely to contain significant amounts of the vitamins necessary for algal growth – notably cyanocobalamin, thiamine, and biotin – and can therefore be used directly. Natural waters should, however, be freed of their vitamin content by treatment with activated charcoal (Stein 1973, Chp. 25, III.E.1).

3. Sterilization. Suitable techniques for sterilizing culture media have been described in Stein (1973, Chp. 12). It should be noted, however, that synthetic media or natural waters that do not contain added chelating agents or pH buffers should be autoclaved before the addition of the nutrients and micronutrients. After sterilization, the culture vessels containing the medium should be removed from the autoclave while still hot (but below 100 C) and allowed to cool under normal air pressure; this prevents problems attributable to the precipitation of calcium carbonate.

4. Nutrient additions. Nutrients should be added aseptically from separately autoclaved stock solutions, their concentrations being kept to the minimal level that will allow growth to occur for 10–14 days and produce healthy cultures. The required additions, particularly of the micronutrients, will depend on the algal species being grown, but the following enrichments (contained in 100 ml medium) have been found to be satisfactory for many different species of marine phytoplankton: (a) 1 ml of nutrient stock solution (10 mM $NaNO_3$, 1 mM

Na_2HPO_4); (b) 0.05 ml of vitamin stock solution (1.5 μM cyanocobalamin, 1.5 mM thiamine·HCl); and (c) 1 ml of trace metal stock solution (0.18 mM $FeCl_3$, 18 μM $MnCl_2$). The nutrient and vitamin stock solutions and media are made up using distilled H_2O and the trace metal stock solution with 0.05 M HCl.

Experience has shown that it is beneficial to add the trace metal solution after the cultures have been inoculated. If the uptake of one of the micronutrients listed is being studied, it would not, of course, be included in these stock solutions. Silicate additions for the growth of diatoms are not usually necessary, as adequate levels result from the use of glass containers.

C. Trace components and radioactive labeling

In order to minimize the decrease in the volumes of the cultures caused by the sampling procedure, the size of the aliquots withdrawn must be kept as small as possible while maintaining the precision of the measurements. This is practicable provided a suitable radioactively labeled form of the trace component being studied is available. Ideally, γ-emitting isotopes should be used so that direct counting of the radioactivity of samples is possible. Alternatively, a β-emitting label can be satisfactory, though the samples will usually need to be processed before their β activity can be determined; details of sample preparation for β-activity liquid scintillation counting are given by Price (1973) and in a series of technical bulletins (Packard Instrument Co.). A list of suitable radioactive tracers for the more important metals and vitamins is given in Tables 43-1 and 43-2; further information about the availability of radioactive isotopes and compounds may be obtained from Oak Ridge National Laboratory or The Radiochemical Centre.

Should no radioactively labeled form of the trace component be available, concentrations could, in principle, be determined by chemical analysis of the samples, though the culture volumes required to make this feasible are usually prohibitively large.

A separate stock solution of the trace component under investigation is required for each concentration to be studied, the amount in each stock being such that a 100-fold dilution into the culture medium gives the required experimental concentration. The range of concentrations in a set of test cultures should be wide enough to cause the algal growth rates to vary from exponential to the slowest that can be achieved. With a toxic metal, growth can, in fact, be stopped by using lethal concentrations, but, in the case of an essential micronutrient, if the algal requirement for it is small, the stock populations must be grown through several subcultures in a medium containing

Table 43-1. *Radioactive isotopes suitable for trace metal uptake studies*

Element	Mose useful radioisotope	Half-life	Most useful radiation	Chemical form usually available
Antimony	124	60 d	γ	$SbCl_3$ in HCl
Arsenic	74	18 d	γ	H_3AsO_4 in dil. HCl
Cadmium	109	470 d	β	$CdCl_2$ in dil. HCl
Chromium	51	27.8 d	γ	$CrCl_3$ in dil. HCl Na_2CrO_4 in H_2O
Cobalt	60	5.26 y	γ	$CoCl_2$ in dil. HCl
Copper	64	12.84 h[a]	γ	Unprocessed irradiated element or compound $CuCl_2/(NO_3)_2$ in dil. HNO_3
Iron	59	45 d	γ	$FeCl_3$ in dil. HCl
Manganese	54	314 d	γ	$MnCl_2$ in dil. HCl
Mercury	203	47 d	γ	$HgCl_2/(NO_3)_2$/acetate in H_2O or dil. acid
Molybdenum	99	67 h[a]	γ[b]	$Na_2/(NH_4)_2 MoO_4$ in H_2O or alkali
Nickel	63	120 y	β	$NiCl_2$ in dil. HCl
Selenium	75	121 d	γ	Na_2SeO_3/SeO_4 in H_2O or dil. HCl
Silver	110m	253 d	γ	$AgNO_3$ in dil. HNO_3
Tin	119m	245 d	γ	H_2SnCl_6 in HCl
Zinc	65	245 d	γ	$ZnCl_2$ in dil. HCl

[a] For short term only.
[b] Must be counted so that γ radiation from daughter product, technetium 99m, is excluded.

Table 43-2. *Radioactively labeled forms of vitamins*

Vitamin	Label	Half-life	Form usually available
Biotin	[14]C	5760 y	D- or DL-(carbonyl-^{14}C)-biotin
Cyanocobalamin	[57]Co	270 d	Aqueous with 0.09% benzyl alcohol as preservative
Thiamine	[14]C	5760 y	(Thiazole-2-^{14}C)-thiamine·HCl
	[35]S	87.2 d	Thiamine·HCl

low concentrations of the micronutrient until the cells are deficient and thus sensitive to small additions of the rate-limiting component in the test cultures.

Metal stock solutions should usually be made up in dilute acid solution (0.05 N HCl is suitable), but where stronger acid or alkaline solutions are necessary, the volumes added to the test cultures must be kept small so that the pH of the medium is not altered.

Vitamin stocks are usually prepared in aqueous solution, though care should be exercised in autoclaving them (Stein 1973, Chp. 2, III.E.1 and 2).

Radioactive tracers should be of as high a specific activity as possible and added to the stock solutions before sterilization in order to ensure isotopic equilibration; they should not be added separately to the culture medium. As most radioactive labels are not carrier-free, allowance must be made for the amount of nonradioactive material introduced into the concentrations in the stock solutions.

D. Culture apparatus

The stock and test cultures should be grown under the same conditions of constant temperature, which should be controlled to within ± 0.5 C and constant illumination of 3–5 klux (Stein 1973, Chp. 11, II.C). Aeration will provide some stirring but, if additional agitation is necessary, this may be carried out by periodically shaking the cultures manually or, more conveniently, by fixing the culture vessels to an inclined platform rotating about the source of illumination. Davies (1974) has described an apparatus suitable for this purpose.

E. Other equipment

Membrane filters (0.45 μm pore size); non-oral pipetting device; well-type scintillation counter for γ-emitting isotopes or liquid scintillation counter for β-emitters (Nuclear-Chicago, Packard Instrument, Nuclear Enterprises); hemocytometer or, preferably, Coulter particle counter with volume measurement capability (Coulter Electronics; see Stein 1973, Chps. 19 and 22).

III. Methods

A. Test organism

Isochrysis galbana Parke (UTEX 987; CCAP 927/1; MBA Culture I). None of these cultures are bacteria-free; axenic stock cultures can be prepared from them using the method of Droop (1967).

B. Procedure for studying the uptake of mercury
and its effect on growth

    1. Collect and, as soon as possible afterward, filter seawater from an unpolluted area (Stein 1973, Chp. 2, II.D.1 and 2). Store until required in glass containers kept in the dark at low temperature.

Table 43-3. *Composition of working mercury stock solutions in flasks A–H*[a]

Stock	A	B	C	D	E	F	G	H
Stable Hg stock (ml)	0	0	0	0.5	2.0	3.5	8.5	21.0
^{203}Hg stock (ml)	0	0.075	0.15	0.2	0.2	0.2	0.2	0.2
0.1 N HCl (ml)	25	25	24.8	24.3	22.8	21.3	16.3	3.8
Concentration in culture (ng at l^{-1})	0	16.3	32.6	48.4	63.4	78.4	128.2	253

[a] Prepared as described in III.B.5. Distilled water is added to bring each flask up to 50 ml.

2. Establish axenic stock cultures of *I. galbana* in seawater enriched as described in II.B.4. Grow under the temperature and lighting conditions to be used in the measurements.

3. Clean eight culture vessels labeled A–H and all the other necessary glassware.

4. Measure refiltered seawater into culture vessels (200 ml into 250-ml flask or by proportion for larger containers) and close input side of bubbler on each vessel to prevent loss of water during autoclaving.

5. Prepare mercury stock solutions labeled with ^{203}Hg in eight 50-ml flasks marked A–H. The following is given as an example of the composition of such stocks: If the ^{203}Hg solution as supplied contains 1.09 mg Hg and 1.03 mCi of radioactive label in 1 ml of 0.1 N HCl, dilute it to 5 ml using 0.1 N HCl; this gives a concentration of 1.09 mg at. Hg l^{-1} and a suitable level of radioactivity for experimental purposes. Also required is a nonradioactive Hg solution containing 49.8 μg at. Hg l^{-1} as HgCl$_2$ in 0.1 N HCl.

Dispense these solutions into the measuring flasks as shown in Table 43-3 to give the working mercury stocks. The mercury concentrations in the cultures given in Table 43-3 result from the addition of 1 ml of Hg stock solution to 100 ml of medium. The values are here calculated from the concentrations of the solutions used to prepare the working stocks. It is, however, preferable to check the concentrations in the working stocks by analysis.

6. Autoclave (15 min at 120 C) the culture vessels containing the seawater, the nutrient stock solutions (II.B.4), and the working Hg stock solutions A–H. Allow the culture vessels to cool to experimental temperature under normal air pressure.

7. Set up culture vessels to receive uniform illumination and start aeration. The air leaving the cultures must be passed through charcoal filters to trap any volatile mercury present.

8. Determine the cell population of the stock culture, which should be growing exponentially.

9. Add the nutrient enrichments (but not the metals) to each culture vessel, then an inoculum of cells from the stock culture to give an initial population of about 5000 cells ml⁻¹, and finally the trace metal stock solutions. (All additions and samplings should be made aseptically.)

10. Allow the cultures to settle overnight. ·

11. Next day (day zero), introduce the working Hg stock solutions, including the "solvent blank" (A), into the appropriate culture, and remove a 5-ml aliquot of the culture by pipette immediately after each addition, transferring it to a test tube marked to indicate the source of the sample.

12. Determine the radioactivities of the initial samples (not A) by placing the test tubes into a well-type counter set up to detect the γ radiation of ^{203}Hg and, using the same samples, measure the algal populations.

13. The cultures are then sampled daily. Every time, filter a 5-ml aliquot of each culture, except A, using two membrane filters placed one on top of the other for each sample. Separate the filters, fold and drop into test tubes, and measure their radioactivities. The difference between them represents the radioactivity contained by the cells retained on the top filter. Remove a second 5-ml aliquot of each culture for measuring the total radioactivity in the culture medium and the cell population.

14. Retain some of the top filters after measuring their radioactivity; dissolve in 5 ml conc. HNO_3 and recount. The ratio of this count to the original gives a correction factor for the difference in counting geometry between the filters and the liquid samples. Several observations must be made and the mean value of the factors used in the calculations.

15. Continue to sample the cultures until growth ceases.

IV. Sample data

The method of processing the experimental data is illustrated in Table 43-4 using results obtained in an actual experiment with mercury-containing cultures of *I. galbana*. It was calculated that between day 7 and day 9 (Table 43-4) when the mean Hg content of the cells was 7.88 ag at. μm^{-3}, the mean specific growth rate of the culture was $(12.2-9.87) \times 10^5/2\sqrt{(9.87 \times 12.2 \times 10^{10})} = 0.1$ day⁻¹, whereas, in the Hg-free control, it was 0.46 day⁻¹. Using this method, it has been shown that the specific growth rate of *I. galbana* is reduced linearly by

Table 43-4. Data obtained from mercury-containing culture of Isochrysis galbana (see III.B and IV)

Time after Hg addition (days)	Total cell volume (μm^3 ml^{-1})[a]	Samples	^{203}Hg radioactivity in samples (counts/100 sec)			Fraction of day 0 count	Equiv. Hg conc. (ng atom l^{-1})[e]	Hg conc. in cells (ag atom μm^{-3})
			Background corrected	Decay corrected[b]	Geometry corrected[c]			
0	2.56×10^5	5-ml Culture	31 801	31 801	31 801	1.00	37.4	—
7	9.87×10^5	5-ml Culture	13 029	14 449	14 449	0.454	16.9	—
		Top filter	7 073	7 844	6 832 } 6587[d]	0.207	7.73	7.83
		Bottom filter	253	281	245 }			
9	12.2×10^5	5-ml Culture	8 698	9 933	9 933	0.312	11.7	—
		Top filter	8 440	9 638	8 395 } 8253[d]	0.259	9.69	7.93
		Bottom filter	143	163	142 }			

[a] Algal populations are expressed as volume of cell material per milliliter because Hg affects cell size. Changes in cell numbers, therefore, do not accurately reflect the growth rate.

[b] Calculated using $N_0 = N_t e^{0.693t/t_{1/2}}$ where N_t and N_0 are, respectively, the counts on day t and the value corrected to day 0, and $t_{1/2}$ is the half-life of the radiotracer (47 days for ^{203}Hg).

[c] Correction factor is 0.871 for ^{203}Hg in the well counter used for these measurements.

[d] Difference between radioactivity of top and bottom filter.

[e] The stable and radioactive Hg are assumed to behave in the same way so that changes in radioactivity counts corrected for decay, etc., are directly proportional to changes in the mercury concentration.

[f] The units are atto (10^{-18}) gram atom of Hg per cubic micrometer of cell material.

increasing concentrations of Hg in the cell material, growth ceasing at a cellular level of 10 ag at. Hg μm^{-3} (Davies 1974).

V. Discussion

One of the shortcomings of the procedure as described is that it does not differentiate between the fraction of the trace component that has entered the cells and become involved in the intracellular metabolic processes and the fraction that is merely attached, by adsorption or adhesion, to the exterior surfaces of the algae. This is especially important in the case of a micronutrient such as iron, because at the pH of most culture media and in the absence of chelating agents, it is present as particulate hydrous ferric oxide, which cannot easily be separated from the cells. However, treatment of samples of the culture with a suitable chelating agent causes rapid dissolution of any extracellular metal, allowing the intracellular content to be determined by separating the cells from the medium in the usual way. Full details of the method are given by Davies (1970).

The intracellar locations of trace components taken up by algae are at the present time largely unknown, but techniques such as electron microscope autoradiography (Rogers 1973) and electron probe X-ray microanalysis (Marshall 1975) are now providing means for identifying these.

VI. References

Davey, E. W., Gentile, J. H., Erickson, S. J., and Betzer, P. 1970. Removal of trace metals from marine culture media. *Limnol. Oceanogr.* 15, 486–8.

Davies, A. G. 1970. Iron, chelation and the growth of marine phytoplankton. I. Growth kinetics and chlorophyll production in cultures of the euryhaline flagellate *Dunaliella tertiolecta* under iron-limiting conditions. *J. Mar. Biol. Assoc. U.K.* 50, 65–86.

Davies, A. G. 1974. The growth kinetics of *Isochrysis galbana* in cultures containing sublethal concentrations of mercuric chloride. *J. Mar. Biol. Assoc. U.K.* 54, 157–69.

Droop, M. R. 1967. A procedure for routine purification of algal cultures with antibiotics. *Br. Phycol. Bull.* 3, 295–7.

Marshall, A. T. 1975. Electron probe X-ray microanalysis. In Hayat, M. A. (ed.). *Principles and Techniques of Scanning Electron Microscopy*, vol. 4, pp. 103–73. Van Nostrand Reinhold, New York.

Price, L. W. 1973. Practical course in liquid scintillation counting: Preparation of samples. *Lab. Pract.* 22, 110–14, 181–6, 194.

Rogers, A. W. 1973. *Techniques of Autoradiography*. Elsevier, Amsterdam. 372 pp.

Stein, J. R. (ed.). 1973. *Handbook of Phycological Methods: Culture Methods and Growth Measurements*. Cambridge University Press, London. 448 pp.

Section VI

Ion content and transport

44: Measurement of ionic content and fluxes in microalgae

J. BARBER

Department of Botany,
Imperial College of Science and Technology, London SW7 2BB, England

CONTENTS

I	Introduction	*page* 450
II	Test organism and its growth	450
III	Ionic analyses	451
	A. Determination of cell parameters	451
	B. Preparation of samples for analyses	452
	C. Techniques for analyses	453
IV	Measurement of ion fluxes	455
	A. Filtration method	455
	B. Typical influx experiment	456
	C. Efflux measurements	458
	D. Centrifugation methods	459
V	Acknowledgment	461
VI	References	461

I. Introduction

Microalgae are useful organisms for studying ion transport. They are easy to grow in continuous culture, often having fast doubling rates, and are convenient experimental material. Moreover, synchronous cultures can be used to study transport at various stages of the life cycle. The precise methodology of experimentation will depend to some extent on the organisms under study (Barber 1968a, 1968b, 1968c, 1969; Shieh and Barber 1971; Barber and Shieh 1972, 1973; Dewar and Barber 1973, 1974). For simplicity I will refer only to studies with *Chlorella*.

II. Test organism and its growth

The specific examples given throughout this chapter were obtained with *Chlorella pyrenoidosa* (UTEX 252). The alga was maintained on agar slants and photoautotrophically cultured for experiments. The sterile liquid cultures were continuously illuminated and shaken and gassed with a 5% CO_2 + 95% air mixture. The composition of the liquid medium was 5 mM KNO_3, 0.5 mM $Na_2HPO_4 \cdot 2H_2O$, 0.5 mM KH_2PO_4, 2.0 mM $MgSO_4 \cdot 7H_2O$, 0.25 mM $Ca(NO_3)_2 \cdot 4H_2O$, 1 mM KCl, plus 2 ml l^{-1} of Hutner's micronutrient solution (Hutner et al. 1950). Composition of Hutner's solution (in grams per liter): EDTA (disodium salt), 25; $ZnSO_4 \cdot 7H_2O$, 22; H_3BO_3, 10.4; $MnSO_4 \cdot H_2O$, 44; $FeSO_4 \cdot 7H_2O$, 5.0; $CoCl_2 \cdot 6H_2O$, 2.16; $CuSO_4 \cdot 5H_2O$, 1.56; $(NH_4)_6 Mo_7O_{24} \cdot H_2O$, 1.1. Cells were harvested during the exponential phase of growth.

It is vital for ion transport studies to establish the composition of the experimental medium. For most animal cells and for algae collected from their natural habitat, the experimental medium can be readily based on the normal physiological bathing medium. For microalgae that can be grown in a variety of artificial culture media the decision is more difficult. In the case of *Chlorella* the liquid culture medium was chosen as the normal experimental medium (Barber 1968a). In fact, this medium had been slightly modified to include both Na^+ and Cl^-, as these ions, together with K^+, play a major role in

[450]

normal cellular ionic regulation processes. In retrospect, it might have been more satisfactory, from a physiological point of view, to have used a medium more like normal pond water (i.e., Na/K ratio > 1), although, of course, this would have limited the ease of culturing large amounts of material.

Before undertaking any ion transport experiments with laboratory-cultured microalgae, some thought should be given to the nature of the experimental medium to be employed.

III. Ionic analyses

When determining whether a particular ion is passively or actively transported by a cell, its internal concentration must be estimated (see Dainty 1962). This quantity is also needed for calculating effluxes. To estimate the internal concentration, several cell parameters have to be obtained, such as cell volume and cell water content.

A. Determination of cell parameters

The most convenient cell quantity is the packed cell volume. This can be determined by centrifugation of the harvested cells in calibrated tubes for a fixed period of time. For *Chlorella,* 1-ml calibrated tubes or hematocrit tubes were used and spun for 5 or 10 min at about $2200 \times g$ to constant packed cell volume.

Resuspension of the pellets to known cell densities is then possible. For example, for *Chlorella* it was convenient in most experiments to use 5 μl packed cells per milliliter of suspending medium. These standard suspensions can then be used to obtain a number of important cell parameters necessary for calculating internal concentrations and fluxes of ionic species.

1. Average dry weight. Known volumes of packed cells are pipetted into weighed vessels and placed in an oven at 90–95 C until constant weight is obtained. For *Chlorella,* 1 ml packed cells = 163.3 ± 15.6 mg dry weight.

2. Average cell number (N). Using standard cell suspensions, count cells manually using a hemocytometer and microscope, or electronically using a Coulter counter. For *Chlorella,* 1 ml of packed cells = 470 × 10^7 cells.

3. Average cell dimensions. Average values of the cell dimensions can be obtained using a microscope with a calibrated eyepiece. For spherical *Chlorella* cells, the average diameter (d) was 6.5 μm.

4. Cell volume. Using (2) and (3), it is possible to calculate the actual cell volume in any quantity of packed cells assuming the cell has a particular shape. Taking a *Chlorella* cell as a sphere, the total cell volume in 1 ml packed cells is given by $4/3 \ \pi r^3 N = 0.675$ ml. Alternatively, the total cell volume of a known quantity of packed cells can be obtained by using ^{14}C-D-mannitol or any other nonpenetrating or nonbinding compound that can measure the free space in a packed cell pellet (e.g., ^{14}C-inulin). Cells are initially suspended in a medium containing the radioactively labeled nonpenetrating species and then spun down in a calibrated tube in the usual way. After the radioactive supernatant is decanted, it is necessary to remove carefully the final remaining drops of active supernate from the surface of the pellet with tissue paper. If care is taken, the free space in the pellet can simply be determined by washing (by resuspension and centrifugation) with nonradioactive medium and then counting the washings. The sum of these counts can then be compared with a known volume of the original radioactive supernatant. With *Chlorella*, ^{14}C-mannitol was used at an activity of 0.1 μCi ml^{-1} and the estimated cell volume was 0.669 ± 0.041 of the packed cell volume, in good agreement with the calculated value above.

5. *Water content.* Estimation of the water content of the cells is possible by weighing a known wet packed cell volume before and after drying to constant weight in an oven at 95 C. Taking the difference and subtracting the value for extracellular water obtained in III.A.4 gives the cell water content. (Note that the dry weight factor obtained in III.A.1 can be conveniently used to relate the dry weight obtained here with packed cell volume.) For *Chlorella*, 1 ml of packed cells was found to contain 0.627 ± 0.04 ml of cell water, corresponding to about 90% water in the cell. Such a high water content may not be found with other algae; for example, *Anacystis nidulans* was found to contain only 65% water (Dewar and Barber 1973).

B. *Preparation of samples for analyses*

In order to carry out ionic analyses it is usually, but not always (see below), necessary to produce a liquid sample.

1. *Wet ashing.* Digest known packed cell volumes in conc. HNO_3 or $HClO_4$. It is advisable to centrifuge digests to remove particles from the suspension and analyze the supernatant. This avoids blocking of the atomizer on analytical machines.

2. *Dry ashing.* Ash known packed cell volumes in a muffle furnace (500 C for 15–30 min) followed by digestion of residue in a small quantity of 0.1 N HNO_3 (evaporate off liquid before placing in furnace).

C. Techniques for analyses

1. Atomic absorption and emission spectroscopy. The digests should be diluted to a known volume. This will be determined by the particular ion being measured and the range of linear response of the instrument. For *Chlorella,* to obtain dilute digests (5 ml) of about 0.5 mM, it is necessary to take about 30 μl packed cells for K^+ and at least 10 times this quantity for Na^+.

Flame emission photometers are quite suitable for K^+ and Na^+, whereas atomic absorption spectroscopy may be necessary for other cations such as Ca^{2+} and Mg^{2+}. In all cases the calibrating solutions should have concentrations similar to that of the extract.

Simultaneous measurement of two or more elements is now possible. In particular, multielemental analyses on single samples (liquid or dry) can now be done using an inductively coupled radiofrequency plasma emission spectrometer (Applied Research Laboratories or SMI Spectrametrics Inc.).

2. Electrometric titration. This can be used to determine Cl^- levels (e.g., Ramsay et al. 1955). For *Chlorella,* the procedure was to dry down 50–100 μl packed cells in alkaline solution. To this 0.5 ml of 1 N HNO_3 was added, quickly stirred, and neutralized with 1 N NaOH (the time of acid treatment is kept to a minimum to avoid loss of Cl_2). The electrometric titration is carried out by placing 50–100 μl of extract on a Teflon square and slowly adding 1 mM $AgNO_3$ by means of an Agla syringe via a glass tube dipping into the drop. A platinum wire sealed into the syringe can act as a reference electrode; the other electrode can be a silver wire dipping into the extract on the Teflon. Measurements can be made using a pH meter or electrometer suitably backed off with a potentiometer to allow the use of more sensitive scales.

3. Isotopic equilibrium. This is a very convenient method for determining ionic levels in microalgae, especially for ions that are in low concentrations (e.g., Cl^- and Na^+). The method is simply to allow the internal radioactivity to come into equilibrium with an external solution of known specific activity. In some cases it may be desirable to grow the organisms in a radioactive culture medium (e.g., for K^+), but this is not always necessary if the loading time is short (e.g., Cl^- and Na^+ in *Chlorella*). This is also a sensitive technique for detecting changes in internal levels brought about by changing conditions (e.g., a light-dark transition; Fig. 44-1). Separation of the labeled cells from the radioactive medium can be done either by centrifugation or, perhaps more conveniently, by the filtration technique outlined below (IV.A).

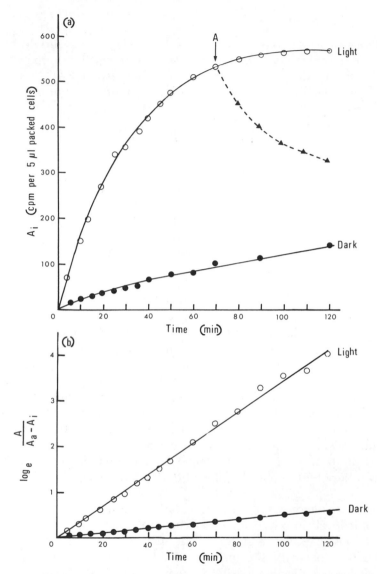

Fig. 44-1. (*a*) Uptake of $^{36}Cl^-$ by illuminated (white circles) and dark-treated (black symbols) *Chlorella* cells. At point *A* some of the illuminated suspension was transferred to a darkened vessel. Temperature, 25 C, white-light intensity, 50 W m^{-2}. (*b*) First-order semilog plot of the data shown in (*a*).

4. Neutron activation. This is a useful nondestructive technique for elemental analyses on wet or dried samples with no need for digestion. The principle of the technique is to irradiate the sample with neutrons and detect the radioactive products by their characteristic energies and lifetimes. Short irradiation times (e.g., 2 min) can be used for

K, Na, Cl, Mg, Ca, and Mn (also for trace metals like V, Cu, Ti, and Al); longer irradiation times extend the list to about 40 different elements. For a useful reference text, see De Soete et al. (1972).

5. Electron probe analysis. Although I know of no reports of the use of this technique with microalgae, it could, in principle, be used to give information on the ionic content and localization of ions in the cells of these organisms. The principle of the technique is to detect characteristic X-ray emissions associated with the irradiation of a particular element by an electron beam. The reader is referred to the book by Reed (1975) on the use of this technique.

6. Ion-specific electrodes. There are several very good ion-specific electrodes on the market that, when calibrated, could be used directly for estimating the ionic content of liquid extracts.

IV. Measurement of ion fluxes

Microalgae are ideal experimental material for studying ionic fluxes. Suspensions of algae in the region of 5 μl packed cells per milliliter will contain many thousands of cells per milliliter and can easily be treated in various ways (e.g., illuminated, subjected to specific inhibitors). Equal aliquots can be taken as a function of time to establish both fluxes and kinetics.

A. Filtration method

This is a very useful approach for radiotracer studies.

1. Procedure for influx studies

   a. Harvest cells, wash with cold distilled water, and resuspend in experimental medium to known packed cell volume. For *Chlorella*, 5 μl packed cells per milliliter is suitable.

   b. Place cells in reaction vessels (lollipop shape is ideal for illumination studies) and pretreat for 20–30 min with the particular condition being studied (e.g., light, dark, inhibitors). Make sure the cell suspension is stirred using a gas stream or magnetic flea and kept at a desired temperature (gas with air or with a test atmosphere such as N_2, CO_2/N_2, CO_2/air). The reaction vessels should be placed in a glass-sided constant-temperature water bath to allow illumination. For dark fluxes, cover vessels with black tape or paint.

   c. Introduce radioisotopes from a suitable stock and at the same time start a clock. The isotope should be at high specific activity and added in small volumes so as not to alter significantly the concentration of its nonradioactive isotope in the suspending medium. For *Chlorella*, the following activities have been used: 0.05–0.1 μCi ml^{-1} for ^{36}Cl and ^{42}K, and 0.1–0.2 μCi ml^{-1} for ^{22}Na.

d. At leisure 1-ml aliquots should be taken with a 1-ml syringe at 5- to 10-min intervals spaced over the duration of the time course. For *Chlorella,* this duration is several hours for K^+ and about 2 h for Na^+ and Cl^-. However, for initial rates there is no need to continue sampling to isotopic equilibrium, and 20–30 min is sufficient.

e. The cells can be separated from the radioactive solution by rapid filtration through Millipore membrane filters clamped in a Millipore filter unit. For anions such as Cl^-, cheaper glass fiber filters are suitable; these are, however, unsuitable for K^+ and Na^+, as they show considerable cation-exchange properties because they are negatively charged. The radioactive solution can be sucked off by means of a suitable pump. It is advisable to inject the 1-ml aliquot of cell suspension into about 20 ml of ice-cold water before sucking down, as this leads to an even cell distribution on the filter. The sedimented cells should then be washed once more with an additional 20 ml of distilled water to remove residual extracellular radioactivity on the filter pad. With *Chlorella,* no loss of internal ions occurred during this washing procedure, but it may be necessary to test for leakage when using other microalgae. The sampling and washing procedure should be as rapid as possible, and membranes with different pore sizes should be investigated. For *Chlorella,* membrane filters with 8-μm pores are the most suitable.

f. Allow the filters with their cell sediments to air-dry under gentle heat from a lamp and then stick them with glue to aluminium planchets for counting. It is very important that each sample contains the same quantity of cells and is evenly distributed on the filter. Poor data often result from taking uneven samples. The thickness of the sedimented cell layer should be kept at a minimum to avoid self-absorption.

g. Take a known volume of the reaction medium (e.g., 50 μl), place on a planchet, dry, and count radioactivity using the same counter as for the assaying of the radioactivity in the cells. Knowing the actual amount of the ion in this standard allows calculation of the specific activity (see below).

h. Count samples using a Geiger counter, ideally with a low background to improve sensitivity. A convenient machine with automatic exchange is the Nuclear-Chicago gas flow counter.

i. Calculate fluxes in terms of amount taken up per unit area per unit time (e.g., moles per square centimeter per second).

B. Typical influx experiment

Figure 44-1*a* shows the results of measuring the uptake of radioactive chloride (^{36}Cl) by an illuminated and a darkened suspension of *Chlorella* cells. The cell suspension was 5 μl packed cells per milliliter, and each point represents the activity in counts per minute recorded for each 1-ml aliquot. The initial Cl^- in the medium was 1 mM and was insignificantly increased on addition of the high-specific-activity ^{36}Cl.

1. Calculation of influxes. The initial slopes of the curves in Fig. 44-1a correspond to 85 and 9 cpm per 5 μl packed cells over the first 5 min for light and dark cells, respectively. For comparison with other cells (algal, plant, and animal), it is necessary to convert these fluxes into more meaningful units (e.g., moles of Cl^- per square centimeter per second).

a. What is the surface area of 5 μl packed cells? Assuming cells to be spheres, SA = $4\pi r^2 \times 5 \times N = 31.2$ cm^2

b. What is the conversion between counts per minute and moles of Cl^- (i.e., specific activity)? A 50-μl standard gave 7000 cpm, and as the external Cl^- concentration was 1 mM, there was 50×10^{-9} mol Cl^- in the standard. Thus 7000 cpm corresponds to 50×10^{-9} mol Cl^-, or 1 cpm = 7.14×10^{-12} mol Cl^-. Therefore, the fluxes are:

$$\text{light:} \quad \frac{85 \times 7.14 \times 10^{-12}}{5 \times 60 \times 31.2} = 0.65 \times 10^{-13} \text{ mol } Cl^- cm^{-2}sec^{-1}$$

$$\text{dark:} \quad \frac{9 \times 7.14 \times 10^{-12}}{5 \times 60 \times 31.2} = 0.68 \times 10^{-14} \text{ mol } Cl^- cm^{-2}sec^{-1}$$

2. Calculation of internal levels. It can be seen that after about 2 h the illuminated cells had come almost into isotopic equilibrium. Thus the internal Cl^- level can be calculated, as it corresponds to about 580 cpm per 5 μl packed cells where 1 cpm = 7.14×10^{-12} mol Cl^-. Using the factors for converting packed cell volume to cell water, this internal level corresponds to 1.32 mmol l^{-1} of cell water, or 1.32 mM. In the case of the dark-treated cells, the final level of activity is lower at isotopic equilibrium. In this experiment samples taken after several hours indicated that isotopic equilibrium occurred at a level corresponding to 320 cpm per 5 μl packed cells or 0.73 mM. The fact that *Chlorella* cells maintain higher levels of Cl^- in the light than in the dark is demonstrated by the net efflux that occurs when some of the illuminated cells containing ^{36}Cl are transferred to a darkened vessel. The drop in internal activity corresponds to a lowering of the internal Cl^- from its light to dark level.

3. Kinetics. *Chlorella* cells seem to act as a single compartment for Cl^- uptake (also true for Na^+ and K^+). Thus, the addition of ^{36}Cl at time zero to cells that are in a steady state gives rise to a rate of increase of internal activity (A_i) corresponding to the following first-order expression:

$$\frac{dA_i}{dt} = aS_o\phi_o - aS_i\phi_i \tag{1}$$

where S is the specific activity in counts per minute per mole, a is the surface area in square centimeters, ϕ is the flux in moles per square centimeter per second, and $_o$ and $_i$ denote outside and inside, respectively. At $t = \infty$ (i.e., at long times), the system is in isotopic equilibrium: $S_o = S_i$ and $A_i = A_\infty$. Integration of equation 1 gives:

$$A_i = A_\infty(1 - e^{-kt}) \tag{2}$$

where k is a rate constant equal to $a\phi/c_iv_i$, c_i is the internal concentration in moles per milliliter, and v_i is the cell volume in milliliters. Rearranging equation 2 gives:

$$\log_e \frac{A_\infty}{A_\infty - A_i} = kt \tag{3}$$

Thus, for a first-order uptake the plot $\log_e A_\infty/A_\infty - A_i$ against time should be linear. As shown in Fig. 44-1b it does follow this relationship when taking $(A_\infty)_{\text{light}} = 580$ cpm and $(A_\infty)_{\text{dark}} = 320$ cpm. From the slopes of the semilog plots the influx rates can be calculated using the relationship:

$$\phi = \frac{kv_ic_i}{a} \tag{4}$$

Thus for the light data, $k = 5.6 \times 10^{-4}\text{sec}^{-1}$, $v_ic_i = 8.9 \times 10^{-7}$ mol Cl^- ml^{-1} packed cells, and $a = 6.24 \times 10^2$ cm^2 ml^{-1} packed cells, so that $\phi_{\text{light}} = 0.8 \times 10^{-13}$ mol Cl^- cm^{-2} sec^{-1}. For the dark rate constant of 8×10^{-5} sec^{-1}, a similar calculation gives $\phi_{\text{dark}} = 0.66 \times 10^{-14}$ mol Cl^- cm^{-2} sec^{-1}. These values correspond closely to those calculated from the initial influxes, and the latter calculations give an alternative approach for flux determinations.

If the cells or the cell suspension behave as a multicompartment system, then more complex kinetic equations should be applied (e.g., see Walker and Pitman 1976).

C. Efflux measurements

Effluxes can be conveniently measured by resuspending cells that have been fully loaded with radioisotope in a fresh inactive medium. A few aliquots of cells should be taken from the loading suspension and filtered in the usual way (also remember to take standards) before centrifugation. This allows calculation of internal ion levels and also serves to check if significant loss of radioactivity occurs during the separation procedure. With *Chlorella,* the separation of the cells from the radioactive medium involved a quick centrifugation followed by a rapid wash with distilled water. After recentrifugation, the cells were resuspended to a standard cell density and aliquots taken, filtered, washed, and counted in the same way as for influx measurements.

The duration of the experiments and the density of the cell suspension should be such as to minimize backflow of tracer. Also, because the efflux of Cl^- and Na^+ from *Chlorella* was significant over short periods (unlike K^+), it was necessary when testing the effect of an inhibitor to pretreat in the dark, as under this condition the loss of radioactivity was slowed and the possibility of tracer backflow reduced. If the above precautions are taken, it can be assumed that the external specific activity (S_o) is essentially zero, and equation 1 reduces to:

$$\frac{dA_i}{dt} = -a\phi_i S_i \tag{5}$$

substituting $S_i = A_i/v_i c_i$ and integrating gives the internal activity (A_i) at any time (t):

$$A_i = A_i^0 \, e^{-kt} \tag{6}$$

where k is the same rate constant as given in equation 2, and A_i^0 is the initial activity of the labeled ion in the cell.

If the cells act as a single compartment, then the plot $\log A_i$ against t yields a straight line and the rate constant k can be obtained from its slope, allowing calculation of the efflux using equation 4.

It should be noted that only k has to be found from the efflux experiment, and it does not matter if internal radioactivity is lost before sampling is begun (e.g., during the separation procedure). However, the steady-state level of the ion in the cell is needed; this as suggested above, can be obtained by counting cells taken directly from the loading suspension. One further point: It is quite possible that a suspension of microalgae may show more complex kinetics for tracer exchange, not because each cell has more than one compartment, but because the suspension itself is heterogeneous, containing, for example, cells at various stages of their life cycle.

D. Centrifugation methods

1. Oil layer separation. As an alternative to the filtering procedure described above, tracer influxes and effluxes could be studied by rapidly centrifuging microalgae through an oil layer of a suitable specific gravity. This can enable rapid separation of cells from their aqueous suspension medium. Although I have not used this technique for microalgae, it has been used successfully in studies of radiotracer uptake by isolated chloroplasts (Heldt 1976).

2. Net fluxes. When *Chlorella* cells are grown in a culture medium high in Na^+ and lacking K^+, the cells become enriched in Na^+. Addition of K^+ to a suspension of Na^+-rich cells results in a significant net efflux of Na^+ and uptake of K^+ (Shieh and Barber 1971; Barber and Shieh 1972, 1973). This type of experiment proved valuable for studying

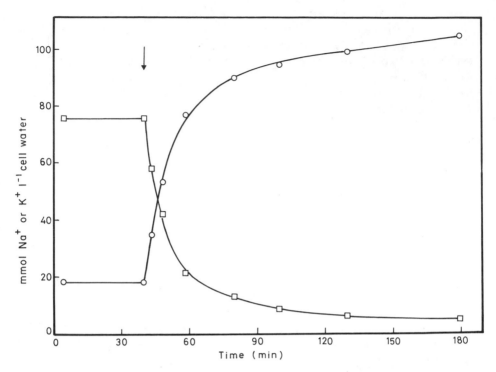

Fig. 44-2. Time course of net Na⁺ extrusion (squares) and K⁺ uptake (circles) induced by injecting 3 mM KCl, as indicated by the arrow, into a suspension of illuminated *Chlorella* cells grown so as to have a high intracellular Na⁺ content. Temperature, 25 C; white-light intensity, 50 W m⁻².

Na^+/K^+ regulation in *Chlorella*. The Na^+-rich medium can be identical to the normal medium given above, but with all the K^+ salts replaced by the corresponding Na^+ salts. A good initial inoculation of normal K^+ cells is usually needed, as growth is relatively slow. The cells can be harvested about 1 week later. For experiments, harvest the cells and resuspend them in the Na^+ culture medium at a density of 10 μl packed cells per milliliter. To follow the net fluxes typically shown in Fig. 44-2, use about 75–100 ml of suspension. Before adding K^+ (e.g., 3 mM KCl), take two or three 7-ml aliquots with a pipette and separate the cells from the medium by rapid centrifugation (discard supernate). Quickly wash pellet with cold distilled water by resuspension and centrifugation. Inject K^+ solution and take further samples with the pipette, centrifuging and washing as rapidly as possible after sampling. The washed pellets can then be digested in hot HNO_3 and the digest carefully made up with water to a final volume of 10 ml. Flame photometry measurements allow estimations of the K^+ and Na^+ contents of the extracts and calculations of the internal concentrations of these ions in terms of cell water, using the conversion factor given earlier.

These net flux curves can be analyzed kinetically and the net K^+ influx and Na^+ efflux calculated (Barber and Shieh 1972). Moreover, the sensitivity of the K^+/Na^+ exchange to temperature, light, and metabolic inhibitors can be tested in the same way as for tracer studies. For illuminated *Chlorella* cells at 25 C, the fluxes were shown to follow the first-order expressions:

$$(K)_t = (K)_{t=0} + Q_K(1 - e^{k_1 t}) \tag{7}$$

where (K) is the internal concentration of potassium at time t, and $Q_K = (K)_{t=\infty} - (K)_{t=0}$, and

$$(Na)_t = (Na)_{t=\infty} + Q_{Na}\, e^{-k_2 t} \tag{8}$$

where (Na) is the internal concentration of sodium at time t, and $Q_{Na} = (Na)_{t=0} - (Na)_{t=\infty}$. The rate constants ($k_1$ and k_2) can be used, as before, for calculating the flux rates.

3. Use of ion-selective electrodes. If net ion fluxes occur that cause changes in ion levels in the suspending medium, these changes may be detected by ion-selective electrodes. I have not used this approach myself, except for H^+ movement (Shieh and Barber 1971), but there is no reason why this technique should not be used for ions such as K^+ and Na^+. In fact, ion-selective electrodes have been used a great deal for studying net ionic movements with various biological tissue, and I refer the reader to the excellent studies with chloroplast suspensions carried out by Hind et al. (1974).

V. Acknowledgment

I would like to thank Martin Luton and Elizabeth Dibb for their help in preparing this chapter.

VI. References

Barber, J. 1968a. Measurement of the membrane potential and evidence for active transport of ions in *Chlorella pyrenoidosa. Biochim. Biophys. Acta* 150, 618–25.

Barber, J. 1968b. The influx of potassium into *Chlorella pyrenoidosa. Biochim. Biophys. Acta* 163, 141–9.

Barber, J. 1968c. The efflux of potassium from *Chlorella pyrenoidosa. Biochim. Biophys. Acta* 163, 531–8.

Barber, J. 1969. Light-induced net uptake of sodium and chloride by *Chlorella pyrenoidosa. Arch. Biochem. Biophys.* 130, 389–92.

Barber, J., and Shieh, Y. J. 1972. Net and steady-state cation fluxes in *Chlorella pyrenoidosa. J. Exp. Bot.* 23, 627–36.

Barber, J., and Shieh, Y. J. 1973. Sodium transport in Na^+-rich *Chlorella* cells. *Planta* 111, 13–22.

Dainty, J. 1962. Ion transport and electrical potentials in plant cells. *Annu. Rev. Plant Physiol.* 13, 379–402.

De Soete, D., Gijbels, R., and Hoste, J. 1972. *Neutron Activation Analysis.* Wiley, New York. 836 pp.

Dewar, M. A., and Barber, J. 1973. Cation regulation in *Anacystis nidulans. Planta* 113, 143–55.

Dewar, M. A., and Barber, J. 1974. Chloride transport in *Anacystis nidulans. Planta* 117, 163–72.

Heldt, H. W. 1976. Metabolite transport in intact spinach chloroplasts. In Barber, J. (ed.). *The Intact Chloroplast,* vol. I, *Topics in Photosynthesis,* pp. 215–34. Elsevier, Amsterdam.

Hind, G., Nakatani, H. Y., and Izawa, S. 1974. Light-dependent redistribution of ions in suspensions of chloroplast thylakoid membranes. *Proc. Natl. Acad. Sci. U.S.A.* 71, 1484–8.

Hutner, S. H., Provasoli, L., Schotg, A., and Haskin, C. P. 1950. Some approaches to the study of the role of metals in the metabolism of microorganisms. *Proc. Am. Philos. Soc.* 94, 152–70.

Ramsay, J. A., Brown, R. H., and Croghan, P. C. 1955. Electrometric titration of chloride in small volumes. *J. Exp. Biol.* 32, 822–9.

Reed, S. J. B. 1975. *Electron Microprobe Analysis.* Cambridge Monograph on Physics. Cambridge University Press, London. 400 pp.

Shieh, Y. J., and Barber, J. 1971. Intracellular sodium and potassium concentrations and net cation movements in *Chlorella pyrenoidosa. Biochim. Biophys. Acta* 233, 594–603.

Walker, N. A., and Pitman, M. G. 1976. Measurement of fluxes across membranes. In Lüttge, U., and Pitman, M. G. (eds.). *Encyclopedia of Plant Physiology,* vol. 2, part A. *Transport in Plants II,* pp 93–126. Springer-Verlag, Berlin.

45: Ion transport and ion-stimulated adenosine triphosphatases

CORNELIUS W. SULLIVAN

*Department of Biological Sciences,
University of Southern California, Los Angeles, California 90007*

CONTENTS

I Introduction   *page* **464**
II Materials and test organism   **466**
  A. Reagents and solutions   466
  B. Test organism   466
III Methods   **466**
  A. Storage of membrane-associated ATPase activities   466
  B. ATPase (EC 3.1.6.3) assays   466
  C. Development of assay conditions   468
  D. Release of ATPases from membranes   470
  E. Enzyme separation by polyacrylamide gel electrophoresis   470
IV **Sample data**
V **Discussion**   **471**
VI **References**   **471**
    475

I. Introduction

This chapter includes criteria and methods for the identification, localization, and isolation of ion-stimulated adenosine triphosphatases (ATPases) that may be part of, or functionally associated with, ion pumps or ion transport systems in microalgae. To begin to understand ion transport in microalgae, one must know ion concentrations outside and inside the cell as well as concentrations within various subcellular compartments (e.g., vacuoles, chloroplasts, and mitochondria), ion fluxes across limiting membranes, and the membrane potentials. These subjects have been addressed and methods outlined for several algae in excellent reviews (Gutknecht and Dainty 1968; MacRobbie 1970).

In biological systems ions can be considered to be acted on by two physical forces: the chemical potential gradient and the electropotential gradient. Together they constitute the electrochemical potential gradient.

$$\bar{u} = d(RT \ln a)/dx + zF \, d\psi/dx \tag{1}$$

In this expression $\bar{u}$ is the electrochemical potential gradient, R is the gas constant, T is the absolute temperature, a is the chemical activity, Z is the algebraic valency, F is the Faraday constant, and ψ is the electrical potential.

If an ion moves "uphill," from a lower to a higher electrochemical potential during its transport through a membrane, the process requires an input of energy, and this transport process is said to be active. The concept of energy-dependent ion transport in algae evolved from the work of Hoagland et al. (1926) and Hoagland and Davis (1929), in which a light-enhanced ion uptake was observed in *Nitella*. Since that time many studies with algae have been directed at the physiological processes and molecular mechanisms involved in ion transport (see reviews by Gutknecht and Dainty 1968; MacRobbie 1970; Hodges 1973; and Higinbotham 1974).

Active transport in algae may be driven by respiration and/or photosynthesis. Although the precise nature of the metabolic coupling into ion transport is unknown, generally anion transport is thought to be coupled directly to electron transfer reactions or to be dependent on some reduced product (MacRobbie 1970); or ATP is involved, and a major component of the system is a membrane-bound ATPase. Only a few ATPase activities have been described in algal membranes (Bonting and Caravaggio 1966; Jokela 1969; Meszes and Erdei 1969; Hemmingsen 1971; Sullivan and Volcani 1974; Sullivan et al. 1974, 1975; Falkowski 1975), and only recently has any evidence for their possible role in ion transport been shown (Sullivan and Volcani 1974, Sullivan et al. 1975, Falkowski 1975).

What characteristics might we expect of an ATPase involved in the transport of an ion (such as Na^+) across algal membranes? The activity should (1) be located in a membrane associated with the transport of Na^+, (2) preferentially catalyze the hydrolysis of ATP compared with other nucleotides, (3) catalyze ATP hydrolysis at a rate dependent on the Na^+ concentration, (4) show specificity for Na^+ compared with other monovalent cations, and although there is no a priori reason to expect it (5) be dependent on the presence of Mg^{2+}, as this has been shown to be a characteristic of all previously described ATPase activities associated with ion transport (Skou 1965).

With these minimal criteria for the ATPase established, the characteristics of the Na^+ transport system should be examined for these same characteristics in whole cells or membrane vesicle preparations. Other parameters can be examined, such as sensitivity to various metabolic or enzyme inhibitors, kinetic constants for Na^+, and kinetics of Na^+ uptake or activation. Then if the Na^+-stimulated ATPase activity is involved in Na^+ transport, a high degree of correlation should be observed among the characteristics tested (Hodges 1973). For instance, Na^+ uptake kinetics might be expected to show a positive correlation with the rate of ATP hydrolysis vs. Na^+ concentration. Note that a positive correlation is not necessarily expected in all comparative studies. Agents that completely block the ATPase are expected to inhibit transport; however, the reverse is not necessarily true. In assessing the degree of correlation, one must remember that, in general, agents or conditions that modify the activity of the ATPase will show a similar quantitative response with respect to the transport activity only if the ATPase component involves the rate-limiting step of the overall transport process.

Methods for examining the first set of criteria for a special case of ion-stimulated ATPase – active linked transport of Na^+ and K^+ in the marine diatom *Nitzschia alba* – are given in this chapter.

II. Materials and test organism

A. Reagents and solutions

All solutions are prepared using glass-distilled deionized water, hereafter referred to as water. All chemicals are reagent grade and were obtained as follows:

1. TRIS-ATP, 45 mM stock solution prepared by neutralizing (pH 7) with a 1 M stock solution of Trizma base (Sigma). This may be stored frozen for up to 3 weeks.

2. Bromphenol blue, hydrazine sulfate, and Nuchar activated charcoal (MC/B Manufacturing Chemists), 100 mg ml^{-1} suspension in water after washing twice in 4 l of 0.5 N HCl, twice in 4 l of water, and collecting on Whatman No. 1 filter paper in a Büchner funnel under vacuum.

3. γ-^{32}P-adenosine-5′-triphosphate, tetratriethylammonium salt, 10 Ci mmol^{-1} (New England Nuclear).

4. Photopolymerization acrylamide kit (Bio-Rad).

5. Solutions for phosphate analysis. Solution 1: 4.4 g $(NH_4)_6Mo_7O_{24} \cdot 4H_2O$ dissolved in 1 l 1.6 N H_2SO_4 and stored at room temperature. Solution 2: 0.5 g $SnCl_2$ and 2.0 g hydrazine sulfate dissolved in 950 ml of 0.6 N H_2SO_4, filtered through a 0.45-μm pore Millipore filter, and stored at 5 C.

B. Test organism

Nitzschia alba (Lewin and Lewin). See Chp. 5 (III.A.1).

III. Methods

Membrane-enriched fractions were isolated as described in Chp. 5.

A. Storage of membrane-associated ATPase activities

The activity of enzymes can be maintained for 1–3 days by the procedures outlined in Chp. 5 (III.D). Never use stored material when initiating a search for a particular enzyme, as it may be labile and may not be detected in stored material. Freshly prepared membranes should always be used as source material, and optimum conditions for storage should be based on the activity of the newly isolated material.

B. ATPase (EC 3.1.6.3) assays

Extensive studies (Sullivan et al., 1974, 1975; Sullivan and Volcani 1974; Okita et al. 1976) have shown that there are basically three classes of ATP hydrolyzing activities associated with diatom membranes: (Mg^{2+})-dependent ATPases that are stimulated by monovalent anions or by monovalent cations, and (Ca^{2+})-dependent ATP hydrolyzing activities that are not further stimulated by additions of monovalent ions. Each activity is assayed by incubation in a medium determined to give a maximum response of that activity. The final concentrations of constituents in the incubation mixtures are 100 mM TRIS-HCl in each and: (1) (Mg^{2+})-ATPase, 4 mM $MgCl_2$, 2 mM ATP, pH 8.8; (2) (Mg^{2+}, Na^+)-ATPase, 2 mM $MgCl_2$, 4 mM ATP, 100 mM KCl, pH 8.5; (3) (Mg^{2+}, K^+)-ATPase, 2 mM $MgCl_2$, 4 mM ATP, 100 mM KCl, pH 8.5; (4) high-salt (Mg^{2+}, Na^+, K^+)-ATPase, 2 mM $MgCl_2$, 4 mM ATP, 100 mM NaCl, 100 mM KCl, pH, 8.5; (5) synergistic (Mg^{2+}, Na^+, K^+)-ATPase, 2 mM $MgCl_2$, 4 mM ATP, 5 mM NaCl, 45 mM KCl or 5 mM NaCl, 350 mM KCl, pH 8.5; (6) (Mg^{2+}, HCO_3^-)-ATPase, 4 mM $MgCl_2$, 2 mM ATP, 50 mM $KHCO_3$, pH 8.2; (7) (Ca^{2+})-ATPase, 2 mM $CaCl_2$, 2 mM ATP, pH 8.8. These assay conditions are used to determine the activity of membrane-bound enzymes in membrane fractions and to assay enzymes released from the membranes by detergent and separated on polyacrylamide gels.

1. Membrane-bound activities. The assay is initiated by the addition of 100 μl, containing 10–50 μg membrane protein, to 900 μl of pre-warmed assay mixture in 12-ml Pyrex tubes. After incubations with shaking at 30 C for 10–30 min, the reaction is stopped by adding 2 ml of 3.5% perchloric acid followed by 0.5 ml of a suspension of activated charcoal. The content of protein per assay and incubation times are adjusted so as not to exceed 10% hydrolysis of the substrate, yet obtain final absorbance readings of 0.4–0.7 above the blank. After thorough mixing, the tubes are placed in an ice-water bath for 30–60 min; charcoal is removed by filtration through 24-mm diam. 0.45-μm membrane filters. The residue containing adsorbed nucleotides is washed with two 1-ml aliquots of ice-cold water, and the filtrates are collected in a 12-ml graduated centrifuge tube. The volume of filtrate is brought to 5 ml with water.

2. Phosphate determination. The phosphate released during ATP hydrolysis is determined by Lindeman's (1958) modification of the Fiske-Subbarow method (1925). To 5 ml of the filtrate is added 3 ml of reagent No. 1, the tube is mixed, then 2 ml of reagent No. 2 is added and mixed immediately. A blue color develops, and after 15 min in the dark at room temperature the absorbance is read at 720 nm. The color is stable for ca. 2 h. From a stock solution of KH_2PO_4 that contains 200 μg P*i* ml^{-1}, appropriate dilutions are made to give 1–20 μg P*i* per 5 ml H_2O, and these are dispensed into assay tubes. Water (5 ml) is used as a blank for the standards and enzyme assays. Phosphate is analyzed as described above for the filtrate, and a standard curve is constructed by plotting absorbancy at 720 nm vs. micrograms of P*i*.

For measuring initial rates during the first 60 sec of the reaction, a more sensitive assay is desired. Without significantly changing the assay procedure described, γ-^{32}P-ATP can be added to the assay to give a final concentration of 0.1–1.0 μCi per assay. The same procedure described above is followed except that the filtrate is collected and 1 ml is added to a scintillation vial with 10 ml Aquasol for radioassay in a liquid scintillation counter.

3. Controls. A control assay must be determined for each variation in the basic assay condition, for each new membrane preparation, and for each new ATP stock solution.

a. Additions. It is particularly important to monitor the effects of any additions to the assay mixture on the color development during phosphate analysis and on the enzyme activity. Inhibitor studies involve dissolution of various agents in 100% ethanol; a final ethanol concentration of 1–10% in the assay mix may significantly decrease or increase the ATPase activity. Thus the standard assay mixture containing the appropriate ethanol concentration must be assayed as a control.

When the effects of anions on ATPase activity are assayed, particular attention must be paid to the effects of their counterions, especially for Na$^+$ or K$^+$, as these ions may stimulate activity dramatically. Care must also be taken not to add monovalent cations to solutions inadvertently through the addition of acids or bases.

C. Development of assay conditions

ATPase assay conditions should yield maximum rates of ATP hydrolysis, and the rate of hydrolysis must be linear with time and amount of enzyme (membrane protein) added. The assay parameters, which are adjusted to maximize the rates of ATP hydrolysis, are the following: pH, substrate level, divalent ion preference and concentration, ATP/Mg^{2+} or ATP/Ca^{2+} molar ratio, and temperature. Because we wish to know the ATPase and ion transport activities under optimum growth conditions, a physiologically realistic temperature is desired. Thus, because the externally measured temperatures are known to be identical with internal temperatures of microorganisms, one might logically choose the growth temperature of the organisms for the assay. The ATPase can be further characterized by determining its response to monovalent cations and anions, nucleotide specificity, kinetic constants, and sensitivity to inhibitors. See Sullivan et al. (1975) and Tables 45-1 and 45-2 for examples.

Table 45-1. *Kinetic constants for ATPase activity associated with various membrane fractions of* Nitzschia alba

Constant	(Mg^{2+})-activity				High-salt (Na^+, K^+, Mg^{2+}) activity			
	Mito.	ER	Golgi	PM	Mito.	ER	Golgi	PM
K_m(mM ATP)	1.1 ± 0.2	0.64 ± 0.1	0.3 ± 0.7	1.3 ± 0.3	0.89 ± 0.5	1.7 ± 0.3	2.6 ± 0.4	0.73 ± 0.2
V(μmol P_i mg^{-1} protein h^{-1})	7.0 ± 0.5	1.7 ± 0.1	0.83 ± 0.1	2.5 ± 0.2	2.5 ± 0.4	30.0 ± 1.9	16.1 ± 1.0	27.3 ± 1.9

Note: Mito., mitochondria; ER, endoplasmic retirulum; PM, plasma membrane.
Source: Sullivan et al. (1975).

Table 45-2. *Effects of anions on the (Mg²⁺)-dependent ATPase activities from membrane fractions of Nitzschia alba*

Anion (100 mM)	Specific activity			Percent of control (Na⁺ or K⁺ salt)		
	Mito.	Golgi	PM	Mito.	Golgi	PM
K⁺ salts						
Cl⁻	3.85	1.10	4.66	100	100	100
Br⁻	4.27	0.85	4.48	90	77	96
I⁻	2.65	0.29	2.22	62	26	48
NO₂⁻	3.05	0.77	4.40	88	70	94
NO₃⁻	3.76	0.83	4.66	71	75	100
HCO₃⁻	10.0	1.68	9.00	234	152	193
SO₄²⁻	7.9	1.77	4.08	185	160	87
Na⁺ salts						
Cl⁻	15.6	9.24	74.22	100	100	100
F⁻	0.97	0.23	0.31	6	2	15
Acetate	8.17	5.23	68.4	52	57	92
None	3.35	0.75	2.48	—	—	—

Note: Mito., mitochondria; PM, plasma membrane
Source: Sullivan et al. (1975).

D. Release of ATPases from membranes

Most ATPases are integral proteins (Singer and Nicholson 1972) and are strongly associated with membranes. However, they can generally be released by one or more of three groups of compounds: detergents, chaotropic agents, and organic solvents.

The nonionic detergent Brij 36T has been shown to be useful for dissociating ATPases from diatom membranes (Okita et al. 1976). The membrane fraction is washed three times by centrifugation at 40 000 × g for 30 min in 20 ml of 20 mM TRIS-HCl, pH 7.4, and resuspended in the same buffer to a concentration of 5 mg protein ml⁻¹. A 10% solution of Brij 36T is added to give a 1:1 (w/w) detergent protein ratio; the mixture is stirred on a magnetic stirrer for 60 min at 4 C. The resulting solubilized membranes are centrifuged at 100 000 × g at 5 C for 60 min. The supernatant fluid is carefully collected with a Pasteur pipet to avoid disturbing the membrane pellet, and the solution is made to 5 mM with 2-mercaptoethanol and 0.4 M sucrose.

E. Enzyme separation by polyacrylamide gel electrophoresis

To discriminate between an ATP hydrolyzing activity and a discrete ATPase protein, physical separation and identification are required.

These can be accomplished by polyacrylamide gel electrophoresis, which separates proteins on the basis of molecular weight (size) and charge.

Gels of 7% polyacrylamide measuring 5×65 mm are prepared and run in TRIS-glycine buffer, pH 8.9, at 2 C according to Gabriel (1971). These running gels are preconditioned by electrophoresis at 2 mA per gel for 5 min before polymerization of the stacking gel. Approximately 200 μg of solubilized membrane protein is loaded on each gel immediately before electrophoresis at 2 mA per gel for 105 min.

1. Localization of ATPase on gel. Immediately following electrophoresis, the gels are rinsed with water, then preincubated for 20 min at 0 C in 100 mM TRIS-maleate buffer, pH 7.8. The gels are rinsed in water and incubated for 20 min at 25 C in the appropriate ATPase assay mix contained in a 12-ml Pyrex test tube. An aliquot of conc. $Pb(NO_3)_2$ is added to give a final concentration of 3.3 mM, the reaction is incubated for 1 h at 25 C, washed overnight in 200 ml water, and the gel developed by submerging in a 1% $(NH_4)_2S$ solution in a 12-ml test tube for 2–5 min. Gels are then removed to water and bands are observed, recorded, or photographed.

IV. Sample data

Tables 45-1 and 45-2 represent (Mg^{2+})-dependent and high-salt (Mg^{2+}, Na^+, K^+)-ATPase characteristics for various membrane-enriched fractions from *N. alba*.

Figures 45-1 and 45-2 and Tables 45-3 to 45-6 show characteristics of the (Mg^{2+})-dependent synergistically stimulated (Na^+, K^+)-ATPase from the 20-fold-enriched plasma membrane fraction of *N. alba* These data demonstrate the minimal requirements of an ATPase activity involved in ion transport. Such an activity as that shown in Fig. 45-1 is thought to represent the presence of a Na^+ pump in the diatom plasma membrane similar to the model suggested by Baker (1966).

Figure 45-3 is a drawing of representative electrophoretic patterns of ion-stimulated ATPase and ADPase activities from solubilized membranes of *N. alba*. This technique allows the separation and enumeration of individual proteins having ATP hydrolyzing ability.

V. Discussion

The data presented here and in previous reports (Sullivan and Volcani 1974; Sullivan et al. 1975; Falkowski (1975) make a strong case for

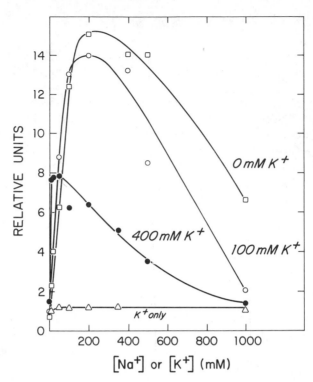

Fig. 45-1. Effects on the (Mg^{2+})-dependent ion-stimulated ATPase activity from *Nitzschia alba* plasma membranes of varied high Na^+ concentrations in the presence of constant K^+ (□, ●, ○) and of varied K^+ in the absence of Na^+ (△). (From Sullivan and Volcani 1974.)

Table 45-3. (Mg^{2+})-*dependence of the* (Na^+, K^+)-*stimulated ATPase from the plasma membrane fraction of Nitzschia alba*

Divalent cation	Na^+/K^+ (mM)	Total ATPase activity (μmol P_i h^{-1} mg^{-1} protein)	Net (Na^+, K^+)-stimulated ATPase activity (specific activity)
None	0/0	0.53	—
	50/0	1.16	0.63
	0/50	0.72	0.19
	25/25	1.16	0.63
	5/45	0.47	−0.06
$MgCl_2$	0/0	2.19	—
	50/0	10.3	8.11
	0/50	3.03	0.84
	25/25	10.3	8.11
	5/45	10.7	8.51

Source: Adapted from Sullivan and Volcani (1974).

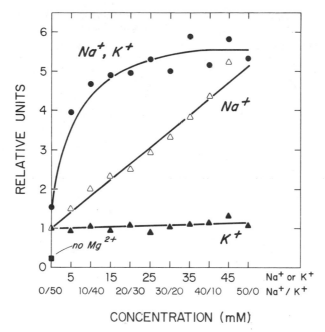

CONCENTRATION (mM)

Fig. 45-2. Synergistic stimulation of *Nitzschia alba* plasma membrane ATPase activity by various molar ratios of Na$^+$ and K$^+$ at constant ionic strength. (From Sullivan and Volcani 1974.)

Table 45.4. *Effects of divalent cations on the (Mg^{2+})-dependent (Na^+, K^+)-ATPase from plasma membranes of Nitzschia alba*

Divalent cations (2 mM)	Total ATPase activity (μmol P$_i$ h^{-1} mg^{-1} protein)	Net (Na$^+$,K$^+$)-stimulated ATPase activity (specific activity)
Mg	6.16	3.97
Mn	4.66	2.16
Ca	2.84	0·
Co	1.97	1.26
Zn	1.16	0
Cu	0.06	0

Source: Adapted from Sullivan and Volcani (1974).

the existence of ATPase systems in algal membranes that have the potential for being involved in the active transport of various monovalent anions and cations.

A plasma-membrane-bound ATPase has been shown convincingly to participate in the coupled transport of Na$^+$ and K$^+$ in animal cells (Skou 1965; Glynn et al. 1971). The association of (Mg^{2+})-dependent

Table 45-5. *Nucleotide substrate specificity of the (Mg^{2+})-dependent (Na^+, K^+)-ATPase from plasma membranes of Nitzschia alba*

Nucleotides (4 mM)	Specific activity of ATPase (μmol P_i h^{-1} mg^{-1} protein)	Net ATPase activity (specific activity)
ATP	6.90	5.50
ADP	0.43	0.38
AMP	0.09	0.09
CTP	1.33	0.87
ITP	1.48	-0.21
GTP	0.19	-0.48
UTP	0.90	0.38

Source: Adapted from Sullivan and Volcani (1974).

Table 45-6. *Monovalent cation specificity of (Mg^{2+})-dependent (Na^+, K^+)-ATPase from plasma membranes of Nitzschia alba*

Monovalent cation	5 mM Na^+ + 45 mM cation		5 mM Na^+ + 350 mM cation		5 mM cation + 350 mM K^+	
	Specific activity[a]	Control (%)	Specific activity[a]	Control (%)	Specific activity[a]	Control (%)
None	3.16[b]	100	3.17[b]	100	3.16[c]	100
Li	3.83	122	3.74	118	3.28	104
Na	—	—	—	—	18.2	576
K	7.10	225	18.1	571	—	—
Rb	5.19	164	3.83	120	3.82	120
Cs	4.44	140	1.09	34	4.55	144
NH_4	5.75	182	10.88	343	4.55	144
Choline	4.20	132	11.49	362	3.46	109

[a] μmol P_i h^{-1} mg^{-1} protein.
[b] Na^+ is present at 5 mM.
[c] K^+ is present at 350 mM.
Source: Adapted from Sullivan and Volcani (1974).

(Na^+, K^+)-ATPase and (Na^+)-ATPase activities with the plasma membrane fraction of *N. alba* is especially interesting, as they could participate in the energy transduction process for Na^+ and K^+ transport systems in diatoms. If so, this would suggest that ATP may be the energy source for cation transport in diatoms, as was suggested for cation transport in *Nitella* (MacRobbie 1965).

Confirmation of this hypothesis is of considerable importance, not only for developing a more thorough knowledge of Na^+ and K^+ transport in algae, but also because of the need to broaden our understanding of the phylogenetic and evolutionary relationships of this important system. In addition, the possible Na^+-dependent cotrans-

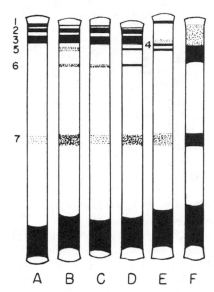

Fig. 45-3. Diagrammatic representation of electrophoretic patterns observed for *Nitzschia alba* solubilized membrane ATPase and ADPase activities on polyacrylamide gels. A Brij 36T solubilized extract of a 20 000 × *g*, 60 min, membrane fraction containing 200 μg protein was layered on each gel. Gels were aligned from the top; the distinct line above band 1 is the interface of stacking and running gels. Numbers 1–7 indicate the relative position of each band. The gels were incubated at the optimal assay conditions described for *A*, (Mg^{2+})-ATPase; *B*, (Mg^{2+}, Na^+)-ATPase; *C*, (Mg^{2+}, K^+)-ATPase; *D*, (Mg^{2+}, Na^+, K^+)-ATPase (high-salt condition); *E*, (Ca^{2+})-ATPase; *F*, (Ca^{2+})-ADPase. (From Okita et al. 1976.)

port of organic solutes (*sensu* Crane 1965; Curran 1965) in heterotrophic and facultatively heterotrophic diatoms may play an integral part in the metabolism and survival of diatoms that utilize organic compounds (see review by Hellebust and Lewin 1977).

VI. References

Baker, P. F. 1966. The sodium pump. *Endeavour* 25, 166–72.

Bonting, S. L., and Caravaggio, L. L. 1966. Studies on Na^+-K^+-activated adenosine triphosphatase. XVI. Its absence from the cation transport system of *Ulva lactuca. Biochim. Biophys. Acta* 112, 519–23.

Crane, R. K. 1965. Na^+-dependent transport in the intestine and other animal tissues. *Fed. Proc.* 24, 1000–6.

Curran, P. F. 1965. Ion transport in intestine and its coupling to other transport processes. *Fed. Proc.* 24, 993–9.

Falkowski, P. G. 1975. Nitrate uptake in marine phytoplankton: (Nitrate, chloride)-activated adenosine triphosphatase from *Skeletonema costatum* (Bacillariophyceae). *J. Phycol.* 11, 323–6.

Fiske, C. H., and Subbarow, Y. 1925. The colorimetric determination of phosphorus. *J. Biol. Chem.* 66, 375–400.

Gabriel, O. 1971. Analytical disc gel electrophoresis. *Methods Enzymol.* 22, 565–78.

Glynn, I. M., Hoffman, J. F., and Lew, V. L. 1971. Some "partial reactions" of the sodium pump. *Philos. Trans. R. Soc. London Ser. B* 262, 91–102.

Gutknecht, J., and Dainty, J. 1968. Ionic relations of marine algae. *Oceanogr. Mar. Biol. Annu. Rev.* 6, 163–200.

Hellebust, J. A., and Lewin, J. C. 1977. Heterotrophic nutrition. In Werner, D (ed.). *Biology of Diatoms,* pp. 169–97. Blackwell, Oxford.

Hemmingsen, B. B. 1971. A mono-silicic acid stimulated adenosinetriphosphatase from protoplasts of the apochlorotic diatom *Nitzschia alba.* Ph.D. thesis, University of California, San Diego. 132 pp.

Higinbotham, N. 1974. Conceptual developments in membrane transport, 1924–1974. *Plant Physiol.* 54, 454–62.

Hoagland, D. R., and Davis, A. R. 1929. The intake and accumulation of electrolytes by plant cells. *Protoplasma* 6, 610–26.

Hoagland, D. R., Hibbard, P. L., and Davis, A. R. 1926. The influence of light, temperature, and other conditions on the ability of *Nitella* cells to concentrate halogens in the cell sap. *J. Gen. Physiol.* 10, 121–46.

Hodges, T. K. 1973. Ion absorption by plant roots. *Adv. Agron.* 25, 163–207.

Jokela, A. C. C. T. 1969. Outer membrane of *Dunaliella tertiolecta:* Isolation and properties. Ph.D. thesis, University of California, San Diego. 123 pp.

Lindeman, W. 1958. Observations on the behaviour of phosphate compounds in *Chlorella* at the transition from dark to light. *Proc. 2nd Int. Conf. Peaceful Uses Atomic Energy* 24, 8–15.

MacRobbie, E. A. C. 1965. The nature of the coupling between light energy and active ion transport in *Nitella translucens. Biochim. Biophys. Acta* 94, 64–73.

MacRobbie, E. A. C. 1970. The active transport of ions in plant cells. *Q. Rev. Biophys.* 3, 251–94.

Meszes, G., and Erdei, J. 1969. (Mg^{2+}-Na^+-K^+)-activated ATPase activity and its properties in *Scenedesmus obtusiusculus* algal cells. *Acta Biochim. Biophys. Acad. Sci. Hung.* 4, 131–9.

Okita, T., Sullivan, C. W., and Volcani, B. E. 1976. Gel electrophoresis of ion-stimulated ATPases from protoplast membranes of the diatom *Nitzschia alba. Plant Sci. Lett.* 6, 129–34.

Singer, S. J., and Nicolson, G. 1972. The fluid mosaic model of the structure of cell membranes. *Science* 175, 720–31.

Skou, J. C. 1965. Enzymatic basis for active transport of Na^+ and K^+ across cell membrane. *Physiol. Rev.* 45, 596–617.

Sullivan, C. W., and Volcani, B. E. 1974. Synergistically stimulated (Na^+, K^+)-adenosine triphosphatase from plasma membrane of a marine diatom. *Proc. Natl. Acad. Sci. U.S.A.* 71, 4376–80.

Sullivan, C. W., Volcani, B. E., Lum, D., and Chiappino, M. L. 1974. Isolation and characterization of plasma and smooth membranes of the marine diatom *Nitzschia alba. Arch. Biochem. Biophys.* 163, 29–45.

Sullivan, C. W., Volcani, B. E., and Lum, D. 1975. Multiple ion-stimulated adenosine triphosphatase activities associated with membranes of the diatom *Nitzschia alba. Arch. Biochem. Biophys.* 167, 437–43.

Section VII

Inhibitors

46: Inhibitors used in studies of algal metabolism

S. S. BADOUR

Department of Botany,
University of Manitoba, Winnipeg, Manitoba R3T 2N2, Canada

CONTENTS

I **Introduction** *page* **480**
II **References** **485**

I. Introduction

Most of the inhibitors listed in Table 46-1 are metabolite analogues, herbicides, antibiotics, chelators, uncouplers, or thiol reagents. The references given in this chapter were selected to represent various studies dealing with the use of inhibitors in algal metabolism. If the reference cited does not include sufficient information on the chemical structure and properties of the inhibitor, Dawson et al. (1969) may be consulted.

With respect to inhibitors of photosynthesis and phosphorylation, the reader is referred to the articles of Losada and Arnon (1963), Good and Izawa (1973), and Metzner (1969). To acquire further knowledge on the chemical properties and action mechanisms of inhibitors of nucleic acid and protein biosynthesis, the concise monograph of Kersten and Kersten (1974) is recommended. The work of Webb (1963) provides detailed information on inhibitors of respiration and related metabolic processes. His comments on their specificity (Chp. 16, vol. I) and his suggestions for planning and reporting inhibition studies (Chp. 17, vol. I) constitute the fundamentals that should be considered when metabolic inhibitors are to be used in phycological research.

Table 46-1. *Inhibitors used in studies of algal metabolism*

Name	Site of inhibition	Organism	Reference
Actinomycin D	DNA-dependent RNA synthesis; also DNA polymerase	*Cyclotella*	Oey and Schnepf (1970)
3-Amino-1,2,4-triazole (amitrole)	Phosphorylase (Q enzyme)	*Chlorella Cyanidium Oscillatoria*	Fredrick (1969)
Amytal	NADH oxidases and energy transfer reactions at the NAD-flavoprotein region	*Prototheca*	Lloyd (1966)

Table 46-1. (*cont.*)

Name	Site of inhibition	Organism	Reference
Antimycin A	Cyclic photo-phosphoryla-tion and reoxidation of reduced cyt *b* in mitochon-dria	*Chlorella*	Tanner et al. (1965)
		Polytomella	Lloyd and Chance (1968)
Arsenite	Respiration (reacts with – SH groups)	*Navicula*	Lewin (1954)
Atebrin	Photosynthetic and oxidative phosphoryla-tion (specific flavin antago-nist)	*Phormidium*	Nultsch (1966)
8-Azaguanine	Protein synthesis (causes transla-tion errors)	*Polytomella*	Lloyd et al. (1968)
Azaserine	Reactions involv-ing transamin-ation	*Scenedesmus*	Barker et al. (1956)
Azide	Fe enzymes (e.g., cytochrome oxidase)	*Anabaena*	Leach and Carr (1970)
Bisulfite	Glycolate dehy-drogenase or oxidase	*Chlorella*	Goulding and Merrett (1966)
Carbon monox-ide	Cytochrome oxi-dase; reversed by light	*Saprospira* *Vitreoscilla*	Webster and Hackett (1966)
Carbonyl cya-nide-*m*-chloro-phenyl hydrazone (CCCP)	Photophosphory-lation (un-coupler)	*Ankistrodesmus*	Urbach and Gimmler (1969)
Chloral hydrate	Protein synthesis	*Chlamydomonas*	McMahon and Blaschko (1971)
Chloramphenicol (D-*threo*-iso-mer)	Protein synthesis on chloroplast ribosomes (70s)	*Euglena*	Davis and Mer-rett (1975)
p-Chloromercuri-benzoate (PCMB)	–SH groups present in cell membrane proteins	*Acetabularia*	Brachet (1975)

Table 46-1. (*cont.*)

Name	Site of inhibition	Organism	Reference
p-Chloromercuri-benzene sul-fonate (PCMBS)	–SH groups present in cell membrane proteins	*Acetabularia*	Brachet (1975)
Chloropromazine	Cytochrome oxi-dase, oxidative phosphoryla-tion, and N_2 fixation	*Anabaena*	Cox (1966)
3-(*p*-Chloro-phenyl)1,1-di-methylurea (CMU)	O_2 evolution step of photosyn-thesis	*Chlorella*	Ogasawara and Miyachi (1970)
Cordycepin	*m*RNA synthesis	*Acetabularia*	Brachet (1975)
Cyanide	Cytochrome oxi-dase, carboxy-dismutase system	*Euglena* *Chlorella*	Sharpless and Butow (1970) Rabin et al. (1958)
Cycloheximide (actidione)	Protein synthesis on cytoplasmic ribosomes (80s)	*Euglena* *Chlamydomonas*	Davis and Mer-rett (1975) McMahon (1975)
Desaspidin	Photophosphory-lation and oxi-dative phosphoryla-tion	*Phormidium*	Nultsch (1969)
Diamox (acetazo-lamide)	Carbonic anhy-drase	*Chlorella*	Graham and Reed (1971)
Dichlone	CO_2 fixation, synthesis of lipids and glu-tamic acid (alters mem-brane permea-bility)	*Chlorella*	Zweig et al. (1972); Sikka et al. (1973)
3-(3,4-Dichloro-phenyl)-1,1-di-methyl urea (DCMU)	O_2 evolution step of photosyn-thesis	*Scenedesmus*	Bishop (1958)
2,4-Dinitro-phenol (DNP)	Oxidative phos-phorylation (uncoupler)	*Chlorella*	Kandler (1958)
Dio-9	Photophosphory-lation and the coupled elec-tron flow	*Chara*	Smith and West (1969)
Diphenylamine	Carotenoid synthesis	*Anabaena*	Ogawa et al. (1970)

Table 46-1. (*cont.*)

Name	Site of inhibition	Organism	Reference
α,α'-Dipyridyl	Fe^{2+} enzymes (iron chelator)	*Golenkinia*	Ellis et al. (1975)
Disalicylidene propanediamine (DSPD)	Reduction of ferredoxin, cyclic photophosphorylation	*Ankistrodesmus*	Urbach and Gimmler (1969)
Fluoride	Mg^{2+}-requiring enzymes, such as enolase and acid phosphatase	*Chlorella*	Sargent and Taylor (1975)
5-Fluorodeoxyuridine	DNA synthesis	*Cyclotella*	Oey and Schnepf (1970)
5-Fluorouracil	Synthesis of chloroplast protein	*Euglena*	Lord et al. (1975)
Germanic acid	Silicate uptake and metabolism	*Cylindrotheca*	Darley and Volcani (1969)
4-(*n*-Heptyl)-hydroxyquinoline-N-oxide (HOQNO)	Electron transport in PS II region, also in mitochondria near cyt *b*	*Hydrodictyon*	Raven (1969)
Hydroxylamine	O_2-evolving system in photosynthesis	*Anacystis* *Scenedesmus*	Cheniae and Martin (1972); Gaffron (1943)
α-Hydroxy-2-pyridine methanesulfonate (α-HPMS)	Glycolate dehydrogenase and glycolate oxidase	*Chlorella* *Anabaena*	Lord and Merrett (1969); Grodzinski and Colman (1970)
Imidazole	Photophosphorylation (uncoupler)	*Nitella*	MacRobbie (1965)
Iodoacetamide	Reactions involving –SH groups	*Chlorella*	Coombs and Whittingham (1966)
Isonicotinyl hydrazide (INH)	Conversion of glycine to serine	*Chlorella*	Pritchard et al. (1961)
Levulinic acid	δ-Aminolevulinic acid dehydratase	*Chlorella*	Beale (1970)
6-Methylpurine	Functional RNA synthesis	*Chlorella*	McCullough and John (1972)
Monofluoroacetate (MFA)	Aconitase	*Phaeodactylum*	Cooksey (1974)

Table 46-1. (*cont.*)

Name	Site of inhibition	Organism	Reference
Nalidixic acid	Chloroplast DNA replication	*Euglena*	Lyman et al. (1975)
Ouabain	Cation transport (the transport ATPases)	*Nitella*	MacRobbie (1962)
o-Phenanthroline	O_2 evolution in photosynthesis (iron chelator)	*Golenkinia*	Ellis et al. (1975)
		Scenedesmus	Gaffron (1945)
Phenylurethane	O_2 evolution in photosynthesis	*Anabaena*	Cox (1966)
Phlorizin	Photophosphorylation and the membrane ATPase	*Hydrodictyon*	Raven (1968)
Puromycin	Protein and RNA synthesis	*Acetabularia*	Bonotto et al. (1969)
Rifampicin	Chloroplast DNA transcription	*Chlamydomonas*	Surzycki et al. (1970)
Rotenone	Flavin region of electron transport chain in mitochondria	*Prototheca*	Lloyd (1966)
Salicylaldoxime	Cyclic photophosphorylation	*Scenedesmus*	Stuart (1971)
SAN 9789	Carotenoid synthesis	*Euglena*	Vaisberg and Schiff (1976)
Selenomethionine	Methionine uptake and metabolism	*Chlorella*	Shrift (1959)
Spectinomycin	Chloroplast protein synthesis at translational level	*Chlamydomonas*	Surzycki et al. (1970)
Tosylarginine methyl ester (TAME)	Trypsin and chymotrypsin-like proteases	*Acetabularia*	Brachet (1975)
s-Triazine(s)	O_2 evolution in photosynthesis	*Ankistrodesmus*	Rau and Grimme (1971)
p-Trifluoromethoxycarbonyl-cyanide phenylhydrazone (FCCP)	Oxidative phosphorylation (uncoupler)	*Anabaena*	Leach and Carr (1970)
Triparanol	Sterol biosynthesis	*Chlorella*	Chan et al. (1974)

II. References

Barker, S. A., Bassham, J. A., Calvin, M., and Quarck, U. C. 1956. Sites of azaserine inhibition during photosynthesis by *Scenedesmus*. *J. Am. Chem. Soc.* 78, 4632–5.

Beale, S. I. 1970. The biosynthesis of δ-aminolevulinic acid in *Chlorella*. *Plant Physiol.* 45, 504–6.

Bishop, N. I. 1958. The influence of the herbicide, DCMU, on the oxygen-evolving system of photosynthesis. *Biochim. Biophys. Acta* 27, 205–6.

Bonotto, S., Goffeau, A., Janowski, M., Vanden Driessche, T., and Brachet, J. 1969. Effects of various inhibitors of protein synthesis on *Acetabularia mediterranea*. *Biochim. Biophys. Acta* 174, 704–12.

Brachet, J. 1975. The effects of some inhibitors of RNA synthesis and proteolysis on morphogenesis in *Acetabularia mediterranea*. *Biochem. Physiol. Pflanz.* 168, 493–510.

Chan, J. T., Patterson, G. W., Dutky, S. R., and Cohen, C. F. 1974. Inhibition of sterol biosynthesis in *Chlorella sorokiniana* by triparanol. *Plant Physiol.* 53, 244–9.

Cheniae, G. M., and Martin, I. F. 1972. Effects of hydroxylamine on photosystem II. II. Photoreversal of the NH_2OH destruction of O_2 evolution. *Plant Physiol.* 50, 87–94.

Cooksey, K. E. 1974. Acetate metabolism by whole cells of *Phaeodactylum tricornutum* Bohlin. *J. Phycol.* 10, 253–7.

Coombs, J., and Whittingham, C. P. 1966. The effect of high partial pressures of oxygen on photosynthesis in *Chlorella*. I. The effect on end products of photosynthesis. *Phytochemistry* 5, 643–51.

Cox, R. M. 1966. Physiological studies on nitrogen fixation in the blue-green alga *Anabaena cylindrica*. *Arch. Mikrobiol.* 53, 263–76.

Darley, W. M., and Volcani, B. E. 1969. Role of silicon in diatom metabolism: A silicon requirement for deoxyribonucleic acid synthesis in the diatom *Cylindrotheca fusiformis* Reimann and Lewin. *Exp. Cell Res.* 58, 334–42.

Davis, B., and Merrett, M. J. 1975. The glycolate pathway and photosynthetic competence in *Euglena*. *Plant Physiol.* 55, 30–4.

Dawson, R. M. C., Elliott, D. C., Elliott, W. H., and Jones, K. M. (eds.). 1969. *Data for Biochemical Research,* 2nd ed. Oxford University Press, London. 654 pp.

Ellis, R., Spooner, T., and Yakulis, R. 1975. Regulation of chlorophyll synthesis in the green alga *Golenkinia*. *Plant Physiol.* 55, 791–5.

Fredrick, J. F. 1969. Effect of amitrole on biosynthesis of phosphorylases in different algae. *Phytochemistry* 8, 725–9.

Gaffron, H. 1943. The effect of specific poisons upon the photoreduction with hydrogen in green algae. *J. Gen. Physiol.* 26, 195–217.

Gaffron, H. 1945. *o*-Phenanthroline and derivatives of vitamin K as stabilizers of photoreduction in *Scenedesmus*. *J. Gen. Physiol.* 28, 269–85.

Good, N. E., and Izawa, S. 1973. Inhibition of photosynthesis. In Hochster, R. M., Kates, M., and Quastel, J. H. (eds.). *Metabolic Inhibitors,* vol. 4, pp. 179–214. Academic Press, New York.

Goulding, K. H., and Merrett, M. J. 1966. The photometabolism of acetate by *Chlorella pyrenoidosa*. *J. Exp. Bot.* 17, 678–89.

Graham, D., and Reed, M. L. 1971. Carbonic anhydrase and the regulation of photosynthesis. *Nature (London) New Biol.* 231, 81–3.

Grodzinski, B., and Colman, B. 1970. Glycolic acid oxidase activity in cell-free preparations of blue-green algae. *Plant Physiol.* 45, 735–7.

Kandler, O. 1958. The effect of 2,4-dinitrophenol on respiration, oxidative assimilation and photosynthesis in *Chlorella. Physiol. Plant.* 11, 675–84.

Kersten, H., and Kersten, W. 1974. *Inhibitors of Nucleic Acid Synthesis.* Springer-Verlag, Berlin. 184 pp.

Leach, C. K., and Carr, N. G. 1970. Electron transport and oxidative phosphorylation in the blue-green alga *Anabaena variabilis. J. Gen. Microbiol.* 64, 55–70.

Lewin, J. C. 1954. Silicon metabolism in diatoms. I. Evidence for the role of reduced sulfur compounds in silicon utilization. *J. Gen. Physiol.* 37, 589–99.

Lloyd, D. 1966. Inhibition of electron transport in *Prototheca zopfii. Phytochemistry* 5, 527–30.

Lloyd, D., and Chance, B. 1968. Electron transport in mitochondria isolated from the flagellate *Polytomella caeca. Biochem. J.* 107, 829–37.

Lloyd, D., Evans, D. A., and Venables, S. E. 1968. Propionate assimilation in the flagellate *Polytomella caeca:* An inducible mitochondrial enzyme system. *Biochem. J.* 109, 897–907.

Lord, M. J., Armitage, T. L., and Merrett, M. J. 1975. Ribulose 1,5-diphosphate carboxylase synthesis in *Euglena.* II. Effect of inhibitors on enzyme synthesis during regreening and subsequent transfer to darkness. *Plant Physiol.* 56, 600–4.

Lord, M. J., and Merrett, M. J. 1969. The effect of hydroxymethanesulphonate on photosynthesis in *Chlorella pyrenoidosa. J. Exp. Bot.* 20, 743–50.

Losada, M., and Arnon, D. I. 1963. Selective inhibitors of photosynthesis. In Hochster, R. M., and Quastel, J. H. (eds.). *Metabolic Inhibitors,* vol. 2, pp. 559–93. Academic Press, New York.

Lyman, H., Jupp, A. S., and Larrinua, I. 1975. Action of nalidixic acid on chloroplast replication in *Euglena gracilis. Plant Physiol.* 55, 390–2.

MacRobbie, E. A. C. 1962. Ionic relations of *Nitella translucens. J. Gen. Physiol.* 45, 861–78.

MacRobbie, E. A. C. 1965. The nature of the coupling between light energy and active ion transport in *Nitella translucens. Biochim. Biophys. Acta* 94, 64–73.

McCullough, W., and John, P. C. L. 1972. The inhibition of functional RNA synthesis in *Chlorella pyrenoidosa* by 6-methyl purine. *New Phytol.* 71, 829–37.

McMahon, D. 1975. Cycloheximide is not a specific inhibitor of protein synthesis *in vivo. Plant Physiol.* 55, 815–21.

McMahon, D., and Blaschko, W. 1971. Chloral hydrate inhibits protein synthesis *in vivo. Biochim. Biophys. Acta* 238, 338–42.

Metzner, H. (ed.). 1969. *Progress in Photosynthesis Research,* vol. 3, pp. 1683–807. Laupp, Tübingen.

Nultsch, W. 1966. Ueber den Antagonismus von Atebrin und Flavinnucleotiden in Bewegungs- und Lichtreaktionsverhalten von *Phormidium uncinatum. Arch. Mikrobiol.* 55, 187–99.

Nultsch, W. 1969. Effect of desaspidin and DCMU on photokinesis of blue-green algae. *Photochem. Photobiol.* 10, 119–23.

Oey, J. L., and Schnepf, E. 1970. Ueber die Auslösung der Valvenbildung bei der Diatomee *Cyclotella cryptica. Arch. Mikrobiol.* 71, 199–213.

Ogasawara, N., and Miyachi, S. 1970. Regulation of CO_2-fixation in *Chlorella* by light of varied wavelengths and intensities. *Plant Cell Physiol.* 11, 1–14.

Ogawa, T., Vernon, L. P., and Yamamoto, H. Y. 1970. Properties of *Anabaena variabilis* cells grown in the presence of diphenylamine. *Biochim. Biophys. Acta* 197, 302–7.

Pritchard, G. G., Whittingham, C. P., and Griffin, W. J. 1961. Effect of isonicotinyl hydrazide on the path of carbon in photosynthesis. *Nature* 190, 553–4.

Rabin, B. R., Shaw, D. F., Pon, N. G., Anderson, J. M., and Calvin M. 1958. Cyanide effects on carbon dioxide fixation in *Chlorella. J. Am. Chem. Soc.* 80, 2528–32.

Rau, I., and Grimme, L. H. 1971. Zum Einfluss verschieden substituierter s-Triazine auf Stoffwechselreaktionen der Grünalge *Ankistrodesmus braunii. Z. Naturforsch.* 26b 919–21.

Raven, J. A. 1968. The action of phlorizin on photosynthesis and light-stimulated ion transport in *Hydrodictyon africanum. J. Exp. Bot.* 19, 712–23.

Raven, J. A. 1969. Effects of inhibitors of photosynthesis and the active influxes of K and Cl in *Hydrodictyon africanum. New Phytol.* 68, 1089–113.

Sargent, D. F., and Taylor, C. P. S. 1975. On the respiratory enhancement in *Chlorella pyrenoidosa* by blue light. *Planta* 127, 171–5.

Sharpless, T. K., and Butow, R. A. 1970. Phosphorylation sites, cytochrome complement, and alternate pathways of coupled electron transport in *Euglena gracilis* mitochondria. *J. Biol. Chem.* 245, 50–7.

Shrift, A. 1959. Nitrogen and sulfur changes associated with growth uncoupled from cell division in *Chlorella vulgaris. Plant Physiol.* 34, 505–12.

Sikka, H. C., Saxena, J., and Zweig, G. 1973. Alteration in cell permeability as a mechanism of action of certain quinone pesticides. *Plant Physiol.* 51, 363–7.

Smith, F. A., and West, K. R. 1969. A comparison of the effects of metabolic inhibitors on chloride uptake and photosynthesis in *Chara corallina. Aust. J. Biol. Sci.* 22, 351–63.

Stuart, T. S. 1971. Hydrogen production by photosystem I of *Scenedesmus:* Effect of heat and salicylaldoxime on electron transport and photophosphorylation. *Planta* 96, 81–92.

Surzycki, S. J., Goodenough, U. W., Levine, R. P., and Armstrong, J. J. 1970. Nuclear and chloroplast control of chloroplast structure and function in *Chlamydomonas reinhardi. Symp. Soc. Exp. Biol.* 24, 13–37.

Tanner, W., Dächsel, L., and Kandler, O. 1965. Effects of DCMU and antimycin A on photoassimilation of glucose in *Chlorella. Plant Physiol.* 40, 1151–6.

Urbach, W., and Gimmler, H. 1969. Effect of disalicylidenepropanediamine and other inhibitors on photophosphorylation and glycolate metabolism in green algae. In Metzner, H. (ed.). *Progress in Photosynthesis Research,* vol. 3, pp. 1274–80. Laupp, Tübingen.

Vaisberg, A. J., and Schiff, J. A. 1976. Events surrounding the early development of *Euglena* chloroplasts. 7. Inhibition of carotenoid biosynthesis by the herbicide SAN 9789 (4-chloro-5-[methylamino]-2-[α,α,α-trifluoro-m-tolyl]-3[2H]pyridazinone) and its developmental consequences. *Plant Physiol.* 57, 260–9.

Webb, J. L. 1963–1966. *Enzyme and Metabolic Inhibitors,* vol. I (1963) 951 pp.; vol. II (1965) 1237 pp.; vol. III (1966) 1028 pp. Academic Press, New York.

Webster, D. A., and Hackett, D. P. 1966. Respiratory chain of colorless algae. II. Cyanophyta. *Plant Physiol.* 41, 599–605.

Zweig, G., Carroll, J., Tamas, I., and Sikka, H. C. 1972. Studies on effects of certain quinones. II. Photosynthetic incorporation of $^{14}CO_2$ by *Chlorella*. *Plant Physiol.* 49, 385–7.

Section VIII

Appendix

List of suppliers

Aldrich Chemical Co., Inc., 940 W. St. Paul Ave., Milwaukee, Wis. 53233

American Instrument Co., Div. of Travenol Labs., Inc., 8030 Georgia Ave., Silver Spring, Md. 20910

Amersham/Searle Corp., 2636 S. Clearbrook Dr., Arlington Heights, Ill. 60005

Applied Research Laboratories, P.O. Box 129, Sunland, Calif. 91040

Applied Science Labs., P.O. Box 440, State College, Pa. 16801

Beckman Instruments, Inc., 2500 Harbor Blvd., Fullerton, Calif. 92634

Bellco Glass, Inc., 340 Edrudo Rd., Vineland, N.J. 08360

Belton's Inc., 2518 Shrewsbury Rd., Columbus, Ohio 43221

Bio-Rad Laboratories, 32nd and Griffin Aves., Richmond, Calif. 94804

B. Braun Instruments, 805 Grandview Dr. S., San Francisco, Calif. 94080

Bronwill Scientific, Rochester, N.Y. 14603

Buchler Instruments, Div. of Searle Analytic, Inc., 1327 16th St., Fort Lee, N.J. 07024

Burdick and Jackson Laboratories, Inc., 1953 South Harvey St., Muskegon, Mich. 49422

Calbiochem, Clinical and Diagnostics Div., 10933 N. Torrey Pines Rd., La Jolla, Calif. 92037

Canadian Scientific Products Ltd., P.O. Box 7177, London, Ontario N5Y 4J9, Canada.

Canalaco, Inc., 5635 Fisher Lane, Rockville, Md. 20852

CCAP = Culture Centre of Algae and Protozoa (Cambridge), The Director, 36 Storeys Way, Cambridge, CB3 0DT, U.K.

The Chemical Rubber Co., 2310 Superior Ave., Cleveland, Ohio 44114

Chromatographic Specialities Ltd., P.O. Box 1150, 300 Laurier Blvd., Brockville, Ontario, K6V 5W1, Canada.

Cole-Parmer Instrument Co., 7425 N. Oak Park Ave., Chicago, Ill. 60648

Corning Glass Works, Houghton Pk., Corning, N.Y. 14830

Coulter Electronics Inc., 590 W. 20 St., Hialeah, Fla. 33010

Curtin-Matheson Scientific Co., P.O. Box 1546, Houston, Tex. 77001

Damon/IEC Div., Damon Corp., 300 Second Ave., Needham Heights, Mass. 02194

Difco Laboratories, P.O. Box 1058 A, Detroit, Mich. 48232

Dow Chemical U.S.A., Barstow Bldg., 2020 Dow Center, Midland, Mich. 48640

E. I. du Pont de Nemours & Co., Chemical Dyes and Pigments Dep., Photoproducts Dep., Instrument Products Division, Wilmington, Del. 19898

Eastman Kodak Co., 343 State St., Rochester, N.Y. 14650

E-C Apparatus Corp., 3831 Tyrone Blvd. North, St. Petersburg, Fla. 33709

Eck and Krebs, Scientific Laboratory Glass Apparatus, Inc., 27-11 40th Ave., Long Island City, N.Y. 11101

The Eppley Laboratory, Inc., 12 Sheffield Ave., Newport, R.I. 02840

Esbe Laboratory Supplies, 401 Alness St., Downsview, Ontario M3K 2T8, Canada.

Falcon Plastics, Div. of Beckton-Dickson Co., 2424 S. Sheridan Way, Mississauga, Ontario L52 2M8, Canada.

Farhenfab. Bayer, A. G., 5090 Leverkusen, Germany

Fisher Scientific Co., 711 Forbes Ave., Pittsburgh, Pa. 15219

Gelman Instrument Co., 600 S. Wagner Rd., Ann Arbor, Mich. 48106

General Electric Co., 1 River Road, Schenectady, N.Y. 12345

Gilford Instrument Labs., Inc., 132 Artino St., Oberlin, Ohio 44074

Gilson Medical Electronics, P.O. Box 27, 3000 West Beltline Highway, Middleton, Wis. 53562

Gould Inc., Instrument Systems Div., 3631 Perkins Ave., Cleveland Ohio 44114

Griffin and George Ltd., Ealing Rd., Alperton, Wembley, Middlesex, U.K. (In Canada: 3650 Western Rd., Toronto, Ontario M9L 1W2)

Haake, Inc., 244 Saddle River Road, Saddle Brook, N.J. 07662

Harshaw Chemical Co., 6801 Cochran Rd., Solon, Ohio 44139

ICN Pharamaceuticals Inc., Life Sciences Group, 26201 Miles Rd., Cleveland, Ohio 44128

ISCO (Instrumentation Specialties Co.), 4700 Superior Ave., Lincoln, Neb. 68504

International Equipment Co., 300 Second Ave., Needham Heights, Mass. 02194

Intertechnique Instruments, Inc., Randolph Park W., Dover, N.J. 07801 (In France: 78370 Plaisir, France)

Ishikawa Co., Shiba-3, Minato-ku, Tokyo, Japan

Kontes, Spruce Street, Vineland, N.J. 08360

Ladd Research Industries, Inc., P.O. Box 901, Burlington, Vt. 05401

The London Co., 811 Sharon Dr., Cleveland, Ohio 44145

Mallinckrodt, Inc., P.O. Box 5439, St. Louis, Mo. 63147

Marine Colloids, P.O. Box 308, Rockland, Me. 04841

Matheson-Coleman and Bell. See MC/B Manufacturing Chemists

Matheson Gas Products, P.O. Box 85, 932 Paterson Plank Rd., E. Rutherford, N.J. 07073

MBA = Marine Biological Association Culture Collection, Plymouth, PL1 2PB, U.K.

MC/B Manufacturing Chemists, 2909 Highland Ave., Norwood, Ohio, 45212

E. Merck, D61, Darmstadt, Germany (In U.S.A.: Brinkman Instruments, Inc. Cantiague Rd., Westbury, N.Y. 11590; in Canada: BDH Chemicals, 350 Evans Ave., Toronto, Ontario M8Z 1K5)

Millipore Corp., Ashby Rd., Bedford, Mass. 01730

National Appliance Co., P.O. Box 23008, Portland, Ore. 97223

New Brunswick Scientific Co., Inc., 1130 Somerset St., P.O. Box 606, New Brunswick, N.J. 08903

New England Nuclear, 549 Albany St., Boston, Mass. 02118

Nuclear-Chicago Corp., 2000 Nuclear Dr., Des Plaines, Ill. 60018

Nuclear Enterprises, Inc., 935 Terminal Way, San Carlos, Calif. 94070 (In U.K.: Sighthill, Edinburgh, EH11 4E4)

Oak Ridge National Laboratory, P.O. Box X, Oak Ridge, Tenn. 37830

Packard Instrument Co., Inc., 2200 Warrenville Road, Downers Grove, Ill. 60515

Permutit Co., E 49 Midland Ave., Paramus, N.J. 07652

Pharmacia Fine Chemicals, Div. of Pharmacia, Inc., 800 Centennial Ave., Piscataway, N.J. 08854

Pierce Chemical Co., P.O. Box 117, Rockford, Ill. 61105

PolyScience Corp., 6366 Gross Point Rd., Niles, Ill. 60648

Prolabo, 12 Rue Pelée, Paris XIe, France (In Canada: Casgrain and Charbonneau, 455 St. Laurent Blvd., Montreal, Quebec H2Y 2Y8)

The Radiochemical Centre, Amersham, Bucks., HP7 9LL, U.K.

Reeve Angel and Co., Inc., 9 Bridewell Pl., Clifton, N.J. 07014

Research Products International Corp., 2692 Delta La., Elk Grove Village, Ill. 60007

Rohm and Haas Co., Independence Mall West, Philadelphia, Pa. 19105

SAI Technology Co., 4060 Sorrento Valley Blvd., San Diego, Calif. 92121

Sartorius-Membranfilter Gmbh., 34 Göttingen, P.O. Box 143, Germany

Schleicher and Schuell, Inc., 543 Washington St., Keene, N.H. 03431

Schwarz/Mann Div., Becton, Dickinson & Co., Mountain View Ave., Orangeburg, N.Y. 10962

Searle Analytic, Inc., 2000 Nuclear Dr., Des Plaines, Ill. 60018

SGA Scientific, Inc., 735 Broad St., Bloomfield, N.J. 07003

Sharples-Stokes Div., Penwalt Corp., 955 Mearns Rd., Warminster, Pa. 18974

Sigma Chemical Co., P.O. Box 14508, St. Louis, Mo. 63178

SMI Spectrametrics, Inc., 204 Andover St., Andover, Mass. 01810

Ivan Sorvall, Inc., Norwalk, Conn. 06856

Charles Tennant & Co. (Canada), Ltd., 34 Clayson, Weston, Ontario M9M 2G8

Thermolyne Corp., 2555 Kerper Blvd., Dubuque, Iowa 52001

Thermovac Industries Corp., 41 Decker St., Copiague, N.Y. 11726

Arthur H. Thomas Co., P.O. Box 779, Vine St. at Third, Philadelphia, Pa. 19105

Turner Associates, 2524 Pulgas Ave., Palo Alto, Calif. 94303

Ultrasonic Systems, Inc., 405 Smith Street, Farmingdale, N.Y. 11735

Union Carbide Corp., Chemicals and Plastics Division, Charleston, W. Va. 25303

Unitec, Inc., 678 Lofstrand Lane, Rockville, Md.

United States Biochemical Corp., P.O. Box 22400, Cleveland, Ohio 44122

UTEX = Culture Collection of Algae, Department of Botany, University of Texas, Austin, Tex. 78712

Utility Chemical Co., 145 Pearl St., Patterson, N.J.

Varian Instrument Div., 611 Hansen Way, Palo Alto, Calif. 94303

VirTris Co., Inc., Route 208, Gardiner, N.Y. 12525

Waring Products Div., Dynamics Corp. of America. Route 44, New Hartford, Conn. 06057

Waters Associates, Inc., Maple St., Milford, Mass. 01757

Whatman, Inc., 9 Bridewell Pl., Clifton, N.J. 07014

Wheaton Scientific, 1000 N. 10th Street, Milville, N.J. 08332

Worthington Biochemical Corp., Halls Mills Rd., Freehold, N.J. 07728

Yeda Research and Development Co., Ltd., Rehovot, Israel

Yellow Springs Instrument Co., Inc., P.O. Box 279, Yellow Springs, Ohio 45387

Carl Zeiss, Inc., 444 Fifth Ave., New York, N.Y. 10018

Subject index

Acetate assimilation, 324–5
Acid phosphatase, assay, 265–6
Acid phosphatase, extracellular, 263
Active transport: ions, 383, 465; organic substrates, 383
ADP determination, 202
ADP/O ratios, 332–3
Adenylate energy charge, calculation of, 203
Agar: analysis, 126–8; content in seaweeds, 128–9; enzymatic hydrolysis, 126; extraction and fractionation, 124–6; precipitation, 125; structure, 110–11
Aldolases: assay, 241; in macroalgae, 240–4
Alginic acid: analysis, 145–7; content in seaweeds, 148; extraction, 145–6; structure, 144
Allophycocyanin: spectrophotometric determination, 73–4
Amino acids: analysis, 193; chromatography, 277–9
Ammonia analysis, 374–5
Ammonium sulfate fractionation: biliproteins, 75–6; enzymes, 241–2, 248
AMP determination, 202
3,6-Anhydrogalactose determination, 114, 116
Antigen–antibody complex, 340–2
Antimycin, optical standardization, 331
Arginine incorporation, 343–5
Artificial seawater medium, 41
Ash content of seaweeds, 187
Ash determination, 184
Ashing, dry, 452
Ashing, wet, 452
ATP determination, 202
ATP utilization, 375
ATPases, ion-dependent: assays, 258–9, 466–70; cation specificity, 474; criteria for involvement in transport, 465; gel electrophoresis, 470–1, 475; kinetic constants, 469; membrane-bound 49–54, 456, 465; nucleotide specificity, 474; release from membranes, 470

Bacterial and fungal contamination, test for, 356, 381
Bile pigments: isolation, 77–8; structure, 72
Biliproteins: gel filtration, 76–7; molecular weights, 74; purification, 74–7; quantitative analysis, 73–4; spectral properties, 74; types and function, 72
Branching isozymes, 246–51
Bromophenols, 158, 167, 172–4, 177
Brushite, 75

Cadmium, metallic for nitrate reduction columns, 405–6
Carbohydrates: anthrone method, 96; hydrolysis, 137; phenol-sulfuric acid method, 96–7, 136–7; total in microalgae, 192
Carbonic anhydrase, 236
Carotenes, 66–9
Carotenoids: analysis, 60, 64–8; chromatography, 66–8; extraction, 64–5; in crude extracts, 65; specific extinction coefficients, 68
Carrageenans: chemical analysis, 114, 116–18; content of different types, 120; enzymatic degradation, 122; extraction and fractionation, 112–14; gel strength, 123; immunochemical technique, 122; infrared analysis, 114–15; moisture content, 120; NMR spectroscopy, 123; polyanionic functions, 122–3; precipitation, 113; structure, 110–11; viscosity measurements, 118–20
Carrageenophytes, separation of nuclear phases, 123–4
Catalase, 34–6
Cell cycle, 339, 354–5
Cell volume determination, 421, 451–2
Centrifugation: density gradient, 6, 8–9, 18, 34–5, 44–7; differential, 19, 28, 44–5; zonal, 7
Chloride, electrometric titration, 453
Chlorophylls: analysis, 60–4, 307–8; chromatography, 63–4; extraction,

Chlorophylls (*Cont.*)
61–2; in crude extracts, 62–3; specific
extinction coefficients, 64; trichromatic equations, 62
Chloroplasts: density gradient centrifugation, 6, 8–9; isolated, properties, 9–
13; isopyknic sedimentation, 7; membranes, preparation, 306–9; polypeptides and proteins, 11–13; stromal
fraction, 11; zonal sedimentation, 7
Chromatin: fractions, 21–2; isolation,
17, 19; RNA polymerase activity, 21
Chromatography: identification
methods, 279; solvent systems, 276
Chrysolaminarin, *see* β-1,3-glucans
CO$_2$ absorption and determination res-
piratory, 273, 322
Co-chromatography, 279
Column chromatography: biliproteins,
75–6; phenols, 163, 165; pigments,
66–7
Combustion of tissues, 273–4
Compartmentation, nucleic acid precur-
sors, 357
Concanavalin A, 54, 83
Concanavalin-Sepharose for purifying
phosphorylase isozymes, 252
Coomassie Brilliant Blue, 92
Culture media, 437–8, 450
Cytochromes, 75
Cytochrome oxidase, 35, 330

DEAE-Sephadex fractionation, crude
agar, 125
Detergent solubilization of enzymes,
257, 470
Disruption of cells: Braun cell homogenizer, 27–8; detergent and pepsin, 16;
enzymic digestion, 16, 75, 235, 273;
freezing and sonication, 73; freezing
and thawing, 17–18, 74; French
pressure cell, 8, 16, 212, 373; glass
homogenizer, 36, 220, 257; grinding
with sand, 33, 74, 240; lysis with
lauryl sarcosine, 82; osmotic lysis, 41,
43–4, 257; sonication, 17–18, 73, 85,
228, 257, 308, 372; Waring Blendor,
16; Yeda press, 44, 308
Disruption of macroalgae: Dangoumou
ball mill, 112, 163; grinding with sand,
248; Waring Blendor, 248; Wiley mill,
121
Dissolved organic carbon analysis, 393
Dithionite oxidation, 375–6
DNA: content, 355; isolation, 82–3; precipitation, 83
DNA analysis: diphenylamine reaction,
83–4; fluorometry, 84; in microalgae,
192; UV spectrophotometry, 83
Dry weight determination, 451

Electron microscope autoradiography,
445
Electron probe X-ray microanalysis, 445
Endoplasmic reticulum, 53
Enzyme purification: ammonium sulfate
fractionation, 241–2, 248; detergent
solubilization, 257; gel electrophoresis,
248–52; gel filtration, 241–2
Euchromatin, 19, 21–2
Exclusion chromatography, 340
Excretion of organic substances, *see* extracellular release
Extracellular products: dissolved organic
carbon, 393; molecular size estimation, 392–3, 395
Extracellular release, 389: methods for
determining, 390–7; nature of products, 390, 393–4; time course for,
394, 396–7
Extraction procedures: agars, 124–6; alginic acid, 145–6; biliproteins, 73–5;
carotenoids, 64–5; carrageenans, 112–
13; chlorophyll, 61; enzymes, 212,
220, 228, 235, 240, 248, 257; fractionation of labelled cells, 322–3; fucoidan, 153; β-1,3-glucans, 136; lipids,
101, 191; low-molecular-weight components, 190–1; nucleic acids, 82–3,
85–6; nucleotides, 200; phenolic compounds, 160, 167; photosynthetic
products, 274; pigments, 112, 124;
seaweed polysaccharides, 112, 124,
145, 153

Fatty acids, 99
Fatty acids, gas-liquid chromatography,
105–6
Ferredoxin, 75
Fiber: content of seaweeds, 187; deter-
mination, 184–5
Firefly lantern extract, 201
Floridean starch synthesis, 246
Fractionation of labeled cells, 322–3
Freeze drying of cells, 227
Fucoidan: analysis, 153–5; content of
seaweeds, 155; structure, 152
Fungal contamination, test for, 356

Galactans, 110
Galactan sulfates, 110
Gas chromatography: ethylene determination, 369
Gas-liquid chromatography: fatty acids,
105–6; phenols, 164–6, 170–2, 175;
sugars, 117, 138
Gas sample storage, 371
Gel electrophoresis: biliproteins, 77; enzyme separation, 248–52, 470–1, 475;
proteins, 11–13; RNA, 86–8

Gel filtration: aldolase, 241–2; biliproteins, 76; extracellular products, 395
Germanic acid (⁶⁸Ge) uptake, 428–33
β-1,3-Glucans: content in diatoms, 138–9; determination, 136–8; structure, 134
Glucose analysis by glucose oxidase method, 137–8
Glucose uptake, 320, 323–4, 326: rapid sampling procedure, 385; time course, 384–6
Glycerolipids, 99
Glycolate: dehydrogenase, 35–6, 236; determination, 393; extracellular release, 395–6
Glycosidic lipids, 100
Glucosyltransferases: detection by protein staining, 250; detection by substrate staining, 250; gel electrophoresis, 248–51; properties, distribution, 246
Glyoxysomes, 36
Golgi from diatoms, 40, 42, 53
Gomori technique, modification, 250
Growth medium, effect on nucleic acid profiles, 355
L-Guluronic acid, 144

H₂ evolution, 375
Harvesting of cells: centrifugation, 17, 33, 43–4, 212, 235, 307, 422–3; continuous centrifugation, 82, 85, 227, 256; filtration, 136
Heterochromatin, 19, 21–2
Heterocyst isolation, 372
Heterotrophic uptake, 380; rapid sampling procedure, 385–6; regulation by environmental conditions, 383–4; specificity, 383
Hill reaction, 305: assay, 309–12; equation, 309
Hydroxylapatite, 75

Iodine determination, 185–7
Iodine content, seaweeds, 187
Inhibitors, metabolic, list, 480–4
Inhibitors, photophosphorylation, 312
Inhibitors, respiratory, 331–3
Ion-exchange chromatography: biliproteins, 76; soluble photoassimilates, 275
Ion fluxes of microalgae, 449, 455–61: centrifugation methods, 455–9; efflux measurements, 458–9; filtration methods, 455–8
Ion transport, *see* ion fluxes of microalgae
Ion-stimulated ATPases, 463–75: *see also* ATPases, ion-dependent
Ionic content, 449: atomic absorption, 453; electron probe analysis, 455;

emission spectroscopy, 453; ion-specific electrodes, 455, 461; isotopic equilibrium, 453; neutron activation, 454–5
Inorganic carbon determination, 272
Inorganic phosphate determination, 258–9
Intercellular space, 326
Irradiance, measurement, 290
Isocitric lyase, 36
Isotopic dilution, 339

Keto-acid determination, 229
Kieselguhr paper, 63–4, 67–8
Kinetic constants, organic substrate uptake, 383

Laminaran: analysis, 135, 137–8; content in seaweeds, 138–40
Light source for photosynthesis, 290
Lipids: determination of total, 102, 191–2; extraction and purification, 101; identification, 102–4; thin-layer chromatography, 102; transesterification, 104–5
Liquid scintillation counting, 274, 322
Low-molecular-weight components: acid soluble, 191, 193; extraction, 274, 322–3
Luciferase, 198–204
D-Luciferin, 198

Macromolecular components, quantitative determination, 189–94
Malate dehydrogenase, 35–6
Mannitol, analysis, 183–4
Mannitol content in seaweeds, 187
D-Mannuronic acid, 144
Mehler reaction, 309
Membranes: characterization, 47–9; composition, 47–9; enzymes for identification, 49–52; Golgi, 40, 42, 53; identification, 52–3; isolation from diatoms, 39–47; morphology, 42, 47; plasma, 53; purification, 256–8
Membrane-bound proteins, release, 470
Mercury: radioactive stock solutions, 442; uptake data, 443–4; uptake of ²⁰³Hg, 441–3
Microbodies, isolation, 32–5
Micronutrients, uptake, 435–41
Mitochondria: isolation, 26–8; respiration, 329; respiratory control, 29
Moisture content, polysaccharides, 120

NADH, optical standardization, 331
Nitrate determination, 406–7
Nitrate reductase: assay, 219, 221; phytoplankton, 218

Nitrate reduction columns, 402–3, 405–6
Nitrate uptake, 401, 405–8
Nitrite determination, 220–1
Nitrogenase: assay methods, 374–6; preparation and purification, 372–4; reactions catalyzed by, 364
Nitrogen fixation, 363
Nitrogen fixation methods: acetylene reduction, 369–72; fixation of $^{15}N_2$, 365–9; growth on N_2, 365
Nuclear membrane, 54
Nuclei: DNA and protein content, 21; isolation of intact, 16–21; structure of isolated, 20
Nucleic acids, 81: labeling artifacts, 358–9; synthesis, precautions, 353, 358–9
Nutrient medium, *Chlorella*, 26–7

O_2 in water, 331–2
Organic substrates, uptake, 379
Osmotic lysis, 234–6
Oxygen electrode, 330: *see also* polarographic measurements
Oxygen exchange: automatic recording, 289; in photosynthesis and respiration, 291–2, 300–1

Paper chromatography: phenols, 163–5, 170; pigments, 63–4, 67–8; soluble photoassimilates, 275–9
PEP carboxylase, *see* phosphoenol pyruvate carboxylase
Periodate oxidation, 183
Peroxisomes, 36
Phenolic compounds: analysis, 161–76; chromatographic data, 164, 168–9, 171, 173–4; chromatography and identification, 163–6, 168–74; extraction, 160, 167; in brown algae, 177; in other algae, 177–8, in red algae, 177; structure, 158–9
Phenylalanine incorporation, 348–9
Phloroglucinols, 158–9, 162–7, 177
Phosphate determination, 413–4, 467–8
Phosphate uptake, 411–7
Phosphatase, in acetone powders, 231
Phosphoenolpyruvate carboxylase: assay, 235–6; in blue-green algae, 234, 236
Phospholipase in acetone powders, 231
Phospholipids, 100
Phosphorylases, 246–7
Photoassimilatory products: fractionation, 274–5; nature of, 274–80
Photophosphorylation, 236, 305: assay 312–14
Photorespiration, 298
Photosynthesis, polarographic methods, 286–95

Photosynthetic electron transport, 312
Photosynthetic rates: field methods, 271–2; laboratory methods, 272–4
C-phycocyanin, spectrophotometric determination, 73–4
Phycocyanobilin, 72, 78
C-phycoerythrin, spectrophotometric determination, 73–4
Phycoerythrobilin, 72, 78
Plasmalemma, *see* membranes, plasma
Polarographic measurements, 285–95: difficulties encountered, 294; Hill reaction, 310–11
Polyacrylamide gel electrophoresis, 248–51
Polysaccharide hydrolysis, 116–17, 137
Pool size, 339, 349: as affecting nucleotide precursor incorporation, 356; nucleotides, 356, 358
Pool swamping, 340, 345–50
Precursor: incorporation, 341, 343, 349; nonspecificity in nucleic acid synthesis, 357–8; pools, 338–9, 343; transport, 356
Proteins: gel electrophoresis, 11–13; staining on gels, 250–1; synthesis, 11–13, 337–50
Protein determination: dye binding, 92–3; Folin phenol method, 340; Lowry method, 265; semimicro Kjeldahl, 28; spectrophotometric, 265; in microalgae, 192
Protein turnover, 339, 350
Protoplasts, from diatoms, 43
Pulse labeling, 338, 340

Radiation, measurement, 290
Radiochromatography, 275
Radioiodination, 340–1
rDNA cistrons, 355
Reducing sugar analysis, fermentation method, 137
Respiration: light-dark transients, 297–303; polarographic methods, 286–95, 297–303
Respiratory assimilation, 317
Respiratory control ratio, 332–4
Ribosomal RNA, 354
Ribulose-bis-phosphate carboxylase: assay, 212–14; phytoplankton, 210, 214
Ribulose-bis-phosphate oxygenase, 215
RNA: content, 355; isolation, 85–6; polymerase, 21
RNA analysis: fluorimetric, 88; gel electrophoresis, 86–8; orcinol method, 88, 193; total in microalgae, 193; UV spectrophotometry, 86

Scintillation counting, 341, 386, 430
Silicalemma, 54
Silicic acid uptake, 427: by diatom cultures, 430; by plankton, 431–2; kinetics, 428, 432
Sodium transport, 465
Solubilization of seaweeds, 273
Specific growth rate of mercury treated cells, 443–4
Spheroplasts, 234–7
Sterilizing labeled compounds, 320
Substrate assimilation, 318
Substrate concentration, effects on transport rate, 382–3
Succinic dehydrogenase, 236
Sucrose gradient: continuous, 34; discontinuous, 46
Sugar analysis: colorimetric, 116; gas-liquid chromatography, 117; paper chromatography, 116; thin-layer chromatography, 127
Sulfate: analysis, 117–8; uptake by microalga, 421–2; uptake by seaweed, 421, 423
Synchronous cultures, 338–9, 343
Synthetic seawater, 40

Thin-layer chromatography: bile pigments, 78; lipids, 102; phenols, 163–4, 170; pigments, 63, 67; soluble photoassimilates, 275–9; sugars, 127
Threonine deaminase: assay, 228–30; in phytoplankton, 226, 230–1
Thylakoid isolation, 8–10
Tonoplast, 54
Toxic metals, uptake, 435–45
Trace metals, radioactive, 439–40
Translocation, 273
Transport: analogues, 326; ions, 455–61; organic substrates, 318, 325
Turnover rates, 338

Uricase, 36
Uronic acid decarboxylation, 145–7

Viscosity measurements, 118–20
Vitamins, radioactivity labeled, 440

Water content of cells, 452

Xanthophylls, 65–9

Author index

*(Italics numbers indicate author citation in full; * indicates chapter author.)*

Aaronson, S., ix, 263*, 265, *266*
Aasen, A. J., 68, *69*
Abelson, P. H., *195*, *327*
Abraham, G. N., 84, *88*
Abraham, S., *55*
Abrams, W. R., *425*
Ackermann, W. W., *90*, 361
Albers, R. W., 256, 259, *260*
Alexander, V., 371, *378*
Al-Hasan, R. H., 393, *397*
Allaudeen, H. S., 226, *231*
Allen, J. R., 354, *359*
Allen, M. B., 365, *376*
Amenta, J. S., *194*
Ames, B. N., 265, *266*
Amy, N. K., 218, *222*
Anderson, D. M. W., 145, 146, 147, *149*
Anderson, J. M., *487*
Andrews, T. J., *215*
Antia, N. J., ix, 225*, 226, 227, *231*, *232*, 240, *244*
Aprille, J. R., 16, 17, *23*
Araki, C., 110, *129*
Armitage, T. L., *486*
Armstrong, F. A. J., *398*
Armstrong, J. J., *487*
Arnon, D. I., 308, *315*, 364, *376*, 480, *486*
Arsenault, G. P., 114, 129, *131*
Asami, S., *244*
Askari, A., 258, *259*
Atkins, C. A., 275, *280*
Atkinson, D. E., 203, *205*
Aurell, B., 247, 249, *253*
Austin, K., 320, *327*
Avron, M., 314, *315*
Azam, F., ix, 427*, 428, 432, 433, *434*

Baardseth, E., 61, *69*, 182, *188*
Babcock, D., *424*
Bächli, P., 135, *140*
Badour, S. S., ix, 209*, 210, 211, *215*, 479*
Bagi, G., *90*
Bahr, J. T., *335*

Bailey, J. M., *253*
Baker, P. F., 471, *475*
Balke, N. E., *259*
Barber, J., ix, 449*, 450, 452, 459, *461*, *462*
Barker, G. R., *89*
Barker, S. A., 481, *485*
Barnet, R. E., 256, *260*
Barry, J. M., *23*
Barry, V. C., 135, *140*
Bartoš, J., 286, *295*
Bassham, J. A., 214, 215, 216, *485*
Bauer, W. R., 355, *359*
Bayen, M., 83, *88*
Beale, S. I., 483, *485*
Beardall, J., 210, *215*
Beattie, A., 134, *140*
Beck, W. S., 241, *244*
Beechley, R. B., 286, *295*
Beevers, H., 32, 36, *37*
Belcher, R., 117, *129*
BeMiller, J. N., *2*
Bendich, A. J., 355, *359*
Bennett, A., 73, 76, 77, *78*
Benson, A. A., *79*
Bergmeyer, H. U., *2*
Berková, E., *295*
Berkner, L., 308, *315*
Bernstein, E., 335, *359*
Bersier, D., 356, 357, *359*
Betzer, P., *445*
Beuhler, R. J., 77, *78*
Beyers, R. J., 293, *295*
Bidwell, R. G. S., 271, 273, 274, 275, *280*, 350, *351*, 356, *361*
Biggins, J., 234, 236, *237*
Bilinski, E., 231, *232*
Birnie, G. D., *89*
Birnstiel, M. L., *89*
Bishop, N. I., 482, *485*
Black, M. K., 420, *425*
Black, W. A. P., 135, *140*, 153, *156*
Blaschko, W., 481, *486*
Blasco, D., 218, *223*
Bligh, E. G., 101, *107*
Bliss, C. I., 404, *408*

Bludstone, H. A. W., 276, 279, *280*
Boardman, N. K., *360*
Boerma, C., 355, *361*
Bogorad, L., 72, 73, 76, 77, *78*
Bolliger, H. R., 67, *69*, *70*
Bolton, E. T., *195*, *327*
Bonner, J., 21, *23*
Bonner, W. D., Jr., 26, 28, *30*, 331, 332, *334*, *335*
Bonotto. S.. 484. *485*
Bonting, S. L., 465, *475*
Booth, C. R., 203, *205*
Borun, G. A., *205*
Bothe, H., 372, *377*
Bourque, D. P., 355, *359*
Boynton, J. E., 355, *359*
Brachet, J., 481, 482, 484, *485*
Bracker, C. E., *260*
Bradfield, G., 421, *424*
Bradford, M., 92, *93*
Bradley, D., *424*
Bradley, M. O., 339, *351*
Braun, R., 356, 357, *359*
Brawerman, G., 6, *13*, *23*, 355, *359*
Bray, G. W., 213, *216*
Breden, E. N., *13*
Briggs, W. R., 273, *282*
Brinkhuis, B. H., 273, *280*
Britten, R. J., *195*, *327*
Brouzes, R., 372, *377*
Brown, D. H., *13*
Brown, M. R., 251, *253*
Brown, P. R., 198, *205*
Brown, R. H., 214, 215, *216*, *462*
Brunk, C. F., 83, *88*
Buchanan, B. B., 215, *216*
Buetow, D. E., ix, 15*, 16, 17, 18, 20, 21, *23*
Bunt, J. S., 318, 324, *326*
Burkholder, P. R., 40, 41, *55*
Burnison, B. K., 274, *280*
Burns, M. W., III, *179*
Burns, R. C., 364, *377*
Burr, A. H., 271, *280*
Burris, R. H., 2, *30*, *205*, 366, 367, 368, 371, 375, 376, *377*, *378*
Burton, K., 83, *88*, *194*
Busby, W. F., *361*
Busch, H., 17, *23*
Butow, R. A., 331, *335*, 482, *487*

Cairns, J., Jr., *398*
Calkins, V. P., 393, *397*
Calvin, M., *79*, *485*, *487*
Cambraia, J., 256, *259*
Cameron, I. L., *361*
Cameron, M. C., 183, *188*
Canelli, E., *416*
Canvin, D. T., 275, *280*

Capel, M. J., *398*
Caperon, J., 405, *408*
Caravaggio, L. L., 465, *475*
Carlucci, A. F., 205
Carpenter, E. J., 402, 405, *409*
Carpenter, J. H., 63, *70*
Carr, N. G., *78*, 481, 484, *486*
Carroll, J., *488*
Carter, C. E., 198, *205*
Carter, J. A., *434*
Čatský, J., *282*
Cattolico, R. A., ix, 81*, 84, 85, 86, *88*, 353*, 354, 355, 356, 357, 358, *359*
Cave, A., *178*
Ceriotti, G., 83, *89*
Chabot, J. F., *359*
Chan, J. T., 484, *485*
Chance, B., 331, 332, *334*, *335*, 481, *486*
Chansang, H., 318, 323, *326*, *327*
Chapman, D. J., 72, 73, 76, *78*, *79*
Chaykin, S., 375, *377*
Chen, L. C.-M., *281*
Chen, P. S., 265, *266*
Cheng, K. H., *237*
Cheniae, G. M., 483, *485*
Chevolot-Magueur, A.-M., 158, *178*
Chiang, K.-S., 357, *361*
Chiappino, M. L., 55, *476*
Chiaroni, A., *178*
Chisholm, S. W., ix, 427*, 428, *433*, *434*
Chopard-dit-Jean, L. H., *70*
Chu, D. K., 214, 215, *216*
Chua, N.-H., *350*
Cifonelli, J. A., 279, *280*
Clark, L. C., Jr., 286, *295*
Clay, W. F., 354, *359*
Coatsworth, J. L., 222, *223*, 402, 408, *409*
Codd, G. A., 210, *216*
Cohen, C. F., *485*
Cohen-Bazire, G., 77, *79*
Cohn, W. E., 198, *205*
Cole, W. J., 77, *79*
Collos, Y., 218, *222*
Colman, B., ix, 233*, 234, 236, *237*, 483, *486*
Colowick, S. P., 2
Conchie, J.; 152, *156*
Connolly, T. N., *232*
Consden, R., 279, *280*
Cook, J. R., 355, *359*
Cooksey, K. E., ix, 317*, 318, 319, 323, 325, 326, *327*, 483, *485*
Coombs, J., *107*, 483, *485*
Cooper, M. F., 358, *360*
Côté, R. H., 152, *156*
Coughlan, S. J., *397*
Counts, W. B., 358, *360*
Cowie, D. B., *195*, *327*

Cox, R. M., 482, 484, *485*
Crabtree, S. J., *205*
Craigie, J. S., ix, xiii, xiv, 109*, *130*, 135, *140*, 157*, 158, 178, *179*, 271, 273, 275, *280*, *281*
Crane, R. K., *475*
Crayton, M. A., 421, *425*
Cremona, T., *244*
Crepsi, H. L., 75, *79*
Criddle, R. W., 210, *216*
Croghan, P. C., *462*
Crosby, W., 129
Cundall, R. B., 123, *129*
Cunningham, D. D., 356, *360*
Curran, P. F., *475*

Dächsel, L., *487*
Dahl, J. L., *260*, *261*
Dahmus, M. E., *23*
Dainty, J., 451, *462*, 464, *476*
Danforth, W. F., 318, *327*, 380, *387*
Darley, W. M., 429, *434*, *483*, *485*
DaSilva, E. J., 177, *178*
Davey, E. W., 438, *445*
Davies, A. G., ix, 435*, 441, *445*
Davies, B. H., 67, 68, *69*
Davies, D. R., 53, *55*
Davis, A. R., 464, *476*
Davis, B., 210, *216*, 481, 482, *485*
Davis, C. H., 250, *252*
Dawson, R. M. C., *2*, 480, *485*
DeBont, J. A. M., 371, *377*
Dedonder, R., 279, *281*
Delieu, T., 311, *315*
Denman, K., 295
Derbyshire, E., 251, *252*
Desai, I. D., 226, 227, 228, 229, *232*
De Soete, D., 455, *462*
Deupree, J. D., *260*
Dewar, E. T., *156*
Dewar, M. A., 450, 452, *462*
Dibb, E., 461
Dische, Z., 154, *156*
Dittmer, J. C., 190, *194*
Dixon, J. F., *260*, *261*
Dobson, R. L., 358, *360*
Dolphin, D., *78*
Doolittle, W. F., 88, *89*
Doshi, Y., 129
Doty, P., 361
Dowd, J. W., 383, *387*
Dowling, E. L., 7, *13*
Doyle, D., 339, *351*
Dreyfuss, J., 420, *424*
Droop, M. R., *327*, 441, *445*
Drozd, J., 371, *377*
Dubin, D. T., 265, *266*
Dubois, M., 96, *97*, 116, *129*, 154, *156*
Dubowik, D. A., *89*

Duckworth, M., 110, 111, 124, 127, 129, *130*, *131*
Dugdale, R. C., 402, *409*
Dulak, N. C., *261*
Duncan, M. J., 271, *280*
Durber, S., 355, *362*
Dutky, S. R., *485*
Dyer, W. J., 101, *107*

Eckhardt, G., *178*
Edelman, M., 83, 84, *89*, *216*, *266*, 358, *360*
Edelstein, T., 129
Eggleton, P., 375, *377*
Ehrenberg, Ł., *90*
Eisenstadt, J. M., 6, *13*, *23*, 355, *359*
El-Hamalawi, A.-R., 88. *90*
Elliot, D. C., *2*, *485*
Elliot, W. H., *2*, *485*
Ellis, R., 483, 484, *485*
Elsden, S. R., *377*
Embree, D. J., 178
Englert, G., *70*
Enns, T., 433
Eppley, R. W., ix, xiv, 205, 217*, 218, *223*, 401*, 402, 405, *409*, 428, 433, *434*
Epstein, H. T., *89*, *360*
Erdei, J., 465, *476*
Er-El, Z., 252, *253*
Erickson, S. J., *445*
Erkenbrecher, C. W., 203, *205*
Ersland, D., 357, *359*
Estabrook, R. W., 330, 331, 332, *334*
Evans, D. A., *486*
Evans, M. C. W., *378*

Falkowski, P. G., ix, 255*, 256, 257, 259, *260*, 465, *471*, *475*
Fambrough, D., *23*
Farkas, G. L., *90*
Farr, A. L., 93, 216, *237*, *266*, *351*
Farr, R. S., 340, *350*
Farrell, K., 210, *216*
Fay, P., 373, *377*
Fedorcsák, I., *90*
Feige, G. B., 275, 276, 277, 279, *281*
Fenical, W., 158, *178*
Ferrari, G., 421, *424*
Filner, P., 421, *424*
Fink, K., 356, *360*
Fischer, E. H., *253*
Fisher, J. D., 259, *260*
Fiske, C. H., 258, *260*, 265, *266*, 467, *475*
Fitzgerald, G. P., *378*
Flamm, W. G., 83, *89*, 358, *360*
Fleck, A., 84, 88, *90*, *190*, *195*
Flemming, M., 135, *140*
Flora, J. B., *89*, *360*

Floyd, G. L., *13*
Fogg, G. E., 295, *296*, 390, 391, 393, *397*, *398*
Folch, J., 101, *107*
Folkes, B. F., 350, *351*
Ford, C. W., 134, *140*
Fork, D. C., 286, *296*
Fox, M., *89*
Fox, S. M., *89*
Frederick, S. E., 32, 33, *37*
Fredrick, J. F., ix, 245*, 246, 247, 248, 249, 250, 251, *252*, *253*, 480, *485*
French, C. S., 4, *79*
Friedman, L., *78*
Friedmann, T. E., 228, *232*
Fries, L., 177, *178*
Fuhs, G. W., *416*
Fujita, Y., 76, 77, *79*

Gaal, O., *90*
Gabriel, O., 471, *476*
Gaffron, H., 483, 484, *485*
Galling, G., 357, *360*
Gallon, J. R., 364, 372, 373, *377*
Gangolli, S. D., 123, *130*
Gantt, E., 72, *79*
Garrett, R. H., 218, *222*
Gavis, J., 405, *409*
Geen, G. H., 276, *283*
Gentile, A. C., 248, *253*
Gentile, J. H., *445*
Gertz, E. W., *261*
Gibbs, M., *13*, 240, 242, *244*
Gibbs, S. P., 84, 85, *88*, 355, *360*
Gibson, W. H., 354, *360*
Gierschner, K., *282*
Gigon, A., 281
Gijbels, R., *462*
Gilles, K. A., *97*, *129*, *140*, *156*
Gillham, N. W., *359*
Gimmler, H., 481, 483, *487*
Givan, A. L., 210, *216*
Glasl, H., *107*
Glazer, A. N., 72, 73, 77, *79*
Glombitza, K.-W., 158, 177, *178*
Glover, J. S., *350*
Glynn, I. M., 473, *476*
Goering, J. J., 402, *409*, 428, 433, *434*
Goffeau, A., *485*
Goldberg, A. L., 358, *360*
Goldman, S. S., 256, 259, *260*
Goldspink, D. F., 358, *360*
Good, N. E., 228, *232*, 480, *485*
Goodenough, U. W., 355, *360*, *487*
Gooderham, K., 354, *359*
Goodman, N. S., 421, 423, *424*
Goodwin, T. W., 60, *69*, *79*
Gordon, A. H., *280*
Gordon, R., 129
Goring, D. A., 123, *130*

Gorman, D. S., 307, *315*
Gough, N., *377*
Goulding, K. H., 481, *485*
Graham, D., 482, *486*
Graham, H. D., 123, *130*
Granger, D., *295*
Grant, N. G., ix, 25*, 27, 29, *30*, 329*, 330, 331, 333, *334*
Grasso, P., *130*
Graves, I. L., *90*, *361*
Greco, A. E., 358, *361*
Green, A., 204, *205*
Green, B. R., 355, *360*
Greenberg, E., 251, *253*
Greenwood, F. C., 340, *350*
Griffin, W. J., *487*
Grimme, L. H., 484, *487*
Grimmler, H., *281*
Grisham, C. M., 256, *260*
Grodzinski, B., 236, *237*, 483, *486*
Groves, S. T., 421, *425*
Gruber, P. J., *37*
Grünes, R., *327*, *388*
Guerrini, A. M., 240, *244*
Guidice, G., 83, 84, *89*
Guillard, R. R. L., 381, *387*, 402, 405, *409*, 415, *416*
Gulmon, S. L., *281*
Gulyás, A., *90*
Gutknecht, J., 464, *476*

Haas, D., *387*
Hackett, D. P., 481, *488*
Hackney, J. F., *260*
Hamilton, J. K., *97*, *129*, *140*, *156*
Hamilton, R. D., 320, *327*
Hanawalt, P. C., 357, *362*
Handa, N., 134, *140*
Hanes, C. S., 279, *281*
Hansen, D., *55*, *260*
Hanson, K. R., 404, *409*
Harding, W. M., 226, *232*
Hardy, R. W. F., 364, 369, 372, *377*
Harrell, C. E., Jr., *179*
Harris, C. E., 242, *244*
Harris, G. P., 287, *296*
Harrison, W. G., 222, 408
Hart, J. W., 421, *424*
Hartree, E. F., 340, *350*
Hase, E., 26, 27, *30*
Haskin, C. P., *462*
Hastings, J. W., 355, *362*
Hatch, M. D., 234, *237*
Hattori, A., 76, 77, *79*, 402, *409*
Haug, A., 61, *69*, 134, 135, 136, 137, 138, 139, *140*, *141*, 144, 148, *149*, 155, *156*, 167, *178*, 182, *188*, 185, 187, *188*, 273, 274, 275, *281*
Hauser, C. R., 331, *334*
Havir, E., *409*

Haxo, F. T., 69
Hayat, M. A., 445
Haystead, A., 364, 372, 373, 374, 377
Healey, F. P., x, 411*, 416
Heath, O. V. S., 271, 281
Heldt, H. W., 459, 462
Hellebust, J. A., x, xiii, xiv, 220, 223, 273, 274, 275, 281, 323, 327, 379*, 380, 381, 384, 387, 390, 393, 394, 398, 475, 476
Hemmingsen, B. B., 41, 55, 256, 260, 429, 433, 434, 465, 476
Henderson, R. E. L., 86, 89
Herbeck, R., 86, 89
Herbert, D., 2
Hershberger, C. L., 354, 360
Hesse, G., 84, 89
Hestmann, S., 140
Hexum, T. D., 261
Hiatt, H. H., 89
Hibbard, P. L., 476
Higinbotham, N., 464, 476
Hilden, S., 259, 260
Hillis, W. E., 166, 179
Hind, G., 461, 462
Hinegardner, R. T., 84, 89
Hirs, C. H. W., 79
Hirsch, C. A., 89
Hirst, E., 140
Hirst, E. L., 140
Hjertén, S., 124, 130
Hoagland, D. R., 464, 476
Hobbie, J. E., 318, 327
Hochster, R. M., 485, 486
Hodges, T. K., 55, 256, 257, 259, 260, 464, 465, 476
Hodson, R. C., x, 419*, 420, 421, 422, 424, 425
Hofer, H. W., 275, 281
Hoffman, J. E., 476
Hoffmann, H.-P., 13
Hokin, L. E., 256, 257, 259, 260, 261
Holby, M., 424
Holden, M., 63, 69
Holleman, J. M., 339, 343, 345, 350
Holman, R. T., 70
Holmes, R. W., 69
Holm-Hansen, O., x, 63, 69, 84, 89, 102, 107, 194, 197*, 198, 204, 205
Holsten, R. D., 377
Holton, R. W., 252, 253
Holzer, H., 232
Hommersand, M. H., 27, 29, 30, 330, 331, 333, 334
Hong, K. C., 130
Hopkins, H. A., 84, 89, 355, 360
Hori, T., 253
Horton, D., 282
Horwitz, W., 166, 178

Hosford, S. P. C., 122, 130
Hoshaw, R. W., 359, 381, 387
Hoste, J., 462
Howell, S. H., 355, 360
Howland, G. P., 358, 360
Huang, R. C., 23
Hubbard, R. W., 83, 84, 89
Huffaker, R. C., 339, 350
Humphrey, G. F., 60, 62, 64, 70
Hunter, W. M., 350
Hutchison, W. C., 88, 90, 190, 194
Hutner, S. H., 450, 462
Hüvös, P., 90
Hyams, J., 53, 55

Ikawa, T., 240, 244, 283
Ingle, J., 190, 193, 194
Ingle, R. K., 237
Isaac, P. K., 215
Isherwood, F. A., 279, 281
Isler, O., 69, 70
Iwanij, V., 343, 350
Izawa, S., 232, 462, 480, 485
Izumi, K., 110, 111, 123, 124, 126, 128, 130

Jackson, E. K., 372, 377
Jackson, W. A., 298, 303
James, A. T., 404, 408
James, T. W., 355, 359
Jamieson, W. D., 178
Janowski, M., 485
Jarvis, P. G., 282
Jassby, A. D., x, 285*, 286, 291, 296, 297*
Jeanjean, R., 420, 425
Jeffrey, S. W., 60, 62, 63, 64, 67, 68, 69, 70
Jensen, A., x, 59*, 60, 61, 62, 64, 66, 67, 68, 70, 135, 137, 138, 139, 141, 140, 166, 179, 187, 188
Jensen, Liaaen, 66, 67, 68, 69, 70
Jerlov, N. G., 271, 281
Jeschke, W. D., 281
Jitts, H. R., 290, 296
Johansson, B. G., 124, 130
John, P. C. L., 483, 486
Johnson, D., 129
Johnson, W. S., 273, 281
Jokela, A. C. C. T., 465, 476
Jolley, E. T., 384, 387
Jones, A. K., 387
Jones, A. S., 117, 129, 130
Jones, K. M., 2, 485
Jones, L. W., 73, 79
Jones, R. F., x, 83, 85, 88, 90, 280, 337*, 339, 343, 347, 349, 351, 355, 359, 360
Jones-Mortimer, M. C., 420, 424
Joy, K. W., 350, 351

Jucker, E., 69, *70*
Jupp, A. S., *486*

Kähr, M., *260*
Kalberer, F., 273, *281*
Kalckar, H. M., 198, *205*
Kalmakoff, J., *232*
Kandler, O., *327*, *388*, 482, *486*, *487*
Kaplan, N. O., *2*
Kapovits, I., *90*
Karl, D. M., x, 197*, 204, *205*
Karlsson, J., *260*
Karrer, P., 69, *70*
Karsten, U., 88, *90*
Kates, J. R., 339, 343, *351*
Kates, M., 100, *107*, *485*
Katona, E., *55*
Katsunuma, T., 226, *232*
Katsurai, T., 73, 74, *79*
Katz, J. J., *79*
Keenan, T. W., *260*
Keir, H. M., *23*
Keller, S. J., *351*
Kemp, J. D., 357, *361*
Kersten, H., 480, *486*
Kersten, W., 480, *486*
Key, J. L., 339, 343, 345, *350*
Khan, M., *107*
Kilham, S. S., 404, *409*
Kim, D. H., 129, *130*
Kirk, J. T. O., 355, *360*
Kirkegaard, L. H., *89*
Klotz, L. C., *359*
Knowles, R., 372, *377*
Knutsen, G., 355, 357, *360*, *361*, *387*
Kobayashi, Y., 76, 77, *79*
Koch, W., *296*
Kochert, G., x, 91*, 95*, 189*
Kofler, M., *70*
Komor, B., *387*
Komor, E., *387*
Komoszyński, M., 256, *260*
König, A., 67, *69*, *70*
Korn, E. D., *55*
Kostetsky, E. Y., 103, *107*
Kowallik, W., 298, *303*
Kratz, W. A., 73, *79*, 365, *377*
Krebs, D., *89*
Krebs, E. G., *252*, *253*
Kredich, N. M., 420, *425*
Kremer, B. P., x, 269*, 271, 273, 274, 276, *281*, *283*
Kripps, R. S., x, 225*, 227, *232*
Krisman, C. R., 250, *253*
Krotkov, G., 88, *90*, *280*
Kuhl, A., 211, *216*
Kuiper, P. J. C., 256, *260*
Küppers, U., *283*
Kurz, W. G. W., 370, *377*

Kycia, J. H., x, 71*
Kylin, A., *260*

Lance, C., 28, *30*
Landymore, A. F., 158, *178*
Lane, M. D., 210, 214, *216*
Lang, K., *70*
Lang, N. J., 373, *377*
Lange, O. L., 287, *296*
Larner, J., 246, *253*
Larrinua, E., *486*
Larsen, B., x, 138, *141*, 143*, *149*, 151*, 152, *156*, 167, *178*, 181*, 182, 185, *188*
Larsen, E., 295
LaRue, T. A., 370, *377*
Laties, G. G.. 332, *335*
Lau, Y. C., *231*, *232*
Laub, D., *232*
Layne, E., 265, *266*
Leach, C. K., 481, 484, *486*
Leake, R. E., 19, *23*
Lean, D. R. S., 392, *398*
Leavitt, C. A., *252*
Lee, K., 393
Lee, R. W., 83, *90*, 355, *360*
Lee, Y. C., *97*
Leech, R. M., 6, *13*
Leedale, G. F., 16, *23*
Lees, M., *107*
Lehninger, A. L., 241, 243, *244*
Leick, V., 83, *88*
Leigh, C., x, 109*, *281*
Lem, N., *107*
Lemberg, R., 73, *79*
Lembi, C. A., 47, 54, *55*
Leonard, N. J., *89*
Leonard, R. T., 49, 53, *55*, 257, *260*
LePage, G. A., 198, *205*
Lesnaw, J. A., *90*
Lessler, M. A., 286, *296*
Letham, D. S., 117, *130*
Leung, S., *107*
Levine, R. P., 307, *315*, 355, *360*, *487*
Lew, V. L., *476*
Lewin, J. C., 218, *222*, 355, *361*, 380, 384, *387*, 475, 476, 481, *486*
Lewin, R. A., *141*, *327*, *362*, 380, *387*, *388*
Lien, S., x, 305*
Lien, T., 355, *361*
Lilley, R. Mc., 215, *216*
Lin, R. I., 194, *195*
Lin, Y.-H., x, 379*
Lindeman, W., 258, *260*, 467, *476*
Lindner, R., *89*
Ling, R., *409*
Linnane, A. W., *360*
Linskens, H. F., 270, 279, *281*, *282*

Liu, M. S., 220, *223*, *387*
Ljones, T., 376, *377*
Lloyd, D., 298, *303*, 330, 331, *335*, 480, 481, 484, *486*
Lloyd, J. B., 137, *141*
Lobban, C. S., 271, 273, *282*
Loeblich, A. R., III, *359*
Loewus, F., 420, 422, *425*
Loftus, M. E., 63, *70*
Loos, E., *387*
Lord, J. M., 210, 214, 215, *216*
Lord, M. J., 483, *486*
Lorenzen, C. J., 63, *69*, *70*, 415, *416*
Lorenzen, H., 211, *216*
Lorimer, G. H., *215*
Losada, M., 480, *486*
Lowry, O. H., *2*, 92, *93*, 212, *216*, 236, 237, 265, *266*, *340*, *351*
Lui, N. S. T., 218, *223*
Lum, D., *55*, *476*
Lüning, K., *282*
Luton, M., 461
Lyman, H., 484, *486*
Lynch, M. J., 17, 18, 19, 20, 21, 22, *23*

MacRobbie, E. A. C., 464, 465, 474, *476*, 483, 484, *486*
Mague, T. H., x, 363*, *378*
Mak, R., *360*
Mandeville, S. E., *79*
Manners, D. J., *140*
Mans, R. J., 338, 341, *351*
Marcelle, R., 271, *282*
Marmur, J., *361*
Marsden, P. F., *13*
Marshall, A. T., *445*
Martin, A. J. P., *280*
Martin, I. F., 483, *485*
Marushige, K., *23*
Maslowski, P., 256, *260*
Matsuda, K., *359*
Matsuka, M., 26, 27, *30*
Matthew, W. T., *89*
Mazliak, P., 100, *107*
Mazumder, R., 226, *232*
McAllister, C. D., *296*
McCandless, E. L., 120, 122, 123, 129, *130*
McCarthy, J. J., 408, *409*
McCullough, W., 483, *486*
McDowell, R. H., 110, *130*, 135, *141*, 149, 152, *156*
McElroy, W. D., 198, 204, *205*
McFadden, B. A., 210, 214, *216*
McLachlan, J., *281*, 290, *296*, 319, 327, 381, *387*
McLaughlin, J. J. A., *327*
McLennan, A. G., *23*
McMahon, D., 481, 482, *486*

Medcalf, D. G., *149*, 152, *156*
Medveczky, N., 416, *417*
Meeuse, B. J. D., 134, 140, *141*
Mehard, C. W., 44, *55*
Mehlitz, A., 279, *282*
Mendiola-Morgenthaler, L. R., *13*
Menzel, D. W., 63, *70*
Merrett, M. J., 210, *216*, 481, 482, 483, 485, *486*
Merrilees, P. A., 101, *107*
Meszes, G., 465, *476*
Metzenberg, R. L., 420, *425*
Metzner, H., 480, *486*, *487*
Meyer, J., 405, *408*
Meyn, T., *424*
Millbank, J. W., 370, *377*
Miller, A. G., 235, *237*
Minas, T., *282*
Mitchison, J. M., 356, *361*
Miyachi, S., 482, *487*
Modi, S. R., 226, *232*
Montgomery, R., *97*
Mooney, H. A., *281*
Moore, S., 193, *195*
Moore, T., 129, 178
Morgenthaler, J.-J., 7, 8, *13*
Morrill, G. A., 86, *90*
Morris, I., 210, *215*, *216*
Moscarello, M. A., *55*
Moulton, D. W., *89*
Mulder, E. G., 371, *377*
Munro, H. N., 84, 88, *90*, 190, *194*, *195*
Myers, J., 73, *79*, 365, *377*
Myklestad, S., x, 133*, 134, 135 136, 138, 139, *140*, *141*, 155, *156*

Nagashima, H., 251, *253*, 271, 275, *282*
Nakabayashi, T., 177, *178*
Nakamura, S., *253*, *282*
Nakatani, H. Y., 462
Nalewajko, C., x, 389*, 390, 392, *398*
Naumova, L. P., 357, *361*
Nauwerck, A., 287, *296*
Neilson, A. H., 380, *388*
Nelson, D. M., *434*
Nelson, E. B., 32, *37*
Ng, R., *360*
Ng Ying Kin, N. M. K., 127, *130*
Nichols, H. W., 319, *327*
Nicholson, G., 470, *476*
Nicholson, N. L., 273, *282*
Nickell, L. G., 40, 41, *55*
Nisizawa, K., x, 239*, *244*, *253*, *282*, *283*
Nissen, P., 421, *425*
Nolan, C., 246, *253*
Nordhorn, G., *283*
North, B. B., 385, *388*
Novelli, G. D., 338, 341, *351*
Novoa, W. B., *253*

Noy, R. J., *378*
Nultsch, W., 481, 482, *486*, *487*
Nutten, A. J., *129*
Nygård, K., 271, *281*

Oaks, A., 350, *351*, 356, *361*
OCarra, P., 72, 73, 77, *79*
O'Connell, K. M., *23*
Oey, J. L., 480, 483, *487*
Ogasawara, N., 482, *487*
Ogawa, T., 482, *487*
Ogur, M., 84, *90*
ÓhEocha, C., 72, 73, 77, *79*
O'Kane, D. J., x, 337*
Okita, T., 54, *55*, 466, 470, 475, *476*
O'Neill, A. N., 152, *156*
Ormeland, R. W., 372, *377*
Ornstein, L., 247, *253*
Osmund, H., xiv
Ozaki, H., *282*
Ozawa, H., *253*

Packard, T. T., 281, 222, *223*
Packer, L., *55*
Padilla, G. M., 17, *23*, *361*
Paech, K., 270, *282*
Paerl, H. W., 198, *205*
Painter, T., *156*
Palaty, V., 259
Pant, A., *397*
Pardee, A. B., 356, *360*, 420, *424*
Parenti, F., 16, *23*
Parker, B. C., 271, *282*
Parson, J. W., 420, *425*
Parsons, T. R., 60, *70*, 258, *261*, 271, 272, *283*, 402, 407, 408, *409*, 413, *417*, 428, 432, *434*
Partridge, S. M., 279, *282*
Pasciak, W. J., 405, *409*
Passonneau, J. V., *2*
Patni, N. J., xi, 263*, 265, *266*
Patterson, G. W., *485*
Paul, J. S., 44, 45, *55*
Paulsen, J. M., 210, 214, *216*
Pearse, J. S., 271, *283*
Peat, S., 135, *141*
Pedersén, M., 177, *178*
Percival, E., 110, *130*, 134, 135, *140*, *141*, *149*, 274, 276, *280*, *282*
Percival, E. G. V., 135, *140*, 152, *156*, *188*
Perdue, J. F, *260*, *261*
Perez, K. T., 274, *280*
Pernas, A. J., 123, *130*
Perry, M. J., 265, *266*, 408
Peterson, L. W., 339, *350*
Phillips, G. O., *129*
Phipps, P. J., *2*
Pierce, R. C., *78*

Pigman, W., *282*
Pitman, M. G., 458, *462*
Plagemann, P. G. W., 356, *361*
v. Planta, C., *70*
Platt, T., 286, 291, 295, *296*, *303*
Pohl, P., 102, *107*
Pollack, M., *13*
Pon, N. G., *487*
Postgate, J. R., 371, *377*
Potier, P., *178*
Powell, J. H., 140, *141*
Preddie, E. C., *244*
Preiss, J., 251, *253*
Prescott, D. M., *2*
Pressman, E. K., *361*
Preston, J. F., *23*
Price, C.A., xi, 5*, 7, 8, *13*
Price, L. W., 439, *445*
Pritchard, G. G., 483, *487*
Provasoli, L., 323, 325, *327*, *462*
Pryce, R. J., 177, *178*

Quark, U. C., *485*
Quastel, J. H., *485*, *486*
Quatrano, R. S., 421, *425*
Quayle, J. R., 320, *327*
Quillet, M., 134, *141*, 276, *282*

Rabin, B. R., 214, *216*, 482, *487*
Rabinowitz, H., 210, *216*
Racker, E., 215, *216*, 241, *244*
Ragan, M. A., xi, 157*, 158, 166, 167, 177, *179*
Raizada, M. K., 226, *232*
Ramakrishnan, T., 226, *231*
Ramsay, J. A., 453, *462*
Ramus, J., 421, *425*
Randall, R. J., 93, *216*, 237, *266*, *351*
Randerath, K., 198, *205*
Rao, V. K. M., 226, *232*
Ratliff, R. L., *362*
Rau, I., 484, *487*
Rauwald, H.-W., *178*
Raven, J. A., 483, 484, *487*
Rawson, J. R. Y., 355, *361*
Raymond, S., 247, 248, 249, *253*
Rebers, P. A., *97*, *129*, *140*, *156*
Reed, L. J., 276, *282*
Reed, M. L., 482, *486*
Reed, S. J. B., 455, *462*
Rees, D. A., 110, 111, 113, 121, *131*
Reich, E., *351*
Reichmann, M. E., *90*
Reid, R. A., *398*
Reimann, B. F., *361*
Reisfeld, A., *216*
Reiss, M. M., 86, *90*
Renfrow, W. B., Jr., 331, *334*
Renger, E. H., 222, 408

Renosto, F., 421, *424*
Reynolds, E. S., 47, *55*
Rhodes, P. R., *359*
Ribbons, D. W., 286, *295*
Richards, F. A., 60, *70*
Riche, C., *178*
Ried, A., 298, *303*
Rifkin, D. B., *351*
Riggs, D. S., 383, *387*
Rigler, F., 416, *417*
Rivera-Ortiz, J. M., 371, 376, *378*
Roberts, P. J. P., *253*
Roberts, R. B., 191, *195*, 322, 323, *327*
Roberts, T. M., *359*
Robinson, G. G. C., 215
Robinson, J. M., *13*
Robinson, R., *377*
Rode, A., 83, *88*
Roels, O. A., 218, *223*
Rogers, A. W., *445*
Rogers, J., 222
Rogers, J. N., 408, *409*
Rold, T. L., *179*
Rosebrough, N. J., *93*, *216*, *237*, *266*, *351*
Rosen, G., 84, *90*
Rosenberg, H., 416, *417*
Rosowski, J. R., 381, *387*
Ross, A. G., *188*
Rotman, B., 265, *266*
Rowlands, D. P., *129*
Rüegg, R., *70*
Ruhland, W., *79*, 270, *282*
Russell, G. K., 240, 242, *244*
Rutschmann, J., 273, *281*
Rutter, W. J., 240, 242, *244*
Ryther, J. H., 381, *387*

Saenger, P., *178*
Sagan, L., 358, *361*
Sagher, D., *216*
Sakshaug, E., 60, 61, *70*, *140*
Salisbury, J. L., 7, 10, 11, *13*
Sandakchiev, L. S., *361*
Sanger, F., 246, *253*
San Pietro, A., 270, *282*, *334*, *335*, 374, *378*
Sargent, D. F., 483, *487*
Saxena, J., *487*
Scaletta, C., *88*
Scarborough, G. A., 47, 54, *55*
Scarsella, A. J., *424*
Schatz, G., 26, *30*
Schell, D. M., 371, *378*
Schiff, J. A., xi, *89*, *360*, 419*, 420, 421, *423*, *424*, *425*, 484, *487*
Schildkraut, C. L., 355, *361*
Schimke, R. T., 339, *351*
Schindler, D. W., 391, *398*

Schjeide, O. A., 194, *195*
Schliselfeld, L. H., *252*
Schmidt, A., 420, *425*
Schmidt, G., 84, *90*
Schmidt, R. R., *89*, *360*
Schmidt, R. V., *398*
Schmitz, K., 271, 273, 274, *281*, *282*
Schneider, P. B., 198, *206*
Schneider, W. C., 84, 88, *90*, 358, *361*
Schnepf, E., 480, 483, *487*
Schonbaum, G. R., 330, *335*
Schotg, A., *462*
Schulze, E. D., *296*
Schürmann, P., 215, *216*
Schwieter, U., 68, *70*
Scor-Unesco, Paris, 60, *70*
Segel, I. H., 420, *424*, *425*
Segovia, Z. M. M., 83, *90*, 358, *361*
Šeliger, H. H., *205*
Senner, J. W., *359*
Šesták, Z., 271, *282*
Šetlík, I., *295*, *296*
Shah, N. M., 393, *398*
Shaltiel, S., 252, *253*
Sharp, J., *89*, *194*
Sharpless, T. K., 331, *335*, 482, *487*
Shaw, D. F., *487*
Shaw, E., *351*
Sheridan, W. F., 357, *361*
Shibata, J., 73, *79*
Shieh, Y. J., 450, 459, 461, *462*
Shrift, A., 484, *487*
Sibley, J. A., 241, 243, *244*
Sieburth, J. McN., 166, *179*
Siegelman, H. W., xi, 71*, 72, 73, 75, 78, *79*
Siekevitz, P., *350*
Siersma, P. W., 357, *361*
Sikka, H. C., 482, *487*, *488*
Simonis, W., *281*
Simpkins, H., 256, *261*
Singer, S. J., 470, *476*
Singh, R. M. M., 232
Siskin, J. E., 357, *361*
Siu, C.-H., 355, *361*
Skou, J. C., 256, *261*, 465, 473, *476*
Slack, C. R., 234, *237*
Slankis, T., *360*
Slater, E. C., 334, *335*
Sloane-Stanley, G., *107*
Sloneker, J. H., 117, *131*
Smestad, B., 276, *280*, *282*
Smetana, K., 17, *23*
Smillie, R. M., 88, *90*, *360*
Smith, F., *97*, *129*, *140*, *156*, 279, *280*
Smith, F. A., 482, *487*
Smith, I., 320, *327*
Smith, I. K., 421, *425*
Smith, K. C., *79*

Smith, R. V., 374, *378*
Smith, W. O., 393, *398*
Sokol, F., *90*, *361*
Solórzano, L., *223*, 408, *409*
Solymosy, F., 86, *90*
Somerfeld, P., *424*
Somogyi, M., 137, *141*
Sonnenblick, E. H., *261*
Sorokin, C., 295, *296*, 321, *327*
Sournia, A., 293, *296*
Spiro, R. G., 96, *97*
Splittstoesser, W. E., 275, *282*
Spooner, T., *485*
Srivastava, L. M., 271, *282*
Ssymank, V., 357, *360*, *361*
Stabenau, H., xi, 31*, 32, 33, 35, 36, *37*
Stahl, E., 102, *107*
Stahl, H., 279, *283*
Stainton, M. P., 392, 393, *398*
Stam, A. C., Jr., 256, *261*
Stambrook, P. J., 356, *361*
Stancioff, D. J., 122, *131*
Stanley, N. F., 122, *131*
Stauffer, J. F., *2*, *30*, 205, *378*
Stearns, J., 205
Steemann Nielsen, E., 272, *283*
Steffensen, D. M., 357, *361*
Stein, J. R., xiii, xiv, *296*, *327*, *387*, 437, 438, 441, *445*
Stein, W. H., 193, *195*
Stephen, W. I., *129*
Stephens, G. G., 384, *388*
Stephens, K., *296*
Stevenson, L. H., *205*
Stewart, P. R., 190, 191, *195*
Stewart, W. D. P., *140*, 210, *216*, *296*, *303*, 364, 370, 372, 374, *377*, *378*, *398*
Stoll, A., 62, *70*
Stone, D. P., 259, *260*
Storey, B. T., *335*
Strange, R. E., *2*
Strehler, B. L., 198, 202, *206*
Strickland, J. D. H., 60, *69*, *70*, 258, *261*, 271, 272, *283*, *296*, 402, 407, 408, *409*, 413, *417*, 428, 432, *434*
Stuart, T. S., 484, *487*
Stuiver, C. E. E., *260*
Sturgess, J. M., 47, *55*
Stutz, E., 83, *90*
Subbarow, Y., 258, *260*, 265, *266*, 467, *475*
Sueoka, N., 339, *351*
Sullivan, C. W., xi, 39*, 42, 45, 46, 48, 49, 50, 51, *55*, 463*, 465, 466, 468, 469, 470, 471, 472, 473, 474, *476*
Sundberg, I., 47, 54, *55*
Surzycki, S. J., 484, *487*
Sutcliffe, W. H., Jr., *89*, *194*
Sutton, D. W., 357, *361*

Svedberg, T., 73, 74, *79*
Swain, T., 166, *179*
Sweeney, B. M., 355, *362*
Swift, H., *361*
Swinton, D., *89*, *360*
Swinton, D. C., 357, *362*

Tabita, F. R., 214, *216*
Tamas, I., *488*
Tan, C. K. *215*
Tanner, W., 326, *327*, 384, *387*, *388*, 481, *487*
Taylor, B. F., 372, *377*
Taylor, C. P. S., 483, *487*
Taylor, Z., *295*
Tempel, N. R., *280*
Teste, J., *178*
Thannhauser, S. J., 84, *90*
Thomas, D. L., 251, *253*
Thommen, H., 65, *70*
Thompson, J. S., *89*
Thompson, T. G., 60, *70*
Tobey, R. A., *362*
Tocher, R. D., 178, *179*, *281*
Tolbert, N. E., 32, 36, *37*, *215*, 287, *29€*
Tominaga, H., 134, *140*
Toribara, T. Y., *266*
Totter, J. R., 198, *206*
Towle, D. W., 271, *283*
Tracey, M. V., 270, *282*
Tracey, T. B., 433
Tregunna, E. B., 259
Trench, M. E., *23*
Trown, O. W., 214, *216*
Tsang, M. L.-S., xi, 419*, 420, *425*
Tsubo, Y., 355, *362*
Tuan, D. Y. H., *23*
Turner, B. C., *79*
Turvey, J. R., 127, *130*
Tyszkiewicz, E., 276, *283*

Ueda, T., *362*
Uesigo, S., 257, *261*
Ullrich, W. R., 416, *417*
Ulrich, F., 36, *37*
Ulsamer, A. G., 47, *55*
Umbreit, W. W., *2*, 28, *30*, 205, 375, *378*
Urbach, W., 481, 483, *487*

Vaisberg, A. J., 484, *487*
Vallée, M., 420, *425*
Van Caeseele, L. A., *215*
Vanden Driessche, T., *485*
Vandrey, J. P., 83, *90*
Vasconcelos, A. C., 6, 7, 11, 12, *13*
Vaskovsky, V. E., 103, *107*
Vaughan, J. H., *88*
Venables, S. E., *486*

Verduin, J., 287, *296*
Vernon, L. P., *487*
Vetter, W., *70*
Vinograd, J., 355, *359*
Vogel, A. I., 161, *179*
Volcani, B. E., 54, *55*, *107*, *361*, 429, *433*, *434*. 465, 466, 471, 472, 473, 474, *476*, 483, *485*
Volk, R. J., 298, *303*
Vollenweider, R. A., 287, *296*
von Stosch, H. A., 134, *141*

Waaland, J. R., *179*
Wagner, G., *425*
Wagner, H., *107*
Walker, D. A., 215, *216*, 311, *315*
Walker, N. A., 458, *462*
Walker, P. M. B., *89*
Wallen, D. G., 276, *283*
Wallentinus, J., 271, *283*
Walter, J. A., *130*
Walters, R. A., 356, *362*
Walton, G. M., 203, *205*
Wang, C. H., 213, *216*
Warner, H., *266*
Warr, J. R., 355, *362*
Watt, A., 231, *232*
Watt, W. D., 391, 393, *398*
Weast, C. R., 331, *335*
Webb, J. L., 480, *488*
Webster, D. A., 481, *488*
Wedding, R. T., 420, *425*
Weglicki, W. B., *261*
Weidner, M., 271, *283*
Weigl, J., 122, *131*
Weinstein, B., 158, 172, *179*
Weistrop, J., *424*
Wells, M. A., 190, *194*
Werner, D., 387, *476*
West, K. R., 482, *487*
Wetzel, M. G., *55*
Whelan, W. J., 137, *141*, 250, *253*
Whistler, R. L., *2*
White, E. H., *205*

Whittingham, C. P., 483, *485*, *487*
Whitton, B. A., *78*, 251, *252*
Wieczorek, G. A., *79*
Wiedemann, E., 62, *70*
Willard, J. M., 242, *244*
Willenbrick, J., 271, 274, *281*, *282*
Williams, G. R., 332, *334*
Williams, J. P., xi, 99*, 101, *107*
Williams, P. M., *107*
Willis, D. L., 213, *216*
Wilson, C. M., 117, *131*
Wilson, P. W., 366, 368, *377*
Wilson, S. B. 334, *335*
Winget, G. D., *232*
Winter, W., *232*
Wold, R., *295*
Wolf, B., 86, *90*
Wolff, D. P., *252*
Wolfrom, M. L., *2*
Wollenberger, A., 88, *90*
Wood, B. J. B., 100, 105, *107*
Wood, W. A., 215, *216*
Woodward, F. N., *156*
Wright, M. G., *130*
Wright, P. L., *55*
Wright, R. T., 318, *327*, 393, *398*
Wu, R., 241, *244*

Yakulis, R., *485*
Yamaguchi, T., 271, 274, 275, *283*
Yamamoto, H. Y., *487*
Yamamoto, L. A., 420, *425*
Yaphe, W., 114, 122, 124, 126, 127, 129, *130*, *131*
Yentsch, C. S., 63, *70*
Yette, M. L., 358, *360*
Yin, H. C., 251, *253*
Yokomura, E., *362*

Zaybin, B. A., 124, *131*
Zderic, J. A., *266*
Zeldin, B., *89*, *360*
Zundel, G., 86, *89*
Zweig, G., 482, *487*, *488*

Taxonomic index

Acetabularia (Chlorophyceae), 357, 481–2, 484
Agardhiella (Rhodophyceae), 122
Alaria (Phaeophyceae), 148, 187
Amphora (Bacillariophyceae), 317–19, 323–4, 326–7
Anabaena (Cyanophyceae), 235, 306, 371–3, 391, 394–6, 402, 481–2, 484
Anacystis (Cyanophyceae), 73, 235, 242, 244, 306, 452, 483
Ankistrodesmus (Chlorophyceae), 481, 483–4
Ascophyllum (Phaeophyceae), 136, 139, 148–9, 155–6, 177, 187

Bacillariophyceae (diatoms), 178
Biddulphia (Bacillariophyceae), 134
Bryopsis (Chlorophyceae), 357

Chaetoceros (Bacillariophyceae), 134–5, 139, 257, 433
Chara (Chlorophyceae), 482
Chlamydomonas (Chlorophyceae), 53, 82, 85, 87–8, 103–4, 210–11, 214, 240, 242, 305, 307, 311–12, 314, 338, 339, 342–3, 347, 349, 354–8, 481–2, 484
Chlorella (Chlorophyceae), 25–9, 82, 84, 251, 306, 326, 330–1, 333–4, 357, 391, 394–7, 420–2, 450–61, 480–4
Chlorogonium (Chlorophyceae), 32–3, 35–6
Chlorophyceae, 178
Chondrus (Rhodophyceae), 112, 115, 120–1, 123, 243, 420–3
Chroococcus (Cyanophyceae), 306
Chroomonas (Cryptophyceae), 77, 227, 230–1
Chrysophyceae, 178
Cladophora (Chlorophyceae), 243
Coccochloris (Cyanophyceae), 235
Cocconeis (Bacillariophyceae), 317–18, 323–4
Codium (Chlorophyceae), 243
Colpomenia (Phaeophyceae), 343
Corallina (Rhodophyceae), 343
Coscinodiscus (Bacillariophyceae), 134
Cyanidium (?), 480

Cyanophyceae, 178
Cyclotella (Bacillariophyceae), 323, 381–6, 480, 483
Cylindrotheca (Bacillariophyceae), 41, 43–4, 483

Dictyota (Phaeophyceae), 243, 357
Ditylum (Bacillariophyceae), 218, 222, 433
Dunaliella (Chlorophyceae), 290, 292–3, 300–2

Ectocarpales, 177
Eisenia (Phaeophyceae), 243
Escherichia (Schizomycophyta), 402
Eucheuma (Rhodophyceae), 115, 121
Euglena (Euglenophyceae), 5–12, 15–17, 20–3, 210, 240, 306, 331, 354, 358, 421, 423, 481–4

Fremyella (Cyanophyceae), 74
Fucaceae, 148
Fucales, 148–9, 156, 177
Fucus (Phaeophyceae), 148, 152, 155–7, 160, 162, 164–5, 167, 170, 177, 187, 271, 278, 421
Furcellaria (Rhodophyceae), 122

Galaxaura (Rhodophyceae), 243
Gelidium (Rhodophyceae), 243
Giffordia (Phaeophyceae), 271, 278
Gigartina (Rhodophyceae), 243
Gigartinaceae, 110, 121, 123
Gloeocapsa (Cyanophyceae), 372–3
Gloiopeltis (Rhodophyceae), 243
Golenkinia (Chlorophyceae), 483–4
Gracilaria (Rhodophyceae), 110, 112, 128–9, 243
Gymnogongrus (Rhodophyceae), 243

Halidrys (Phaeophyceae), 148, 187
Halymenia (Rhodophyceae), 243
Hemiselmis (Cryptophyceae), 227, 230–1
Himanthalia (Phaeophyceae), 148
Hizikia (Phaeophyceae), 242–3
Hydrodictyon (Chlorophyceae), 483–4
Hypnea (Rhodophyceae), 121
Hypneaceae, 110

Ishige (Phaeophyceae), 243
Isochrysis (Haptophyceae), 178, 436, 441–4

Laminaria (Phaeophyceae), 135–6, 139, 148, 185, 187, 271
Laminariaceae, 135
Laminariales, 148–9, 177, 273
Lithodesmium (Bacillariophyceae), 433

Monochrysis = *Pavlova* (Haptophyceae), 290
Monostroma (Chlorophyceae), 178

Navicula (Bacillariophyceae), 178, 391, 395–6, 428–9, 433, 481
Nitella (Charophyceae), 464, 474, 483–4
Nitzschia (Bacillariophyceae), 41–5, 49, 257, 428–30, 432–3, 465–6, 469–75
Nostoc (Cyanophyceae), 371

Ochromonas (Chrysophyceae), 82, 84, 134, 264
Olisthodiscus (Xanthophyceae), 82, 84–6, 354, 357–8
Oscillatoria (Cyanophyceae), 243, 480

Padina (Phaeophyceae), 357
Pelvetia (Phaeophyceae), 148, 187, 271, 278
Petrocelis (Rhodophyceae), 121
Phaeodactylum (Bacillariophyceae), 134, 290, 292, 483
Phaeophyceae, 177, 274
Phormidium (Cyanophyceae), 75, 77, 306, 481–2

Plectonema (Cyanophyceae), 373
Polysiphonia (Rhodophyceae), 157, 160, 167, 171–2, 175, 177
Polytomella (Chlorophyceae), 331, 481
Porphyra (Rhodophyceae), 73, 78, 242–3
Porphyridium (Rhodophyceae), 306, 421
Prototheca (Chlorophyceae), 480, 484
Pseudomonas (Schizomycophyta), 122, 126

Rhodophyceae, 110, 178
Rhodymenia (Rhodophyceae), 73, 246–52

Saccharomyces (Eumycota), 137
Saprospira (Schizomycophyta), 481
Sargassum (Phaeophyceae), 243
Scenedesmus (Chlorophyceae), 306, 391, 395–6, 415, 481–4
Serraticardia (Rhodophyceae), 243, 251
Skeletonema (Bacillariophyceae), 134, 136, 139, 257, 433
Solieriaceae, 110
Spatoglossum (Phaeophyceae), 243
Spirogyra (Chlorophyceae), 32–3, 35–6, 178

Thalassiosira (Bacillariophyceae), 136, 139, 433
Tolypothrix (Cyanophyceae), 75, 77

Ulva (Chlorophyceae), 239, 241–3
Undaria (Phaeophyceae), 243

Vitreoscilla (Schizomycophyta), 481

Xanthophyceae, 178